Lecture Notes in Physics

Springer
Berlin
Heidelberg
New York
Barcelona
Hong Kong
London
Milan
Paris
Singapore
Tokyo

The Editorial Policy for Proceedings

The series Lecture Notes in Physics reports new developments in physical research and teaching – quickly, informally, and at a high level. The proceedings to be considered for publication in this series should be limited to only a few areas of research, and these should be closely related to each other. The contributions should be of a high standard and should avoid lengthy redraftings of papers already published or about to be published elsewhere. As a whole, the proceedings should aim for a balanced presentation of the theme of the conference including a description of the techniques used and enough motivation for a broad readership. It should not be assumed that the published proceedings must reflect the conference in its entirety. (A listing or abstracts of papers presented at the meeting but not included in the proceedings could be added as an appendix.)
When applying for publication in the series Lecture Notes in Physics the volume's editor(s) should submit sufficient material to enable the series editors and their referees to make a fairly accurate evaluation (e.g. a complete list of speakers and titles of papers to be presented and abstracts). If, based on this information, the proceedings are (tentatively) accepted, the volume's editor(s), whose name(s) will appear on the title pages, should select the papers suitable for publication and have them refereed (as for a journal) when appropriate. As a rule discussions will not be accepted. The series editors and Springer-Verlag will normally not interfere with the detailed editing except in fairly obvious cases or on technical matters.
Final acceptance is expressed by the series editor in charge, in consultation with Springer-Verlag only after receiving the complete manuscript. It might help to send a copy of the authors' manuscripts in advance to the editor in charge to discuss possible revisions with him. As a general rule, the series editor will confirm his tentative acceptance if the final manuscript corresponds to the original concept discussed, if the quality of the contribution meets the requirements of the series, and if the final size of the manuscript does not greatly exceed the number of pages originally agreed upon. The manuscript should be forwarded to Springer-Verlag shortly after the meeting. In cases of extreme delay (more than six months after the conference) the series editors will check once more the timeliness of the papers. Therefore, the volume's editor(s) should establish strict deadlines, or collect the articles during the conference and have them revised on the spot. If a delay is unavoidable, one should encourage the authors to update their contributions if appropriate. The editors of proceedings are strongly advised to inform contributors about these points at an early stage.
The final manuscript should contain a table of contents and an informative introduction accessible also to readers not particularly familiar with the topic of the conference. The contributions should be in English. The volume's editor(s) should check the contributions for the correct use of language. At Springer-Verlag only the prefaces will be checked by a copy-editor for language and style. Grave linguistic or technical shortcomings may lead to the rejection of contributions by the series editors. A conference report should not exceed a total of 500 pages. Keeping the size within this bound should be achieved by a stricter selection of articles and not by imposing an upper limit to the length of the individual papers. Editors receive jointly 30 complimentary copies of their book. They are entitled to purchase further copies of their book at a reduced rate. As a rule no reprints of individual contributions can be supplied. No royalty is paid on Lecture Notes in Physics volumes. Commitment to publish is made by letter of interest rather than by signing a formal contract. Springer-Verlag secures the copyright for each volume.

The Production Process

The books are hardbound, and the publisher will select quality paper appropriate to the needs of the author(s). Publication time is about ten weeks. More than twenty years of experience guarantee authors the best possible service. To reach the goal of rapid publication at a low price the technique of photographic reproduction from a camera-ready manuscript was chosen. This process shifts the main responsibility for the technical quality considerably from the publisher to the authors. We therefore urge all authors and editors of proceedings to observe very carefully the essentials for the preparation of camera-ready manuscripts, which we will supply on request. This applies especially to the quality of figures and halftones submitted for publication. In addition, it might be useful to look at some of the volumes already published. As a special service, we offer free of charge LaTeX and TeX macro packages to format the text according to Springer-Verlag's quality requirements. We strongly recommend that you make use of this offer, since the result will be a book of considerably improved technical quality. To avoid mistakes and time-consuming correspondence during the production period the conference editors should request special instructions from the publisher well before the beginning of the conference. Manuscripts not meeting the technical standard of the series will have to be returned for improvement.

For further information please contact Springer-Verlag, Physics Editorial Department II, Tiergartenstrasse 17, D-69121 Heidelberg, Germany

John W. Clark   Thomas Lindenau
Manfred L. Ristig (Eds.)

Scientific Applications
of Neural Nets

Proceedings of the 194th W.E. Heraeus Seminar
Held at Bad Honnef, Germany, 11-13 May 1998

Springer

Editors

John W. Clark
Department of Physics
Washington University
St. Louis, MO 63130, USA

Thomas Lindenau
Manfred L. Ristig
Institut für Theoretische Physik
Universität zu Köln
D-50937 Köln, Germany

Library of Congress Cataloging-in-Publication Data.

Die Deutsche Bibliothek - CIP-Einheitsaufnahme

Scientific applications of neural nets : proceedings of the 194th W.
E. Heraeus Seminar, held at Bad Honnef, Germany, 11 - 13 May
1998 / John W. Clark ... (ed.). - Berlin ; Heidelberg ; New York ;
Barcelona ; Hong Kong ; London ; Milan ; Paris ; Singapore ; Tokyo
: Springer, 1999
  (Lecture notes in physics ; Vol. 522)
  ISBN 3-540-65737-1

ISSN 0075-8450
ISBN 3-540-65737-1 Springer-Verlag Berlin Heidelberg New York

Typesetting: Camera-ready by the authors/editors
Cover design: *design & production*, Heidelberg

SPIN: 10644319   55/3144 - 5 4 3 2 1 0 – Printed on acid-free paper

Preface

The chapters of this volume are based on invited reviews presented at the 194th W.E. Heraeus Seminar, which was devoted to the topic "Scientific Applications of Neural Nets." The workshop was organized to foster communication between scientists who are active in this highly interdisciplinary research area and have made important contributions to the development and implementation of neural-network algorithms suited to the analysis and solution of problems at the leading edge of science.

Recent years have seen a rapid expansion of neural-network applications in many areas of science, including physics, astronomy, geoscience, chemistry, biology, and linguistics. A growing list of interesting examples includes the deployment of neural-network models for event analysis in experimental high-energy physics; star/galaxy discrimination; control of adaptive optical systems; prediction of nuclear properties; fast interpolation of potential energy surfaces in chemistry; classification of mass spectra of organic compounds; protein-structure prediction; analysis of DNA sequences; and design of pharmaceuticals. This intense activity has produced new insights into regularities and mechanisms as well as substantial progress toward the creation of quantitative and reliable predictive tools. At the Heraeus Seminar, speakers discussed applications based on feedforward connectionist systems taught by example, recurrent attractor nets, and self-organizing feature maps, as well as hybrid systems. The presentations ranged over a spectrum of fields:

⋄ Astronomy – Adaptive optical systems have been invented to improve the clarity of images formed by ground-based telescopes from light passing through the turbulent atmosphere of the earth. Neural networks are used to monitor and instruct the operation of these systems by wave-front sensing, reconstruction, and prediction, with the goal of approaching the diffraction limit of resolution.

⋄ Nuclear physics – The production of a host of new nuclei at radioactive-beam facilities has stimulated interest in predictive global models of nuclear properties. As a promising alternative to traditional theory-rich modeling, feedforward neural nets and higher-order probabilistic perceptrons are being applied to statistical analysis of the existing nuclear database. In recent work, notable successes in the development of neural-network predictors have been recorded in the modeling of atomic masses, branching probabilities for different modes of decay, and lifetimes for beta disintegration.

◇ Experimental particle physics – Neural networks offer attractive options for meeting the technical challenges of analyzing the intense data stream from the next generation of high-energy particle accelerators. Of particular interest is the development of a hybrid neural network for determining energy correction factors in the ATLAS calorimeter that is part of the Large Hadronic Collider program at CERN. The design under study involves a recurrent-backpropagation network with nearest-neighbor feedback in the input layer, in conjunction with a competitive network for an event-classification step that precedes the learning process.

◇ Bioinformatics – Understanding the process of protein folding and gaining the power to predict and manipulate protein structure are among the most difficult and important goals in all of science. Their achievement in the coming century will open vast horizons in bioengineering. Neural networks, primarily of the multilayer perceptron class, are making significant contributions toward these goals through statistical prediction of secondary structure, fold class, superficiality, ligand binding sites, distance matrices, and 3D backbone structure based on sequence information. Significant advances in secondary-structure prediction have been realized by invoking evolutionary information embodied in the existing databases to boost the performance of neural networks. In the vital field of drug design, Kohonen networks are emerging as valuable tools for adaptive pattern recognition.

◇ Linguistics – The role of statistics in linguistic analysis has been expanded by the introduction of neural-network techniques. A striking example of this innovative use of computational methods in the basic human sciences is seen in the application of Kohonen self-organizing feature maps to the automated classification of written Slovenian texts and poems according to the styles of the authors.

◇ Information processing – There is a continuing development of new methods for performing sophisticated computational tasks with connectionist systems. Recurrent nets exhibit a complex variety of temporal behavior and are accordingly well suited for processing temporal information. Recurrent networks are also versatile competitors in the arena of combinatorial optimization problems, especially as represented by networks of continuous Potts mean-field neurons. Networks built from chaotic units and hybrid networks that combine supervised and unsupervised learning processes are creating intriguing new opportunities for the modeling of pattern recognition and association.

The seminar provided a forum in which thirty-three invited participants could evaluate the current status of this cross-disciplinary effort and coordinate the planning of future research. The setting of the meeting in the Physikzentrum at Bad Honnef was most conducive to productive discussion and exchange of experience and ideas. During the three days of the meeting, twelve area surveys and six short talks were delivered by internationally recognized experts in their fields, coming both from Europe and the United States. Eight of the review papers appear in this volume. These area surveys, spanning astronomy, nuclear physics, high-energy particle physics, protein structure prediction, linguistics, combinatorial optimization, and neurodynamics, are preceded by a long introduction by John W. Clark. This introduction serves to initiate the non-expert into the field of neural networks as a unifying framework developed for scientific applications of connectionist systems.

The meeting was made possible by the generosity of the W.E. Heraeus Stiftung. We wish to express our sincere gratitude to Dr. V. Schäfer and Mrs. Jutta Hartmann from the Foundation for their splendid efforts in the general organization of the seminar. We are indebted to Henrik Bohr for his help, as a co-organizer, in bringing together scientists from diverse fields of the natural sciences. We benefited from the expert assistance of Michael Barber in the preparation of two of the papers for publication. Finally, we would like to extend our thanks to all speakers, participants, and staff members for their contributions to a successful workshop.

Washington University, St. Louis
Universität zu Köln
January 1999

J. W. Clark
T. Lindenau
M. L. Ristig

Contents

List of Participants

Anderson, Charles H.   cha@shifter.wustl.edu
  Washington University, School of Medicine, 660 South Euclid Ave., St.
  Louis, MO 63110, USA

Bohr, Henrik   hbohr@fysik.dtu.dk
  Center for Biological Sequence Analysis, Bldg. 307, The Technical University of Denmark, DK-2800 Lyngby, Denmark

Clark, John W.   jwc@howdy.wustl.edu
  Washington University, Department of Physics, Campus Box 1105, St.
  Louis, MO 63130, USA

Compiani, Mario   compiani@camserv-unicam.it
  Università di Camerino, Department of Chemical Sciences, Via S. Agostino
  1, I-62032 Camerino MC, Italy

Denby, Bruce   denby@cetp.ipsl.fr
  Université de Versailles/CETP, 10-12 Av. de l'Europe, F-78140 Velizy,
  France

Dittmar, Susan   sdm@thp.uni-koeln.de
  Universität zu Koeln, Institut für Theoretische Physik, Zülpicher Str. 77,
  D-50937 Köln, Germany

Dixit, Vijai V.   dixitvv@slu.edu
  Parks College of Saint Louis University, Department of Science and Mathematics, P. O. Box 56907, St. Louis, MO 63156-0907, USA

Doren, Douglas   doren@udel.edu
  University of Delaware, Department of Chemistry, Newark, DE 19716,
  USA

Fariselli, Piero   farisel@kaiser.alma.unibo.it
  Università di Bologna, Dipartimento di Biologia, Via Irnerio 42, I-40126
  Bologna, Italy

Farnell, Damian J. J.   df@thp.uni-koeln.de
  Universität zu Köln, Institut für Theoretische Physik, Zülpicher Str. 77,
  D-50937 Köln, Germany

Gasteiger, J.   gasteiger@ccc.chemie.uni-erlangen.de
  Computer-Chemie Centrum, Institut für Organische Chemie, Universität
  Erlangen-Nürnberg, Naegelsbachstr. 25, D-91052 Erlangen, Germany

Gernoth, Klaus A.   mccs7kag@thpd1.phy.umist.ac.uk
Department of Physics, UMIST, University of Manchester, P. O. Box 88,
Manchester, M60 IQD, United Kingdom

Hescheler, J.   jh@physiologie.uni-koeln.de
Universität zu Köln, Institut für Neurophysiologie, Robert-Koch-Str. 39,
D-50931 Köln, Germany

Hohlneicher, Georg   ghohln@hp710.pc.uni-koeln.de
Universität zu Köln, Institut für Physikalische Chemie, Luxemburger Str.
116, D-50939 Köln, Germany

Jönsson, Henrik   henrik@thep.lu.se
Department of Theoretical Physics II, University of Lund, Sölvegatan
14A, S-22362 Lund, Sweden

Krisement, Otto   krisement@uni-muenster.de
Institut für Theoretische Physik II, Universität Münster, Carossastr. 21,
D-48161 Münster, Germany

Krogh, Anders S.   krogh@cbs.dtu.dk
Center for Biological Sequence Analysis, Bldg. 208, The Technical University of Denmark, DK-2800 Lyngby, Denmark

Kürten, Karl E.   kuerten@acpx.exp.univie.ac.at
Universität Wien, Institut für Experimentalphysik, Boltzmanngasse 5,
A-1090 Wien, Austria

Lindenau, Thomas   tl@thp.uni-koeln.de
Universität zu Köln, Institut für Theoretische Physik, Zülpicher Str. 77,
D-50937 Köln, Germany

Lorenz, Soenke   slorenz@fhi-berlin.mpg.de
Fritz-Haber-Institut der Max-Planck-Gesellschaft, Faradayweg 4-6, D-14195 Berlin, Germany

Mavrommatis, Eirene   emavrom@atlas.uoa.gr
University of Athens, Physics Department, Div. of Nuclear and Particle
Physics, Panepistimioupoli, GR-15771 Athens, Greece

McGuire, Patrick   mcguire@as.arizona.edu
Center for Astronomical Adaptive Optics, Steward Observatory, University of Arizona, Tucson, AZ 85721, USA

Mladenka, Alice   mlad@cent.gud.siemens.co.at
Technische Universität Wien, Institut 114, Wiedner Hauptstr. 8-10, A-1040 Wien, Austria

Peterson, Keith   Keith_Peterson@post.wesleyan-college.edu
Wesleyan College, Department of Chemistry & Physics, 4760 Forsyth
Road, Macon, GA 31210, USA

Ristig, Manfred L.   ristig@thp.uni-koeln.de
Universität zu Köln, Institut für Theoretische Physik, Zülpicher Str. 77,
D-50937 Köln, Germany

Rost, Burkhard   rost@embl-heidelberg.de
EMBL, Meyerhofstr. 1, D-69012 Heidelberg, Germany

Seixas, João   joao.seixas@cern.ch
    CFIF(Lisbon)/CERN, CERN TH. Div., CH-1211 Geneve 23, Switzerland
Söderberg, Bo   bs@thep.lu.se
    Department of Theoretical Physics, University of Lund, Sölvegatan 14A,
    S-22362 Lund, Sweden
Tafill, Friedrich   fritz@synapse.iaee.tuwien.ac.at
    Institut für Elektronik, Technische Universität Wien, Gusshausstr. 27e,
    A-1040 Wien, Austria
Toussaint, Marc   mt@thp.uni-koeln.de
    Universität zu Köln, Institut für Theoretische Physik, Zülpicher Str. 77,
    D-50937 Köln, Germany
Vilela Mendes, Rui   vilela@alf4.cii.fc.ul.pt
    Grupo de Fisica Matemàtica, Complexo Interdisciplinar, Univ. Lisboa,
    Av. Gama Pinto 2, Pt-1699 Lisboa Codex, Portugal
Volk, Daniel   dv@thp.uni-koeln.de
    Universität zu Köln, Institut für Theoretische Physik, Zülpicher Str. 77,
    D-50937 Köln, Germany
Zupan, Jure   jure.zupan@ki.si
    National Institute of Chemistry, Hajdrihova 19, SLO-1000 Ljubljana,
    Slovenia

Neural Networks: New Tools for Modelling and Data Analysis in Science

John W. Clark

McDonnell Center for the Space Sciences and Department of Physics, Washington University, St. Louis, MO 63130, USA

Abstract. To provide a primer for the study of scientific applications of connectionist systems, the dynamical, statistical, and computational properties of the most prominent artificial neural-network models are reviewed. The basic ingredients of neural modeling are introduced, including architecture, neuronal response, dynamical equations, coding schemes, and learning rules. Perceptron systems and recurrent attractor networks are highlighted. Applications of recurrent nets as content-addressable memories and for the solution of combinatorial optimization problems are described. The backpropagation algorithm for supervised training of multilayer perceptrons is developed, and the utility of these systems in classification and function approximation tasks is discussed. Some instructive scientific applications in astronomy, physical chemistry, nuclear physics, protein structure, and experimental high-energy physics are examined in detail. A special effort is made to illuminate the nature of neural-network models as automated devices that learn the statistics of their data environment and perform statistical inference at a level that may approach the Bayesian ideal. The review closes with a critical assessment of the strengths and weaknesses of neural networks as aids to modeling and data analysis in science.

1 Introduction

To the physics community, the initial attraction of neural networks or "connectionist" systems lay in their novel properties as dynamical and statistical systems. Neural networks are systems of neuron-like units (called simply "neurons") that store information in the connections between the units. In the language of physics, the neurons may be thought of as particles, while the weighted connections between the units may be associated with the interactions between these particles. The neuronal units exhibit varying degrees of activity, and the activation of a given neuron in turn promotes ("excites") or suppresses ("inhibits") the activity of any neuron to which it extends a connection. An excitatory [inhibitory] connection or "synapse" is represented by a positive [negative] weight parameter. The dynamics of the neural system is generally nonlinear and dissipative, with the possibility of chaotic motion as well as flow to fixed points and terminal cycles. Mixtures of negative and positive weights lead to the phenomenon of frustration well known in spin systems, opening the possibility of a large number of equilibrium configurations and a complex thermodynamic phase structure. Indeed, a fruitful correspondence may be established between highly simplified neural-network models and infinite-range spin-glass systems.

The spin-glass analogy stimulated many condensed-matter theorists to visit the field of neural networks in the 1980s. In the aftermath of this involvement, the prevailing sense of the field among physicists is dominated by the theoretical developments resulting from the exploitation of advanced techniques of statistical physics. Outside the physics community, such developments are often regarded with the admiration warranted by their high level of theoretical virtuosity. However, this body of work is also received with considerable reservation. The neuroscientist points to the remoteness of the neuronal and network models from biological reality, and the engineer, to the limited applicability of the results derived.

Perhaps the most extraordinary feature of living neural systems, imitated by their artificial counterparts, is their ability to adapt and to learn. In general, the neuronal interactions change with time, depending on the states recently occupied by the system. Thus, as the network experiences various stimuli, knowledge can be stored in the network connections and their weights, for later retrieval in some information processing task. The search for optimal learning rules has been another productive endeavor of theoretical physicists.

In the main, the contributions of physicists to neural-network research have been rather abstract and have addressed the simplest model systems. However, the situation is evolving. Some theorists with biological inclinations have remained in the field, have learned more neuroscience, and are now making serious attempts to understand the ways that real brains encode and process information. Such forays are easy to defend: the study of the dynamics of existing material systems – whatever their nature – lies in the rightful domain of physics.

In addition to the role of physics in understanding neural networks, one can turn things around and ask about possible roles for neural networks in physics itself, and in creating new physics. The answer to this question – which can just as well be framed for other sciences like astronomy and chemistry – becomes easier when one realizes that artificial neural networks, however complicated and refined, are basically just another set of tools for doing statistical analysis and statistical inference. It then seems natural to exploit the pattern-recognition capabilities of these systems, for example, in the classification of astrophysical images, and especially in the analysis of high-energy physics experiments where the event rate and multiplicities can overwhelm conventional techniques. Thus neural nets can be employed in automated processing of experimental data. Another application is more theoretical: neural nets can be used to construct global statistical models of properties of complex physical, chemical, astronomical, and biological systems. With suitable training schemes, neural-network models can capture the regularities of the existing information on a given class of systems and make specific predictions for unknown properties of individual members of this class. Over the last decade, scientific applications of both kinds – nominally "experimental" and "theoretical" – have grown at a rapid pace.

This volume of the Lecture Notes in Physics series presents a snapshot of recent progress and current efforts in the highly interdisciplinary research area of scientific applications of neural networks. This lead chapter is intended to serve as a "neural-network primer," designed to make the technical chapters that follow more accessible to a general scientific readership. In the limited space available, it will not be possible to give a comprehensive survey of the tremendous range of neural-network techniques that can be brought into play. For presentations in depth as well as broader perspectives of connectionist research, the reader is directed to the excellent texts of Müller and Reinhardt (1990), Hertz, Krogh, and Palmer (1991), and Haykin (1999); and to the review articles of Lippmann (1987), Cowan and Sharp (1988), and Clark (1991). Another very good source is the handbook on neural computation compiled by Arbib (1995).

Neural-network research has always been like a two-edged sword, the one edge being applied to modeling the computational behavior of neurobiological systems, and the other edge to solving real-world problems of pattern recognition, classification, nonlinear nonparametric regression, and optimization. The jury is still out on whether either or both is or will be a "cutting edge." At any rate, we are concerned here only with the second role of neural networks, extending the term "real world" to include science (strangely excluded in some definitions).

In this real-world role, neural networks are realizations of artificial intelligence within the Neurobiological Paradigm (Clark 1991): the adaptation of principles of natural intelligence in the design of machines or algorithms that can perform sophisticated information-processing tasks. Three principal developmental thrusts may be identified by the following sets of key phrases:

(a) Supervised learning by example; layered feedforward networks (perceptrons); backpropagation learning algorithm; classification and regression tasks;

(b) Content-addressable memories; recurrent attractor networks; Hopfield model; combinatorial optimization; image reconstruction;

(c) Unsupervised Hebbian learning; principal component analysis; unsupervised competitive learning; self-organized feature maps; Kohonen nets; vector quantization.

It should be noted that there can be some blending of these general efforts. For example, limited kinds of recurrence are allowed in some multilayered systems taught by example (as in Chapter 4), while perceptrons, Hopfield nets, and Kohonen maps can all perform pattern recognition. The blurring of putatively distinct categories is quite evident in Chapter 9, where hybrid learning schemes and networks with complex nodes are studied.

All three of the developmental thrusts (a)–(c) have been exploited in scientific applications. The first two will be treated in this review. Supervised learning by example, utilizing the backpropagation learning algorithm, will receive the most attention, simply because this technique and variations upon it have seen the most use, by far, in scientific (and other) problems. Hopfield-type attractor nets will also be emphasized, both because of their novel applications in scientific pattern-recognition and optimization tasks and their importance in "physics culture." Some of the original scientific applications (in astronomy, nuclear physics, chemistry, and high-energy physics) will be adopted as examples in discussing the basic techniques.

Self-organizing feature maps (Kohonen 1989,1997) – a dominant theme of thrust (c) – are considered in Chapter 7, where Kohonen nets are used for clustering and classification in a linguistic analysis of works in the Slovenian language. Other promising applications of Kohonen nets may be found in high-energy physics (Lönnblad et al. 1991a) and in problems related to chemometrics and drug design (Li et al. 1993, Gasteiger et al. 1994, Holtzgrabe et al. 1996, Schuur et al. 1996, Gasteiger 1998). Self-organizing feature maps have a special attraction because their behavior is presumed to mimic the development of pattern-recognition capabilities of cortical areas in the brain. Some innovative recent theoretical work in the area of unsupervised learning is described in Chapter 9. The books of Hertz et al. (1991) and Haykin (1999) contain accessible and authoritative accounts of unsupervised Hebbian and competitive learning, principal component analysis, and self-organizing maps. In addition, the monograph of Zupan and Gasteiger (1993) includes a succinct introduction to Kohonen networks, set within a discourse on chemical applications of neural nets.

Sections 2 and 3 introduce the fundamentals of neural-network modeling: neuronal state variables, dynamics, architecture, and learning rules. These aspects are exemplified in terms of the simplest neural-network models: the McCulloch-Pitts (1943), Little (1974), Hopfield (1982), and Cowan-Hopfield (Cowan 1967,1970, Hopfield 1984) models. The dynamical behaviors of these models are taken up in Sect. 2, while Sect. 3 presents a taxonomy of learning rules. We then turn to issues of computational implementation. Sect. 4 deals with the uses of recurrent attractor networks, specifically of Hopfield and Cowan-Hopfield types, as content-addressable memories and as vehicles for the solution of combinatorial optimization problems. The application of such networks to track reconstruction in high-energy physics is highlighted. Sect. 5 focuses on multilayer feedforward networks as tools for pattern recognition and function approximation (regression). Under supervision by a "teacher," such networks demonstrate the capacity to learn by example. The most popular supervised learning algorithm – backpropagation – is derived explicitly. In Sect. 6, the theoretical development of Sect. 5 is illustrated by applications to discrimination between star and galaxy images on sky survey plates; to modeling of potential energy surfaces of chemical systems; to distinction of stable from unstable nuclides and assignment of spin and parity to nuclear ground states; and to prediction of protein secondary and tertiary structure from the primary amino-acid sequence. Sect. 7 explores connections of neural networks with Bayesian probability theory, thereby providing a formal background for Sec. 8, which presents a general appraisal of the utility of neural-network approaches to data analysis and model-building in science. Strong emphasis is placed on the fact that techniques for classification and function approximation based on artificial neural nets fit comfortably within the enlarged repertoire of methods for statistical inference and nonparametric, nonlinear regression that has become available in the last quarter century. To say that neural nets are "just" approximating Bayes inference may disappoint some advocates by lifting the illusory magic aura that has been cast over these systems by the assertion that they are simulating natural intelligence. The irony is that living neural nets must also be performing Bayesian inference at some level, since evolution would dictate that the organism optimize its processes for estimation and decision in a world filled with uncertainty (Anderson 1994,1996, Anderson and Van Essen 1994). For more extensive discussion of neural networks within the context of statistics and probability theory, the reader may consult Smith (1993) and Cherkassky et al. (1994).

2 Neural Network Models

For the physicist, interest in neural-network models lies mainly in their dynamical and statistical behavior; for the engineer, in their problem-solving or computational abilities; and for the neurobiologist, both in their computational aspects and their value in simulating the activity of living nerve nets. In this section, the most basic neural-network models will be introduced and their dynamical properties will be examined. Computational and statistical aspects will be touched upon occasionally, in preparation for a later preoccupation with these aspects.

Neural-network models may be classified in several ways, and in particular according to (i) the nature of the dynamical variables used to describe the activity of the model neurons from which they are constructed, (ii) the dynamical laws obeyed by these variables, and (iii) the pattern of interactions or couplings between the basic units. The *activity variables* characterizing the state of the model system may be *discrete* ("digital") or *continuous* ("analog"). Likewise, the system may evolve in *discrete* or *continuous time*.

In the case of a network operating in discrete time, the states of the individual neurons may be updated simultaneously (*synchronous* or *parallel* dynamics) or one at a time (*asynchronous* or *sequential* dynamics), or in a more complicated manner. The time development may be *deterministic* or *probabilistic*.

The simplest and oldest neural-network models are *digital* in nature and share important abstract features with a system of Ising spins (Cragg and

Temperley 1954, Little 1974). These models include the original McCulloch-Pitts (McP) model (McCulloch and Pitts 1943), extensions of it by Caianiello (1961) and Little (1974), and the celebrated Hopfield (1982) model. Each neuron of the N-unit system is described by a dichotomic state variable σ_i that assumes the value $+1$ when neuron i is "firing" or "on" and -1 when it is "silent" or "off." The state of the system is specified by one of the 2^N "firing patterns" $\{\sigma_1 \ldots \sigma_N\}$, and this state evolves on the vertices of an N-dimensional hypercube. It is often convenient to work with the alternative binary dynamical variable $S_i = (\sigma_i + 1)/2$, which takes the value 1 when the neuronal unit is active and 0 when it is silent. The total synaptic input to model neuron i is then typically expressed as a linear superposition $\sum_j V_{ij} S_j$ of input signals from neurons j in the network, and each neuron is assigned an activation threshold V_{io}. The real square matrix $V = (V_{ij})$, where i and j have ranges $1, \ldots N$, defines the synaptic couplings – the interactions – between the model neurons. If the effect of j on i is excitatory [inhibitory], the coupling V_{ij} is positive [negative], and if j extends no connections to i, the associated matrix element V_{ij} is zero. In a formally analogous system of Ising spins, which can only point "up" ($\sigma_i = +1$) or "down" ($\sigma_i = -1$), the quantity $v_{ij} = V_{ij}/2$ represents the spin-spin coupling strength between spins i and j, while $h_{io} = \sum_j V_{ij}/2 - V_{io}$ represents an external magnetic field affecting spin i.

A complementary theme of neural modeling is played out in the *analog* domain: the digital activity variable S_i is replaced by an analog state variable that ranges continuously from 0 to 1, and which may be updated either in continuous or discrete time. In deterministic formulations where an underlying neurophysiological substrate is being modeled, this analog variable is usually viewed as a normalized instantaneous firing frequency (idealizing the operational notion of a firing rate calculated over a short time interval). In probabilistic descriptions, the analog variable replacing S_i is interpreted as a firing probability. Analog models offer certain technological advantages and, potentially, certain advantages in realistic biological simulation. As we have seen, the digital models have their corresponding spin systems. Similarly, an electrical circuit corresponding to a given continuous-time analog model can usually be formulated and can often be realized in hardware.

Finally, there is the matter of architecture, or "wiring diagram," which is specified by saying which directed connections exist between the neuronal units. Extreme examples of neural-network architectures are seen in perceptrons and Hopfield nets. In a *perceptron*, the units are arranged in an ordered series of layers, the first layer providing an input interface and the last generating the output of the network. Connections extend only in the forward direction, usually from one layer to the next. There are no backward connections, hence no feedback; also there are no lateral connections, hence no recurrence. In a fully coupled *Hopfield net*, every neuron is connected to every other neuron, in both directions, forward and backward, with symmetrical weights. This is the archetypal *recurrent* network, where information can flow in loops.

The performance of some assigned computational task – classification, function approximation, content-addressable memory, combinatorial optimization – hinges critically on architecture, and on judicious choices of the nonvanishing couplings or connection weights V_{ij} or a judicious choice of "learning rule" by which appropriate weights are realized adaptively. The nature of such rules will be considered in the next section. Our main concern in this section is neurodynamics, without regard to specific interaction parameters. Given a neural network of one type or another, we are interested in what dynamical motions are accessible to it, and how is it likely to behave.

2.1   McCulloch-Pitts Model

The first systematic mathematical analysis of assemblies of neuron-like units was carried out in the pioneering 1943 paper of McCulloch and Pitts (McP). The "formal neurones" introduced by these authors are made in the image of real neurons, but the image is very crude: a neuron is represented by a binary, "on-off" device obeying a threshold logic. Thus the binary variable S_i, or the Ising variable σ_i, may be assigned to neuronal unit i. A generic model neuron receives inputs from other neurons, and possibly from itself and sources external to the N-neuron assembly. The total synaptic input to a given neuron is represented in the form $\sum_j V_{ij} S_j$. This model incorporates two basic properties of living neurons: (i) the all-or-none character of the action potential (a neuron fires an action potential, or "spike," if and only if its total stimulus exceeds a certain threshold value) and (ii) linear superposition of incoming stimuli (which holds as a first approximation).

A key assumption of the McP formulation of neurodynamics is that the delay time τ for direct transmission of information, in the form of excitation or inhibition, from the upstream (presynaptic) neuron j to the downstream (postsynaptic) neuron i is *universal*, i.e., independent of the pair ij. Additionally, it is assumed that the stimulus received at time t decays to a negligible value by time $t + \tau$. The network then necessarily evolves synchronously on a regular time grid with spacing τ, according to the simple threshold dynamics

$$S_i(t) = \Theta\big(F_i(t)\big) , \tag{1}$$

where $t = n\tau$ (n being an integer or zero) and

$$F_i(t) = \sum_{j=1}^{N} V_{ij} S_j(t - \tau) - V_{io} \tag{2}$$

is the total stimulus in excess of threshold, or "firing function." The step function $\Theta(x)$ delivers the value 1 if x is positive or zero, and otherwise delivers 0. The state of the system at time t is uniquely determined by the

8   John W. Clark

state one time step earlier, at $t-\tau$. Changes of the firing state or firing pattern $\{\sigma_i\}$ of the assembly of model neurons can occur only at times $n\tau$, and the states of all individual neurons are updated simultaneously. The model is easily extended to incorporate a refractory period, in analogy with the dead time manifested by real neurons after firing an action potential (Harth et al. 1970, Anninos et al. 1970, Anninos 1972, Clark et al. 1985).

Since the model is purely deterministic, and since the number of system states is finite, the motion must inevitably relax to a fixed point or a terminal cycle. In general, there will be a transient period during which no state is repeated, but eventually some state must be revisited, after which time the system stays in the same state or is locked into periodic motion. The model therefore defines a mapping (ordinarily, a many-to-few mapping) from initial states to final operating conditions. These terminal modes may be regarded as "attractors" of the dynamical flow (see Fig. 1). For prescribed thresholds V_{io}, the menu of fixed points and terminal cycles is governed by the matrix (V_{ij}) and can be altered by adjustment of this matrix, systematically or otherwise.

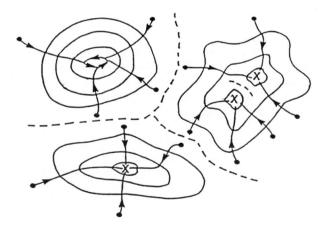

Fig. 1. Highly schematic view of dynamical flow toward attractors (two fixed points and a terminal cycle) in the state space of a neural network. A dashed line indicates a separatrix between areas ("basins") of attraction. For N McCulloch-Pitts neurons, the state space will actually consist of the 2^N vertices of an N-dimensional hypercube and the system will hop from corner to corner.

Such features make the system potentially useful as a repository of content-addressable memories. McCulloch and Pitts (1943) were able to show, more generally, that networks of binary threshold neurons, with synchronous updating, can in principle carry out any computational operation that can be given a precise logical formulation (see also Cowan and Sharp (1988) and Cowan (1990)). In this sense McP networks may be called *computationally universal*. Simple examples of functions that may be computed by *single* McP neurons are: x AND y, x OR y, and NOT x. However, a single McP neuron cannot compute the logical function x AND NOT y, also called XOR (exclusive OR) – although of course nets built from McP neurons can do so. This fundamental limitation had a pivotal role in the history of perceptrons (Rosenblatt 1958,1962, Minsky and Papert 1969, Cowan and Sharp 1988, Clark 1991). (We shall return to this point in Sect. 5.)

The dynamical behaviors of the McP model and its close relatives have been studied by many authors (e.g. Farley and Clark 1954, Caianiello 1961, Harth et al. 1970, Anninos et al. 1970, Anninos 1972, Grondin et al. 1983, Clark et al. 1985,1989, Personnaz et al. 1986, Littlewort et al. 1988, Kürten 1988a-e,1989) using both analytical methods and computer simulation. The examples that have received the most attention involve randomly chosen couplings, but there are also quite extensive investigations of the effects of learning (or antilearning) rules on dynamical behavior.

2.2   Little Model

The Little (1974) model extends the McCulloch-Pitts model to allow for threshold noise (Clark 1988, Bressloff and Taylor 1990, Bressloff 1991,1992, Hertz et al. 1991): a neuron is given some chance to fire even when the threshold test is not met, and some chance not to fire even if the threshold is surpassed. The updating rule (1) is replaced by a conditional probability for firing ($\sigma_i = +1$) or not firing ($\sigma_i = -1$):

$$\rho_i\big(\sigma_i(t)|\{\sigma_j(t-\tau)\}\big) = \Big\{1 + \exp\big[-\beta_i\sigma_i(t)F_i(t)\big]\Big\}^{-1}, \qquad (3)$$

where F_i is given by Eq. (2) and the firing pattern at time $t-\tau$ is known to be $\{\sigma_j\}$. The amount of noise afflicting neuron i is measured by a "temperature" parameter β_i^{-1}, normally taken to have the same value $T = \beta^{-1}$ for all neurons. The larger this parameter, the more likely it is that the neuron i will deviate – unpredictably – from strict threshold logic. The interpretation of $1/\beta$ as a temperature is motivated by the observation that (3) is of precisely the same form as the spin-update probability arising in the Glauber dynamics of an Ising spin system at temperature $T = 1/k_B\beta$, where k_B is Boltzmann's constant (Glauber 1963; see also Sect. 4.2). In the latter problem, σ_i is the orientation of spin i and F_i is to be interpreted as the local magnetic field at i. (It must be kept in mind that the analogy is purely formal and that the parameter β_i^{-1} in (3) does not represent a real physical temperature, as measured, say, in a piece of neural tissue or in a silicon implementation.)

The states of all N neurons are to be updated simultaneously, as in the McP model. Indeed, the McCulloch-Pitts model is obtained as the zero-temperature limit ($\beta^{-1} \to 0$) of the Little model, assuming no F_i is ever *exactly* zero. The particular functional form $g_L(u) = [1 + \exp(-\beta u)]^{-1}$ adopted by Little for the dependence of the firing probability ρ_i on the firing function $u = F_i$ is called the logistic function. This "sigmoidal" form occurs frequently in neural models.

The statistical mechanics of the Little model (and related stochastic models) has been developed in considerable detail by Peretto (1984), Clark (1988), Bressloff and Taylor (1990), and Bressloff (1991,1992).

2.3   Hopfield Model

As in the McCulloch-Pitts design, the Hopfield (1982) model imposes a sharp threshold condition for neuronal firings, but unlike the older model, it implements an asynchronous dynamics. It is imagined that each neuron updates its state at random times, with some prescribed mean waiting time. In practice, a neuron is picked at random, and its state is updated; then another neuron is picked at random and its state updated; and so on. At a given updating event, the neuron administers the threshold test based on the *current* firing states of the other neurons, taking no account of synaptic or other delays. The neuron adjusts its state (or leaves it alone) to conform with the condition

$$F_i \sigma_i \geq 0 , \tag{4}$$

where F_i is evaluated as in Eq. (2) but with $\tau = 0$. The general dynamical behavior implied by asynchronous updating has been discussed by Grondin et al. (1983). A crucial feature emphasized by Hopfield is that if the synaptic couplings are *symmetrical*, and the diagonal couplings v_{ii} are zero (or non-negative), then the energy function that we would naturally define for the system from the spin analogy, i.e.

$$E = -\frac{1}{2} \sum_{i,j} v_{ij} \sigma_i \sigma_j - \sum_i h_{io} \sigma_i , \tag{5}$$

never increases. (Unless otherwise stated, sums over i and j run from 1 to N.) Indeed, this energy function decreases in stages as individual neuronal states are altered, until a local minimum is attained. Such local minima correspond to fixed points characterized by the satisfaction of condition (4) *for all i.* These fixed points are termed "stable," in the sense that no further changes of neuronal states can occur under the one-neuron-at-a-time sequential-update dynamics of the model. Terminal cycles and chaotic behavior are ruled out. On the other hand, for *generic* couplings (v_{ij}), the function defined by (5) does not have any especially useful properties – it no longer serves as a

Lyapunov function for the system, and limit cycles are both possible and common. Due to the random updating sequence, the system can also exhibit a chaotic wandering in state space, which is not possible for the synchronously updated McCulloch-Pitts model.

From the biological point of view, Hopfield's assumption of symmetrical couplings is quite unrealistic. Nevertheless, the introduction of this assumption has been invaluable in providing a firm foothold from which fundamental progress could be made toward understanding the collective computational abilities of neural networks.

The proof of *Hopfield's energy theorem* is elementary. Let the evolution of the net proceed one step, by updating the kth "spin" from σ_k to σ'_k, for a change $\Delta\sigma_k$ that can be $+2$, -2, or 0. The corresponding change in E is

$$\Delta_k E = -\frac{1}{2} \sum_{j \neq k} v_{kj} \sigma_j \Delta\sigma_k - \frac{1}{2} \sum_{i \neq k} v_{ik} \sigma_i \Delta\sigma_k - h_{ko} \Delta\sigma_k \,. \tag{6}$$

Renaming the dummy index i in the second term on the right as j, setting $v_{jk} = v_{kj}$, and recalling the definition of F_k, we may rewrite (6) as

$$\Delta_k E = -\Big(\sum_{j \neq k} v_{kj} \sigma_j + h_{ko} \Big) \Delta\sigma_k = -F_k \Delta\sigma_k + v_{kk} \sigma_k \Delta\sigma_k \,. \tag{7}$$

By the updating rule, $F_k \geq 0$ implies $\Delta\sigma_k \geq 0$, while $F_k < 0$ implies $\Delta\sigma_k \leq 0$. Hence the first term of the last expression for $\Delta_k E$ is never positive. The second term is either $-2v_{kk}$ or 0. Consequently

$$\Delta_k E \leq 0 \quad \forall\, k \,. \tag{8}$$

Since the function E is bounded below, the system must wind down to some local minimum of the E "surface" defined on the space of network states $\{\sigma_i\}$, i.e. on the vertices of the N-dimensional hypercube.

For such minima, Eq. (4) must hold for all i; otherwise it would be possible to reduce the energy still further by flipping a single "spin," and this spin would eventually be tested by the updating procedure. However, if v_{kk} were larger than $|\sum_{j \neq k} v_{kj} \sigma_j + h_{ko}|$ for some set of $\sigma_{j \neq k}$, then both $\sigma_k = +1$ and $\sigma_k = -1$ would satisfy (4). This ambiguity is undesirable in content-addressable memory applications, since it can create additional unwanted stable states in the vicinity of a desired attractor (Hertz et al. 1991). Such behavior can be ruled out by taking $v_{kk} = 0$, as is generally done (Hopfield 1982).

A network whose dynamics is governed by a non-increasing energy function like (5) may operate efficiently as a content-addressable memory. To realize this application one needs to manipulate the couplings $v_{ij} = V_{ij}/2$ so that chosen memories, represented by particular firing patterns $\{\sigma_i^{(\mu)}\}$, are stored as stable fixed-point attractors. How this may be done, in an approximate sense, is one of the subjects of Sect. 4, where we shall revisit some of the main theoretical results on associative memory storage that have been obtained during the last sixteen years, with the Hopfield model as a basis for exact analysis.

It is worth mentioning that if the connection matrix $V = (V_{ij})$ is symmetric and *positive semi-definite*, the above energy theorem holds for the synchronous dynamics of the McP model as well as the asynchronous updating scheme of the Hopfield model (Psaltis and Venkatesh 1988). The restriction to positive semi-definite connection matrices is stronger than nonnegativity of the diagonal couplings. Does there exist a Lyapunov function for the McP model for the case of *arbitrary* symmetric V? The answer is yes. Taking *normal thresholds* $V_{io} = \frac{1}{2} \sum_j V_{ij}$ for simplicity, one can show that the function

$$E_M = - \sum_i |\sum_j v_{ij} \sigma_j| \tag{9}$$

never increases under parallel updating (Psaltis and Venkatesh 1988). An important distinction from the asynchronous case is that cyclic modes of period 2 are allowed along with fixed points, even with symmetrical couplings. In particular, note that if the system alternates between some state and its mirror image ($\{\sigma_i\} \leftrightarrow \{-\sigma_i\}$), the quantity E_M remains unchanged. Expression (9) is known as the Manhattan norm.

The dynamics of the Hopfield model, as described above, is actually stochastic by virtue of the random timing or order of updating events. Some researchers may prefer to update the neurons in a prearranged order, thus removing this element of uncertainty. Either way, one has the option of introducing a noise temperature in the sense of the Little model and permitting the firing decision itself to be probabilistic. Equations (3) and (2) would again be invoked in updating the state of a selected neuron, but with τ reduced to zero. In some circumstances, the presence of a moderate amount of noise may actually enhance performance in memory retrieval (Hertz et al. 1991). Moreover, this ingredient is crucial to the success of the Boltzmann machine (Ackley et al. 1985, Hinton and Sejnowski 1986) in approximating specified input-output mappings.

The commonality of the threshold condition imposed in Hopfield and McP models implies that Hopfield and McP nets with the same couplings have the same fixed points. If one *only* cares about these fixed points (as repositories of memories) and does not care *how* the net gets to them in the course of time, the two models are equivalent.

2.4   Cowan-Hopfield Model

To show the robustness of his findings for the collective computational properties of asynchronous nets of two-state neurons, Hopfield (1984) formulated a continuous-time analog model that displays similar properties in associative memory storage and recall. Subsequently, Hopfield and Tank (1985,1986) used this type of attractor network in an innovative approach to combinatorial optimization tasks, notably the traveling-salesman problem. Actually, other authors, including Cowan (1967,1970,1990) and Cohen and Grossberg (1983), studied the same model (or more general models) at earlier times. It is probably fairest to call it the Cowan model.

To set up the Cowan or Cowan-Hopfield model in an intuitive fashion, we may appeal once again to well-known properties of real neurons, but now place the emphasis on analog behavior and corresponding analog variables. Following Hopfield (1984), let u_i be the transmembrane potential of a model neuron i, and let a_i denote its firing frequency, or pulse rate. It is convenient to measure the latter variable in units of the maximum firing rate, given by the inverse of the absolute refractory period r_i of neuron i. The equation governing the motion of u_i is taken as

$$C_i \frac{du_i}{dt} = -\frac{1}{R_i} u_i(t) + \sum_j T_{ij} a_j(t) + I_i(t) \,, \quad i = 1, \dots N \,, \qquad (10)$$

where the parameters C_i and R_i, respectively, are an input capacitance and a transmembrane resistance assigned to the cell membrane of model neuron i. This equation describes the charging of the cell membrane by input currents from other cells, represented by the sum on the right, and by an additive input current I_i from the environment. The roots of (10) in the Eccles equation (Eccles 1957,1964) have been traced by Cowan (1990). The parameter T_{ij} determines the conductance between the output a_j of neuron j and the cell body of neuron i and clearly plays the role of a synaptic efficiency, or synaptic coupling. The output a_i of cell i is related to its membrane potential u_i through a "sigmoidal" voltage-to-pulse-rate conversion function,

$$a_i(t) = g_i\big(u_i(t)\big) \,, \qquad (11)$$

where $g_i(u)$ is supposed to be monotonic in u on the real line $[-\infty, +\infty]$ with asymptotes $g_i(-\infty) = 0$ and $g_i(+\infty) = 1$. Functions of this type are a staple of mathematical accounts of neuronal activity (cf. Eq. (3)). They are called "sigmoids" (because they are shaped vaguely like an "S") or "squashing functions" (because they squash the whole real line down to the interval $[0, 1]$), or "transfer functions" (because they transfer the neuron's input to its output). Using relation (11) to eliminate the activity variables a_j in favor of the u_j, Eq. (10) gives a coupled set of ordinary nonlinear differential equations for the N potential variables $u_i(t)$.

Hopfield chose $g_i(u) = (1/\pi) \tan^{-1}(\pi\gamma_i u/2) + 1/2$ for the squashing function in (11), whereas Cowan used $g_{Li}(u) = [1 + \exp(-2\gamma_i u)]^{-1}$, the logistic function that we have already encountered in the Little model. The "gain" parameter γ_i measures the slope of $g_{Li}(u)$ at $u = 0$. Almost universally, this parameter is assumed to be independent of the neuron index, i.e., $\gamma_i = \gamma$. Some formulations employ an analog activity variable $2a_i - 1 \equiv \xi_i \in [-1, +1]$, corresponding to the spin variable σ_i, in place of $a_i \in [0, 1]$, which corresponds to the binary variable S_i. In such treatments $\xi_i = \tanh(\gamma_i u_i)$ is a popular choice for the function that transforms the membrane potential into the activity variable; this is consistent with the logistic choice for $g_i(u)$ made by Cowan and Little.

Eqs. (10) and (11) define the simplest kind of continuous-time, leaky-integrator, nonlinear analog model. It is an analog model because the dynamical variable a_i characterizing the activity of neuron i has a continuous range: the response of the neuron is "graded." To understand the appellation "leaky integrator," consider the equation of motion (10) for a particular neuron i, and imagine first that the transmembrane resistance R_i is infinite. Then the model cell performs as a perfect integrator of input currents induced by the environment and by the presynaptic cells j:

$$ u_i(t) = C_i^{-1} \int_{t_o}^{t} \Big[\sum_j T_{ij} a_j(t') + I_i(t') \Big] dt' + u_i(t_o) . \tag{12} $$

Suppose, on the other hand, that the inputs are shut off, but R_i is finite. Then (10) reduces to the equation for the discharge of a capacitor through a resistor; the voltage across the membrane drops exponentially due to the existence of finite leakage currents. Generally, both integration and leakage are in play.

The Cowan-Hopfield model has a nonlinear soft-threshold response described by the sigmoidal squashing function $g_i(u_i)$. The neuronal activity increases continuously from nearly zero to nearly unity as the transmembrane potential passes through a "threshold" value, with a rate of increase in the transition region controlled by the gain parameter γ_i. This threshold happens to be zero for all of the choices of $g_i(u)$ indicated above, but could be shifted to an arbitrary value u_{io} by the replacement $u \to u - u_{io}$ in the argument of g_i.

Within the Cowan-Hopfield description of neuronal function, the response itself is not expressed in terms of individual action-potential spikes or their absence. Instead, a short-term average over spiking activity is implied. Accordingly, thinking in terms of the neurobiological substrate, this model utilizes *frequency coding* (Clark 1991). A more precise description, on a finer time scale, could be achieved by allowing the neuron to fire action potentials at the precise times the membrane potential reaches the actual ignition threshold. This is the basic idea of "integrate and fire" models (Knight 1972).

Comparing analog and digital Hopfield models, the variable a_i may be likened to S_i, and the correspondence grows sharper as the gain parameter γ_i increases and the sigmoid $g_i(u)$ turns into a step function. Moreover, in the steady state, u_i has a form similar to the function F_i representing the superthreshold excitation of neuron i in the discrete model. If the synaptic couplings are symmetrical, $T_{ij} = T_{ji}$, one can again identify an energy function E that never increases under the dynamics,

$$E = -\frac{1}{2}\sum_{i,j} T_{ij} a_i a_j - \sum_i I_i a_i + \sum_i R_i^{-1} \int_o^{a_i} g_i^{-1}(x)dx . \qquad (13)$$

The proof follows easily in a few steps: just calculate dE/dt and make use of the symmetry property, the differential equation (10), and the connection (11). Thus a Lyapunov function can be defined and the system cannot oscillate. All attractors are steady-state solutions of (10)–(11) corresponding to local minima of E, and the network can function properly in associative memory storage and recall, provided the number and location of the fixed points can be adequately controlled. We note that in the high-gain limit, the energy function defined by (13) has the same form as that constructed for the digital Hopfield model, Eq. (5).

Hopfield and Tank (1986) have made the interesting point that Eqs. (10)–(11) may be thought of as the "classical" equations of neurodynamics, in roughly the same sense that Newton's laws are the equations of classical mechanics, since their formulation ignores (i) propagation time delays and (ii) the "quantal" nature of action potentials and of neurotransmitter release at synapses.

The reader may turn to Chapter 9 for deeper investigations of neurodynamics that focus on (i) recurrent continuous-state–continuous-time networks of a broad class containing the Cowan-Hopfield model, (ii) unsupervised learning in generalized networks, and (iii) information processing by chaotic networks. The occurrence and implications of chaos in neural systems has been addressed by many authors (see, for example, Kürten and Clark (1986), Clark (1991), and references cited therein). For a thorough mathematical description of how real neurons work, see Johnston and Wu (1995).

3   Learning Rules

Artificial neural networks may be required to carry out such information-processing tasks as associative memory storage and recall, optimization, adaptive learning from examples, and image processing. Having decided what kind of model neurons will be used for a given task and what kind of dynamical laws will be obeyed by their activities, one must address two further design considerations:

- o An appropriate architecture – the disposition of nodes and the pattern of connections – must be specified.
- o The weights of the connections must be determined through a suitable learning rule.

The term "learning rule" is used in a broad sense. Most simply, the rule may just be an explicit, "one-shot" assignment of weights to the existing connections of the chosen architecture. Such *nondynamical* rules, not involving any subsequent changes of the weights, are intended to endow the system with some prescribed information-processing ability by original design. At one end of a spectrum of nontrivial *dynamical* possibilities, the weights may be altered stepwise under the direction or influence of some external agent, so as to modify, enhance, or correct the performance of the net. This is the case of *supervised learning* ("learning with a teacher" or "learning with a critic"): there is active, purposive intervention of the environment in the determination of the network's synaptic interactions, in response to current network behavior. At the other end of the spectrum, the temporal development of the synaptic couplings is governed by some autonomous dynamical law, changes in the couplings being dependent only on the states recently visited by the system, and made without reference to externally ordained "targets" for performance. Processes of this kind are naturally referred to as *unsupervised learning* or *self-organization*, since the environment has only a passive role as a provider of stimuli.

The foregoing discussion assumes the perspective in which architecture is separate from, and unaffected by, the learning process: one does not envision the creation of new neurons and/or new connections, or the elimination of old ones. More generally, the "wiring diagram" as well as the weights may be allowed to evolve with experience – and indeed this phenomenon does occur in the biological setting, where it is usually regarded as an aspect of neural development rather than learning. There have been some noteworthy efforts aimed at optimizing the architectures of artificial neural networks:

(a) Through schemes for growing new neurons and/or new connections, as in the cascade-correlation (Fahlman and Lebiere 1990) and tiling (Mézard and Nadal 1989) algorithms, or
(b) Through pruning of neurons and/or connections of lesser importance, as in weight decay (Hinton 1986), optimal brain damage (Le Cun et al. 1990b), and the exhaustive procedure employed by Gernoth et al. (1993) and Gernoth and Clark (1995a).

3.1   Hebbian Learning

In keeping with the Neurobiological Paradigm, it is fitting to seek guidance from ideas and findings on the neurophysiological basis of learning in living neural networks. The basic guide is the principle enunciated by Donald Hebb (1949):

> When an axon of cell A is near enough to excite cell B and repeatedly or persistently takes part in firing it, some growth process or metabolic change takes place in one or both cells such that A's efficiency, as one of the cells firing B, is increased.

A modern formulation of Hebb's original proposal for synaptic facilitation might read as follows (Herz et al. 1989):

Firing of the postsynaptic neuron triggers enhancement in just those synapses that are eligible to change by virtue of a time window set by presynaptic activity.

A variety of experimental results (Kelso et al. 1986, Malinow and Miller 1986, Gustafsson et al. 1987, Brown et al. 1988, Bonhoefter et al. 1989, Bliss and Collingridge 1993) have given strong cumulative evidence for the existence, in mammals, of Hebbian synapses in this more restricted sense. Even so, it is clear that the situation is much more complicated than that envisioned by Hebb. The Hebb rule, as such, refers only to excitatory synapses, and only to the case that activity of the presynaptic cell is followed by (and presumably related causally to) activity of the postsynaptic cell. Thus, the nature of synaptic modifications (enhancement versus suppression of synaptic strength) is left open for cases in which the synapse is inhibitory and in which either the presynaptic cell or the postsynaptic cell is silent. Moreover, a significant departure from the Hebbian theme involving correlated presynaptic and postsynaptic activity may be found in the learning model proposed by Alkon et al (1990) to explain the experimental findings on conditioning in the nudibranch mollusc *Hermissenda crassicornis* (Alkon 1984,1988). It is hypothesized that learning involves the association of two simultaneous input events, without regard to the current state of the postsynaptic cell. For a readable survey of important modern discoveries that are revealing the cellular and molecular nature of biological learning mechanisms, see Kandel (1992). Experiments on biological preparations ranging from the primitive "brain" of the sea snail *Aplysia californica* to the rat hippocampus, are informing us of the learning rules that are actually implemented in living nervous systems of both invertebrates and vertebrates. The observed neuronal correlates of the learning process include activity-dependent facilitation in *Aplysia* and long-term potentiation (LTP) in the hippocampus.

While it is tempting to delve further into the biology of learning, our aim here is much more modest, namely to establish a useful taxonomy of local learning rules.

3.2   Formal Classification of Local Learning Rules

The learning theory literature is vast and no attempt will be made at a comprehensive survey. For extensive sets of references, see Clark (1991), Hertz et al. (1991), and Watkin et al. (1993).

Following Rumelhart et al. (1986a), we take the original idea of Hebb as a starting point. Let A_i be the "activation," or, more properly, the excitation, of neuron i due to all its inputs (this would be simply $\sum_j V_{ij} S_j - V_{io}$ in any of the discrete models of Sect. 2, and $u_i - u_{io}$ in the continuous model); and let O_j be the output of neuron j (given by S_j in the discrete models, a_j in the continuous models). The change in the coupling strength V_{ij} is then ordinarily expressed as a product of two functions f_1 and f_2,

$$\Delta V_{ij}(t^+) = f_1\big(A_i(t), T_i(t)\big) f_2\big(O_j(t), V_{ij}(t)\big), \tag{14}$$

where the notation t^+ indicates that the updated coupling is to be effective at a time shortly after t and T_i represents a possible teaching input to neuron i. The original Hebb rule does not involve a teacher and assumes the simple form

$$\Delta V_{ij}(t^+) = \eta A_i(t) O_j(t), \tag{15}$$

where η is a learning-rate parameter. Normally, the learning rate η is taken to be positive, but in general it might be negative, in which case one speaks of *antilearning*. In the so-called *delta rule*, T_i provides a target for the "activation" A_i:

$$\Delta V_{ij}(t^+) = \eta[T_i(t) - A_i(t)]O_j(t). \tag{16}$$

This algorithm, also known as the Widrow-Hoff learning rule (Widrow and Hoff 1960, Widrow 1962, Sutton and Barto 1981), is a generalization of the *Perceptron learning rule* (Rosenblatt 1962, Minsky and Papert 1969, Cowan and Sharp 1988, Clark 1991). Another specialization of (14),

$$\Delta V_{ij}(t^+) = \eta A_i(t)[O_j(t) - V_{ij}(t)], \tag{17}$$

proposed by Grossberg (1976), leads to a type of competitive learning. Commonly, the "activation" A_i in (14) and (15) is replaced by the output O_i, resulting in a symmetrical version of the Hebb rule which, for the discrete-state models of Sect. 2, reads

$$\Delta V_{ij} = \eta S_i S_j \tag{18}$$

(with appropriate time arguments for the neuronal state variables). At the same level of simplification and specialization, the asymmetrical rule

$$\Delta V_{ij} = \eta S_i \sigma_j \tag{19}$$

has been examined by Peretto (1988), who, citing the work of Rauschecker and Singer (1981), maintains that it has a stronger neurophysiological basis than rule (18). As will be seen in Sect. 4, the symmetrical rule (Anderson 1970, Cooper 1973, Hopfield 1982)

$$\Delta V_{ij} = \eta \sigma_i \sigma_j \tag{20}$$

provides the basis of countless studies of the statistical physics of memory in attractor neural networks (Sompolinsky 1988, Amit 1989). Inspection of (18)–(20) suggests another asymmetrical rule,

$$\Delta V_{ij} = \eta \sigma_i S_j, \tag{21}$$

which completes a quartet of very simple learning algorithms. Computer simulations (Witt and Clark 1990) have shown that this last example has a strong stabilizing effect in consolidation of classically conditioned responses of asynchronously updated networks of threshold neurons containing multiple feedback loops. We shall call (18) the "true" Hebb rule; (19), the postsynaptic asymmetrical rule (since it comes into play only when the postsynaptic neuron i is active); (20), the Anderson-Cooper-Hopfield (ACH) rule; and (21), the presynaptic asymmetrical rule.

The learning rules discussed above are all *local* in character – local both in space and in time. More specifically, the change in the synaptic coupling V_{ij} depends only on quantities referring to the presynaptic neuron j and the postsynaptic neuron i, and only on the values taken by these quantities within a short time window. Thus the implementation of these rules does not require global information about the state of the network, a clear advantage in both natural and synthetic contexts.

Peretto (1988) has made a systematic study of the memory storage abilities of networks with connectivity prescribed by local learning rules. In his treatment, the small causal delay $t^+ - t$ implied in Eq. (14) shrinks toward zero. The learning process is regarded as "incremental," which in the standard jargon means that changes of the V_{ij} are made additively, resulting in superposition of different memories in the coupling matrix. (Another possibility (Clark et al. 1985) is to implement a multiplicative learning rule, in which V_{ij} is changed by a factor $1 \pm \delta$ at each update, with $0 \leq \delta < 1$.) Based on scaling and mean-field arguments, Peretto concludes that in the task of encoding a maximal number of stable memories as fixed point attractors, the ACH rule (20) is the most effective member of the general class of learning local rules

$$\Delta V_{ij} = N^{-1}(A\sigma_i\sigma_j + B\sigma_i + C\sigma_j + D), \tag{22}$$

which includes the quartet (18)–(21) as special cases (A, B, C, and D being adjustable constants). Another criterion for optimization of learning rules, namely minimization of the relative entropy of the desired probability distribution of successor states evaluated with respect to the actual dynamical distribution, has been introduced by Qian et al. (1991). These authors find that the preferred rule, among (18)–(21), depends on the choice adopted for the immutable threshold of generic neuron i. If V_{io} is chosen, the best performance is obtained with the ACH rule (20) or the presynaptic asymmetrical rule (21) (a "tie"). If, instead, $V'_{io} = -h_{io} = V_{io} - \sum_j V_{ij}/2$ is chosen as the fixed threshold, then the ACH rule is the winner. (Peretto makes the latter choice, as is conventional in analyses that invoke spin-glass ideas and techniques.)

Alternative classifications of local learning rules, possibly more convenient depending on the application, have been given by Palm (1982) (who defines a linear vector space of local learning rules) and Clark et al. (1985) (who present a catalog of plasticity tracks). Adopting Palm's formulation, Dayan and Willshaw (1991) have solved the problem of learning-rule optimization based on the criterion of maximum signal/noise, with results that may be compared with those of Peretto and Qian et al.

4   Computational Properties of Attractor Networks

We now turn to some prominent information-processing applications that draw upon the models discussed in the preceding sections. In this section we consider two examples involving recurrent networks with highly specialized couplings, designed in the one case (i) to allow storage and recall of a system of memory vectors, and in the other (ii) to permit approximate solution of a combinatorial optimization problem. In the former case, the system relaxes (ideally) to a fixed point associated with the input stimulus, among a number of fixed points representing the stored memories. In the latter, it relaxes (ideally) to the network state that yields a global minimum of an objective function measuring deviations from the optimal solution.

4.1   Hopfield Attractor Model for Associative Memory

The discrete Hopfield model of associative, content-addressable memory (Hopfield 1982) has three salient features, relative to earlier neural models. The first two, considered already in Sect. 2.3, are the asynchronous dynamics and the symmetry of the couplings. To complete the specification of the Hopfield model, the particular symmetrical learning rule (20) is applied in a one-shot additive manner, a term of the form (20) being included for each member μ of a set of n "nominated" memory patterns $\sigma^{(\mu)} = \{\sigma_i^{(\mu)}\}$ that are to be stored by the system. Thus the synaptic couplings $v_{ij} = V_{ij}/2$ (for $i \neq j$) are assigned the strengths

$$v_{ij} = \frac{1}{N} \sum_{\mu=1}^{n} \sigma_i^{(\mu)} \sigma_j^{(\mu)} \,, \tag{23}$$

where the prefactor $1/N$ (alternatively, $2/N$) has been introduced to ensure regularity when going to the limit of a large system (thermodynamic limit $N \to \infty$). This ansatz is often referred to – incorrectly – as the Hebb rule; due to its lineage it is more properly called the Anderson-Cooper-Hopfield (ACH) rule (cf. Anderson 1970; Cooper 1973). To avoid the ambiguity noted in Sect. 2.3, the self-couplings v_{ii} are customarily set to zero. It is also customary to assume normal thresholds, i.e. $V_{io} = \sum_j v_{ij}$ or equivalently $h_{io} = 0$. (There is an alternative formulation (Hopfield 1982, Bruce et al. 1987) based on the S_i variables, in which zero thresholds V_{io} are assumed. The distinction is basically the same as that discussed by Qian et al. (1991) in connection with the optimization of learning rules.) In general, a fully connected architecture is implied, and there is no separation of neurons into definite excitatory and inhibitory subsets.

From inspection of the ansatz (23) for the connectivity matrix, we see that the memories μ are superposed and distributed. The weight $v_{ij} = v_{ji}$ of each reciprocal synapse ij is determined by a superposition of information from *all* of these memories, and information about a given memory is dispersed over all of these synapses. A natural question is: how many memories can we pack into the neural system before the superposition becomes too dense and recall becomes confused by cross-talk? In other words, what is the storage capacity of the Hopfield associative memory? Much theoretical and computational labor has been devoted to this issue. The remainder of this subsection outlines different approaches to an answer, which together reveal the emergence of collective computational properties with growing network size.

Since the neuronal interactions are symmetric with all $v_{ii} = 0$, Hopfield's energy theorem is in force and the only attractors for the dynamics are fixed points satisfying $\sigma_i F_i \geq 0$, for all i, at which points the energy E of Eq. (5) reaches local minima. To each such minimum there belongs a basin of attraction, consisting of the set of all states from which the system flows to that attractor under the dynamics of the model. If the network is to function properly as a content-addressable memory, its fixed points should closely resemble the nominated patterns $\sigma^{(\mu)}$, thus allowing recall of these memories with relatively few errors. The following arguments make it plausible that the desired closeness is attainable. For the time being, attention is restricted to the case of zero temperature, i.e. we do not yet modify the Hopfield asynchronous threshold dynamics by a probabilistic firing decision with noise temperature β^{-1}.

Let $\sigma = \{\sigma_i\}$ be an arbitrary point in the state space of the system. A suitable measure of the "closeness" of this point to the nominated patterns μ is provided by the overlap function (or order parameter)

$$m_\mu(\sigma) = \frac{1}{N} \sum_{i=1}^{N} \sigma_i \sigma_i^{(\mu)} , \qquad (24)$$

which is recognized as a normalized scalar product of the vectors σ and $\sigma^{(\mu)}$. Substituting the learning rule (23) into the energy function (5), we find

$$E(\sigma) = -\frac{N}{2} \sum_{\mu=1}^{n} [m_\mu(\sigma)]^2 + \frac{1}{2}n . \qquad (25)$$

Now imagine (as is frequently supposed) that the firing patterns to be stored are *uncorrelated* or *random*, meaning that in any given pattern μ each neuron has an equal chance of being "on" ($\sigma_i^{(\mu)} = +1$) or "off" ($\sigma_i^{(\mu)} = -1$), and take N to be large. Consider two possibilities for the "input" σ (Domany 1988):

(a) A typical state, namely a random firing pattern, but not one of the nominated memory states. The terms in the overlap then perform a random walk with the expected (large-N) result $m_\mu(\sigma) = \pm 1/\sqrt{N}$, for all μ. In this case, $E(\sigma) \simeq 0$.

(b) One of the chosen memory states, say $\sigma^{(\kappa)}$. There will be a single large overlap, $m_\kappa(\sigma) = 1$, but the $m_\mu(\sigma)$ with labels $\mu \neq \kappa$ are again of expected value $\pm 1/\sqrt{N}$. Thus we estimate that $E(\sigma) \simeq -N/2$.

The result approached in case (b) is much lower than the (nearly zero) energies of typical states obtained in case (a). If the nominated memory states happen to be mutually orthogonal instead of uncorrelated, we have $E = -(N - n)/2$ *exactly* in case (b) without regard to the sizes of N and n. Generally these results indicate that the memory states nominally stored by the ACH algorithm may usefully approximate stable states – local energy minima – of the Hopfield network, provided n is sufficiently small compared to N.

To continue in a similar vein (Keeler 1986), let us feed the network one of the nominated memories, $\sigma^{(\kappa)}$, and see (i) whether it qualifies as a *fixed point* of the network dynamics and (ii) if so, whether it qualifies as an *attracting* fixed point. The fixed point will be an attractor of range d (where $d \geq 1$) if, under a perturbation altering the state from $\sigma^{(\kappa)}$ in any d single-neuron firing bits, the dynamics of the system acts to drive the state back to $\sigma^{(\kappa)}$. The relevant firing function takes the explicit form

$$F_i(\sigma^{(\kappa)}) = \frac{1}{N} \sum_{\mu=1}^{n} \sum_{j=1}^{N} \sigma_i^{(\mu)} \sigma_j^{(\mu)} \sigma_j^{(\kappa)} = \sigma_i^{(\kappa)} + \frac{1}{N} \sum_{\mu \neq \kappa} \sum_{j=1}^{N} \sigma_i^{(\mu)} \sigma_j^{(\mu)} \sigma_j^{(\kappa)} . \quad (26)$$

(The restriction of (23) to $j \neq i$ has been released, for simplicity, since we are primarily interested in large N.) The first term on the extreme right of Eq. (26) is the "signal" and the last term collects the "noise" arising from the presence of the other stored patterns. If the noise can be neglected, it is obvious that $\sigma^{(\kappa)}$ is a fixed point, since F_i and $\sigma_i^{(\kappa)}$ have the same sign, for all i. (We note that $\sigma^{(\kappa)}$ is in fact an eigenvector of the coupling matrix (v_{ij}) with eigenvalue unity.) Now, let the initial state be displaced from $\sigma^{(\kappa)}$ by d bits (change d of the $\sigma_i^{(\kappa)}$), and consider the generic F_i for this perturbed state. The signal term is then given by $(N - 2d)\sigma_i^{(\kappa)}$. The sign of F_i remains the same, and hence the state $\sigma^{(\kappa)}$ qualifies as an attracting fixed point, provided only that $N > 2d$.

If mutually orthogonal patterns are taken for the nominated memories, i.e., $\sigma^{(\mu)} \cdot \sigma^{(\nu)} = \delta_{\mu\nu}$, all $\mu, \nu = 1, \ldots n$, then the noise term is exactly zero. But suppose the nominated memories are chosen randomly, in the sense defined earlier. Then the noise term evidently has zero mean, and, again thinking in terms of a random walk, its rms value is $[(n - 1)/N]^{1/2}$. Thus, the signal-to-noise ratio will be large as long as the number n of nominally stored patterns is much less than the number of neurons N. From this kind of analysis, we can see why the system works as a content-addressable memory, and why it is robust against damage (Denker 1986). Suppose $N = 10^4$ and $n = 10^2$. If the initial state σ matches one of the nominated memory states $\sigma^{(\kappa)}$ in just *half* of its bits (half the individual neuronal states), the signal term is cut in half and the signal-to-noise ratio is only reduced from 10 to 5. And if *half* the synapses are cut out, the signal term is diminished by a further factor 2, but the residual signal-to-noise ratio remains relatively high at 2.5, high enough that further iteration through the nonlinear threshold condition usually converges rapidly to the relevant fixed point.

From the above considerations based on energetics, stability against perturbation, and signal-to-noise ratio, we expect that recall will be reliable, with few errors, as long as one does not attempt to store too many memory patterns. Computer experiments show that if one does try to build in too many memories, the corresponding basins of attraction will begin to distort one another, and *spurious memory states*, hybrids of the nominated patterns, will be created (Hopfield, 1982). It may also be noted that if a particular state $\sigma^{(\kappa)}$ is stored using the rule (23), then so is the mirror-image state $-\sigma^{(\kappa)}$ with the signs of all σ_i variables reversed. This feature follows from the "up-down" or "particle-hole" symmetry of the model, expressed in the invariance $E(\sigma) = E(-\sigma)$.

The limits of useful memory storage based on the ACH ansatz (23) have been elucidated in the elegant theoretical papers by Amit, Gutfreund, and Sompolinsky (1985a,b,1987a,b). This work exploits the analogy between the Hopfield model and an infinite-range spin glass (see also van Hemmen (1986) and Bruce et al. (1987)). A spin glass is a type of magnetic material in which mixed ferromagnetic and antiferromagnetic bonds compete for the attention of individual spins, giving rise to frustration and hence to the possibility of a large number of equilibrium states. Appealing to existing theories of spin glasses, particularly the so-called SK model (Kirkpatrick and Sherrington 1978), it is found that the equilibrium statistical mechanics of the Hopfield (or Little) model with symmetrical ACH couplings is *exactly soluble* in the thermodynamic limit. The key point is that mean-field theory becomes exact by virtue of the infinite range of the interactions entering the Hamiltonian of the Hopfield model.

The analyses of Amit et al. reveals some intriguing features, and indeed some surprises (Sompolinsky 1988). In the case where the number n of nominated firing patterns is kept finite as N goes to infinity (Amit et al. 1985a), there will be $2n$ degenerate ground states, which correspond exactly to the proposed memories and their mirror images. Even at rather small n there are, however, additional local minima of E corresponding to the spurious memories encountered in the computer simulations. These spurious memories are characterized by overlaps m_μ of order unity with more than one nominated pattern μ. They proliferate very rapidly with N as one attempts to pack in more memories. This behavior is in accord with independent arguments of Weisbuch and Fogelman-Soulié (1985) to the effect that the number of firing patterns which can be stored *perfectly* by (23) (meaning that they remain exact fixed points of the dynamics in the presence of the noise term) scales like $N/(2\ln N)$ for large N.

What happens if one tries to store a number of patterns $O(N)$, and N is large? The analysis of Amit et al. (1985b,1987a), based on replica-symmetry theory and other spin-glass techniques, shows that in spite of the instability of the nominated patterns in this case, the system can still function acceptably as a content-addressable memory, provided the load parameter $\alpha = n/N$ remains below a critical capacity $\alpha_c \simeq 0.138$. As the nominated memories are destabilized, new local minima appear in the energy landscape, corresponding to "surrogate" patterns that differ from the desired ones in relatively few bits. For $\alpha < \alpha_1 \simeq 0.052$, these new states are stable. At α_1 there is a first-order phase transition, and in the region $\alpha_1 < \alpha < \alpha_c$ the surrogate states are only metastable with respect to the spurious memory states, although they may be protected by high energy barriers. In the nearly saturated regime corresponding to finite load n/N, the spurious states no longer have significant overlaps with the nominated patterns and resemble spin-glass states. As α_c is approached, there is an abrupt and total loss of memory. We may identify the load domain $0 < \alpha < \alpha_1$ with a "ferromagnetic" (F) phase, the range $\alpha_1 < \alpha < \alpha_c$ with a mixed "ferromagnetic + spin-glass" (F+SG) phase, and the range $\alpha > \alpha_c$ with a pure "spin-glass" (SG) phase.

If the nearly saturated system is elevated to a finite temperature, the attendant thermal noise will not destroy the recall process (thanks to the energy barriers around the surrogate states) until one reaches a critical temperature which decreases monotonically with the load α. However, recall errors are increased, and the critical parameters α_1 and α_c of the two "phase boundaries" between F and F+SG and between F+SG and SG phases are decreased in magnitude. Both α_1 and α_c shrink to zero as the temperature is increased to some limiting value beyond which the behavior of the neuronic system is totally confused by thermal agitation ("ergodic"). In the magnetic analogy, the latter condition corresponds to a paramagnetic phase. These various features of the phase diagram of the Hopfield content-addressable memory network (see Fig. 2) have been documented in computer simulations.

Increased memory-storage capacity and/or facility with correlated (i.e., non-random) memory patterns can be achieved with "more advanced" learning rules. The *nonlocal* pseudo-inverse rule (Kohonen 1989, Personnaz et al. 1986, Hertz et al. 1991) stores N linearly independent firing patterns, while $O(N^2/(\ln N)^2)$ highly correlated, sparsely-coded patterns can be stored using an algorithm that was introduced three decades ago by Willshaw et al. (1969). In a brilliant contribution, Elizabeth Gardner (1988) carried the use of attractor dynamics in neural-network theory to a higher plane of abstraction: she treated the synaptic couplings v_{ij} as dynamical variables governed by a relaxational dynamics based on a suitably constructed energy function. A given computational problem, i.e. a given information-processing task, is to be solved by finding an appropriate set of synaptic couplings v_{ij}, which are subject to specific constraints imposed by the nature of the problem. Gardner's energy function serves as a (nonnegative) cost function that measures the violation of these constraints by a chosen set of couplings. The solution of the problem is accordingly reduced to a search for the (zero) minimum, or minima, of the Gardner energy, commonly using gradient-descent techniques.

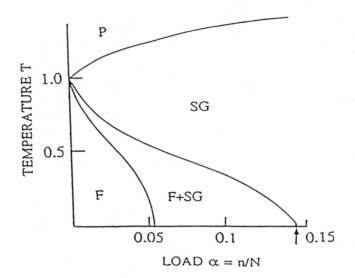

Fig. 2. Equilibrium phase diagram for the Hopfield discrete-time model, in the temperature (T) – load (α) plane (without replica symmetry breaking). The load is defined by $\alpha = n/N$, where N is the number of neurons and n the number of nominally stored memories. The different phases are classed as ferromagnetic (F), spin-glass (SG), mixed (F + SG), and paramagnetic (P). Retrieval phases are F and F + SG.

Two simple implementations of this idea for the associative-memory problem yield results that are essentially equivalent to the Perceptron and Adaline (or Widrow-Hoff) learning rules (Rosenblatt 1962, Widrow and Hoff 1960, Widrow 1962). For random patterns, the theoretical maximum (Cover 1965) of $2N$ stored memories can be attained by this procedure, and larger capacities are possible for correlated memories.

Extensive references to this fascinating area of theoretical research may be found in the review by Clark (1991), the books by Müller and Reinhardt (1990), Hertz et al. (1991), and Peretto (1992), and the issue of Journal of Physics A dedicated to the memory of Elizabeth Gardner (Sherrington, 1989).

4.2 Attractor-Net Solution of Optimization Problems

Hopfield (1984) has demonstrated that the Cowan-Hopfield continuous model (10)–(11) possesses content-addressable memory properties comparable to those of the discrete model studied in Sect. 4.1. It also has other uses: it can be adapted to attack the graph-bisection problem, the traveling-salesman problem, routing problems, scheduling problems, track-reconstruction in experimental high-energy physics, and other important problems involving combinatorial optimization subject to constraints. With symmetrical interactions T_{ij}, the dynamics necessarily drives the energy function E of the model, generally of form (13), toward some local minimum. Accordingly, the strategy is to arrange matters so that the minima of this function correspond to acceptable solutions of the given problem, with deeper minima corresponding to better solutions. The couplings T_{ij} should thus be chosen to incorporate the constraints inherent in the problem, so that E plays the role of a suitable "cost function."

A corresponding strategy can be implemented based on a stochastic version of the *discrete* Hopfield model introduced in Sect. 2.3 (Hertz et al. 1991, Peretto 1992). We shall later return to this possibility.

The Hopfield-Tank Algorithm for the Traveling-Salesman Problem. The most famous combinatorial optimization task is the traveling-salesman problem, in which M cities are to be visited once each in a continuous tour, such that the total distance traveled is a minimum. In the approach of Hopfield and Tank (1985,1986) based on the Cowan-Hopfield model, M^2 neurons are placed in an M-by-M array: each row corresponds to a particular city and the column number indicates the order in a tour. The solution arrived at by the network may be read off from the firing pattern of the neurons in this array, after the system has relaxed under the dynamics prescribed by Eqs. (10)–(11). The cost function constructed by Hopfield and Tank, of the form prescribed by (13) in the high-gain limit, is characterized by:

o Bilinear terms in the neuronal outputs a_i, weighted by the distances between the corresponding cities so as to penalize longer hops;

o Inhibitory components in the T_{ij}, which suppress tours that violate the syntax of the problem by appearance of any city more than once, or by appearance of two cities in the same position;

o A term that incurs a syntactic cost unless exactly M neurons are strongly activated, i.e., unless all cities are visited and all positions occur in a tour.

Three adjustable parameters were introduced to characterize the strengths of the three constraints.

Although the Hopfield-Tank procedure does not generally lead to the *best* solution, for M not too large it is possible to obtain good solutions in a remarkably short time – provided the constraint parameters appearing in the cost function are carefully tuned. Hertz et al. (1991) offer a more robust formulation (involving only one independent parameter), but the tours generated are not quite as good. Despite further improvements on the Hopfield-Tank approach (e.g. Peterson and Söderberg 1989), the best conventional methods for solving the traveling-salesman problem (e.g. Lin and Kernighan 1973) still seem to be superior.

"Hard" Optimization Problems; Stochastic Search and Simulated Annealing.
As in the foregoing example, many problems encountered in science and
technology can be cast as optimization problems in which some cost function $C(X)$ is to be minimized in the space of configurations X, subject to
certain constraints (Bounds, 1987). These problems are considered "hard" if
the number of computational steps s required for their solution grows rapidly
with the size M of the problem, faster than a polynomial in M. There is a
class NP of problems (NP standing for "non-deterministic polynomial") with
the property that the validity of a proposed solution can be tested in polynomial time. Problems of polynomial complexity evidently belong to this class,
but there is also a subclass of much nastier problems, called NP-complete.
For these problems, there is no known algorithm whose worst-case complexity
is bounded by a polynomial in M. The NP-complete problems are convertible to one another by an algorithm requiring at most $s' \sim M^p$ steps, with
p some integer. The implication is that if a deterministic algorithm can be
found that solves one such problem in polynomial time, they *all* can be solved
in polynomial time. However, it is generally believed that the time required
to solve such problems scales *exponentially* with the size M. The traveling-salesman problem and most spin-glass models are NP-complete. Indeed, the
traveling-salesman example is the archetype of a very important subclass of
NP-complete problems with discrete variables – combinatorial optimization
problems – in which the "best" order must be determined for a finite number
M of objects. The number of possible arrangements evidently grows like the
combinatorial factor $M!$.

The cost function involved in an optimization problem may have many local minima, corresponding in general to approximate solutions, and one seeks
the deepest of these, the global minimum or "ground state." Straightforward
search algorithms such as gradient descent (see Sect. 5.3) have the drawback
that they can get stuck in local minima. An intuitively appealing remedy is
provided by the idea of *simulated annealing* (Kirkpatrick et al. 1983). Following the simplest Monte Carlo search algorithm (Metropolis et al. 1953), a
random walk with small step size can be made in the state space $\{X\}$, a given
move being accepted if and only if the cost function is lowered. Evidently this
technique, like a gradient-descent search, will lead to a local minimum, since
all moves are "downhill." To avoid getting trapped in such a minimum, one
may introduce some degree of (simulated!) thermal agitation, measured by
a temperature parameter $T = \beta^{-1}$: if a trial move increases the cost function $C(X)$ by $\Delta C(X) > 0$, it is accepted with probability $\exp[-\beta\Delta C(X)]$.
As before, if the cost function is decreased by the move, it is accepted with
probability unity. Although there is a bias toward low-cost states, there now a
chance of jumping to a higher value of the cost function and thus ascending a
barrier and moving on to another, perhaps lower, minimum. In simulated annealing, the temperature is changed with time according to some annealing
schedule. At high temperatures, equilibrium is attained rapidly, the lower-

cost states being only slightly favored; at low temperatures, the bias toward lower-cost states is strong, but the system takes a long time to settle down. Generally, the most efficient strategy is to start from a high temperature and then slowly reduce it in stages. At some stage during the process, one should reach the best compromise between the absolute rates of transitions and the ratio of cost-lowering and cost-raising moves (Hinton and Sejnowski 1986). If all goes well, the system will first find a coarse-grained minimum on the cost surface agitated by the thermal noise, and thereafter refine its choice in a stepwise fashion, possibly discovering the absolute minimum – or at least a good local minimum – within a reasonable budget of computer time.

For the content-addressable memory task studied in Sec. 4.1, one actually *wants* the system to get stuck in a local minimum of the energy function, namely that minimum corresponding to the memory being suggested by the stimulus.

Mean-Field Optimization. Returning from these general considerations to optimization by neural networks, let us now take up the other option mentioned above, and use the original Hopfield model, with binary neurons updated asynchronously in random sequence. However, in the spirit (but not the letter) of the Monte Carlo procedure described above, we introduce a stochastic updating patterned after that of the Little model (Eq. (3)). The probability that neuron i, when updated, assumes state $\sigma_i = \pm 1$ is given by

$$\rho(\sigma_i = \pm 1) = g_L(\pm h_i) = \left[1 + \exp(\mp 2\beta h_i)\right]^{-1} \tag{27}$$

where $h_i = \sum_j v_{ij}\sigma_j + h_{io} = F_i/2$ may be extracted from the energy function E of Eq. (5). It is easily checked that changes of the overall configuration σ of the network that decrease the energy E are more probable than those that increase it. In the low-temperature limit where $\beta = 1/T$ goes to infinity, the original hard-threshold discrete Hopfield dynamics is regained and energy changes cannot be positive. However, escape from local minima is possible at finite temperatures. A simulated-annealing schedule can be adopted for adjustment of the temperature parameter $1/\beta$ as the system relaxes under the stochastic algorithm.

It has been pointed out by Peterson and Anderson (1987,1988) and others (see Hertz et al. (1991)) that for a discrete Hopfield system with many degrees of freedom σ_i and high connectivity (most $v_{ij} \neq 0$), it is not actually necessary to play out the computationally expensive stochastic "hill-climbing" procedure sketched above. Under these conditions, it can be good approximation to apply mean-field theory, deriving a set of deterministic nonlinear algebraic equations

$$\langle \sigma_i \rangle = \tanh\left(\beta \langle h_i \rangle\right) = \tanh\left(\beta \left[\sum_j v_{ij}\langle \sigma_j \rangle + h_{io}\right]\right) \tag{28}$$

for the average values of the variables σ_i in thermal equilibrium or the steady state. The corresponding equations for the average values $\langle S_i \rangle$ of the binary variables $S_i = (\sigma_i + 1)/2$ are of exactly the same form as one obtains for the steady-state values of the firing rate variables a_i in the continuous Hopfield (or Cowan-Hopfield) model with logistic squashing function. (In establishing this connection, v_{ij} can be identified with $R_i T_{ij}$, apart from a constant that can be absorbed in the parameter β.) Moreover, the dynamical equation (10) of the Cowan-Hopfield model may be used to approach the steady state. In this sense, the Hopfield-Tank procedure is equivalent to a mean-field treatment of the relaxation of the stochastic discrete model. We observe that some benefits of stochasticity in escaping local minima may be lost, but an annealing process can still be carried out by manipulation of the parameter β. Further information on mean-field optimization is provided in Chapter 8.

Any discussion of the efficacy of neural-net approaches to NP-complete problems should be accompanied by a word of caution (Müller and Reinhardt 1990). Under plausible assumptions, it has been demonstrated by Bruck and Goodman (1988) that there can exist no network of polynomial size in M capable of solving the traveling-salesman problem to a preselected accuracy. Even so, Peterson and Söderberg (1989) have proposed a promising approach utilizing a network built from neurons modeled after Potts spins (see Chapter 8). Potts spins can have any integral number of states, in contrast to the two possibilities ("up" and "down") for Ising spins. The use of M-state Potts neurons thus allows each city of the traveling-salesman problem to be represented by a single neuron, whereas M Ising neurons are needed. The dimensionality of the configuration space is reduced from 2^{M^2} to M^M, a considerable economy.

Track Reconstruction by Neural Networks. Neural networks were first introduced into high-energy physics by Denby (1988) and Peterson (1989), who proposed that connectionist systems might provide efficient algorithms for track reconstruction. Their proposals were motivated by the prodigious event rates and very high multiplicities that will be seen in the next generation of high-energy particle accelerators, as represented by the Large Hadronic Collider at CERN. The hope was that rapid and reliable on-line track-finding could be achieved in practice, by implementing the neural-network parallel processing in hardware using VLSI or optical technologies. It will be instructive to outline the original idea in some detail, following the presentation given by Peterson, the formulation of Denby being very similar. (See also Denby and Linn (1990) and Stimpfl-Abele and Garrido (1991).)

Restricting the treatment to two dimensions for the sake of simplicity, the problem is to join M space points (where detector signals are located) with a number of continuous, smooth tracks. The tracks must not bifurcate. No more than one directed line segment can enter or leave a given point, and no point can belong to more than one track. Exactly $M(M-1)$ directed line segments $i \to j$ may be drawn between the M signals $i, j = 1, \ldots, M$. With each directed segment $i \to j$ we associate a binary neuron, labeled ij. The state variable S_{ij} of neuron ij takes the value 1 if the segment $i \to j$ is part of a track, and the value 0 if it is not. The configuration of the neural network is then a posed solution of the track construction problem.

The next step in setting up a neural-network optimization algorithm for this problem is to choose a suitable energy function (or cost function). The two primary considerations are the lengths of the segments and the angles between adjacent segments. The angle θ_{ijl} between segments $i \to j$ and $j \to l$ is defined by $\cos\theta_{ijl} = \mathbf{r}_{ij} \cdot \mathbf{r}_{jl}$, where $\mathbf{r}_{ij} = \mathbf{r}_j - \mathbf{r}_i$ and $\mathbf{r}_{jl} = \mathbf{r}_l - \mathbf{r}_j$. The energy is constructed as a quadratic form in the variables S_{ij}, as is consistent with the standard energy expression (5) in terms of the σ_i (or σ_{ij}) variables. The bilinear term is of the form $-\frac{1}{2}\sum_{ijkl} v_{ijkl} S_{ij} S_{kl}$. We note that since the neurons have two site labels, the synaptic weights must have four. The weights v_{ijkl} have two parts, one accounting for "costs" and the other for "constraints." The cost portion, taken as

$$V_{ijkl}^{(\text{cost})} = \delta_{jk}\frac{\cos^m\theta_{ijl}}{r_{ij} + r_{jl}}, \tag{29}$$

where m is an odd exponent, is designed to discourage long segments and large relative angles. In particular, angles θ exceeding $\pi/2$ incur a strong penalty.

The constraint portion, $V_{ijkl}^{(\text{constraint})}$, has two purposes and thus again divides into two terms, $V_{ijkl}^{(1)}$ and $V_{ijkl}^{(2)}$. The first term has the task of suppressing the occurrence of tracks that bifurcate. In neuronal language, this means that two S variables with one overlapping index should not be activated in the same network configuration. Simultaneous activation of such variables is suppressed by choosing

$$V_{ijkl}^{(1)} = -\frac{1}{2}\alpha_1\left[\delta_{ik}(1 - \delta_{jl}) + \delta_{jl}(1 - \delta_{ik})\right], \tag{30}$$

where α_1 is a Lagrange parameter. The purpose of the second term, $V_{ijkl}^{(2)}$, is to prevent the number of active neurons from deviating greatly from the number of signals M. As seen below in the final form (31) for the energy, it accomplishes this through a global inhibition that increases with the magnitude of $\sum_{kl} S_{kl} - M$.

We arrive at the expression

$$E = -\frac{1}{2}\sum_{ijkl}\delta_{jk}\frac{\cos^m\theta_{ijl}}{r_{ij} + r_{jl}}S_{ij}S_{kl} + \frac{1}{2}\alpha_1\sum_{l\neq j}S_{ij}S_{il} +$$
$$+ \frac{1}{2}\alpha_1\sum_{k\neq i}S_{ij}S_{kj} + \frac{1}{2}\alpha_2\left(\sum_{kl}S_{kl} - M\right)^2 \tag{31}$$

for the total energy, where α_2 is a second Lagrange parameter, associated with the global constraint. The final step in establishing the algorithm is to derive mean-field equations for the average values of the S_{ij} variables, obtaining

$$\langle S_{ij} \rangle = \tanh \left\{ \beta \left[\sum_l \frac{\cos^m \theta_{ijl}}{r_{ij} + r_{jl}} - \alpha_1 \left(\sum_{l \neq j} \langle S_{il} \rangle + \sum_{k \neq i} \langle S_{kj} \rangle \right) - \alpha_2 \left(\sum_{k \neq l} S_{kl} - M \right) \right] \right\}.$$

(32)

The problem may be simplified by appealing to the fact that two distant signals have low likelihood of being joined by a single segment. Thus, attention may be restricted to "legal segments," defined to have lengths below some appropriate cutoff distance R_c. The dimension of the system (32) is correspondingly reduced.

Peterson (1989) solved the reduced set of mean-field equations iteratively based on the parameter choices $\beta = 1$, $\alpha_1 = \alpha_2 = 1$, and $m = 7$. Good convergence, reliable track construction, and parametric robustness were found for test cases involving small numbers of signals and tracks. Similar tests, with generally similar results, have been described by Denby (1988). In some instances, spurious segments appear in the final solutions, violating the edict against bifurcations. However, the offending segments may be removed by a "greedy heuristic procedure," invoking the criterion of minimal increase of the cost component of the energy (Peterson and Anderson 1988).

While widely recognized as a very innovative step, the Denby-Peterson algorithm has since been superseded by more powerful algorithms for inferring a nonintersecting set of curves from a complex set of points. As indicated in Chapter 8, track reconstruction is one of a large class of optimization problems that involve both discrete combinatoric assignment and parametric fitting. Such *parametric assignment problems* can be attacked by a hybrid approach based on deformable templates. The approaches designed by Gyulassy and Harlander (1991) and Ohlsson et al. (1992) are representative of elastic tracking algorithms that are robust with respect to noise and measurement error and extend tracking capabilities to high track densities. Gyulassy and Harlander discuss a possible neural-network implementation of their algorithm.

This section on computational applications of recurrent attractor networks should not be concluded without mentioning an important body of theoretical work in protein chemistry that exploits techniques and insights gained from spin-glass theory and related neural-network models (Stein 1985, Bryngelson and Wolynes 1987, Friedrichs and Wolynes 1989,1990, Bryngelson et al. 1990, Sasai and Wolynes 1990, Bohr and Wolynes 1992, Bohr et al. 1993b, Bohr 1998). A working description of the protein folding process – one which permits computational determination of native 3D structure from sequence information – must infer a reliable free-energy functional and somehow deal with the fact that this functional will have many local minima. On the one hand, the problem of pattern recognition involved in folding to the preferred final configuration is closely analogous to that solved by Hopfield associative memories; and, on the other, the problem of avoiding local minima during the dynamical process by which folding is simulated is closely analogous to that solved by stochastic optimization based on Hopfield-type neural networks. These analogies become most apparent when binary-valued residue contact variables are used to specify the 3D structure (Bohr and Wolynes 1992, Bohr 1998).

5   Multilayer Perceptrons: Backpropagation Learning

Multilayer feedforward neural networks made up of neuron-like units with analog response are able to learn by example and solve complex input-output mapping problems when trained with a suitable algorithm for stepwise adjustment of connection weights. Most applications of neural networks are of this type. The ubiquity and success of multilayer perceptrons may be attributed on the one hand to the versatility of the feedforward architecture, and on the other to the availability of a training procedure – "backpropagation" – that is very easy to implement on the current generation of workstations and PCs. This section provides a review of the general principles of supervised learning (more specifically, learning with a teacher) and includes a derivation of the backpropagation algorithm. Ingenious applications of backpropagation and related supervised learning procedures now abound in engineering, business, medicine, and other endeavors. In this section, focused exemplifications will be confined to cases of direct *scientific* interest.

5.1   Supervised Learning by Example

As shown in Fig. 3, we consider layered feedforward networks containing an input layer, an output layer, and one or more intermediate or "hidden" layers. The intermediate layers of neurons are called hidden because they have no direct communications with the outside world. For input, hidden, and output neurons, we adopt the standard labels k, j, and i, respectively. Generic neurons are labeled m, m'. Every neuron extends a forward connection to each neuron in the next layer, but there are no lateral or backward-going connections in the system. (In general, forward connections could skip layers, but that is not considered here.)

The input and output patterns are represented by the collective activities $\{a_k\}$ and $\{a_i\}$ of the neurons in the input and output layers, respectively. In some problems it is advantageous to employ binary (i.e., McCulloch-Pitts) units to encode the incoming data in the input layer. While granting the special status of input units, the state of a *generic* unit of the network will be characterized by an analog activity variable a_m that ranges between 0 and 1. As in the continuous Cowan-Hopfield model, the activity of neuron m is taken to be a smooth, nonlinear function of its stimulus. In turn, the stimulus felt by neuron m is a linear superposition $\sum_{m'} V_{mm'} a_{m'}$ of the signals it receives from other neurons m' sending connections to it, where $V_{mm'}$ is the weight of the connection from m' to m. At this point it is convenient to change the sign convention for the neuronal threshold, working instead with a *bias* parameter: henceforth, V_{mo} will denote *minus* the threshold of neuron m. The bias of m may be treated as just another weight, through the introduction of an auxiliary unit o that is always maximally active ($a_o = 1$) and extends a connection to m having the bias V_{mo} as its weight (see Fig. 3). The stimulus to neuron m, *in excess of threshold*, is then given by

$$F_m = \sum_m V_{mm'} a_{m'} \tag{33}$$

with $m = o$ included in the sum.

The differentiable nonlinear function $g(F_m)$ that transforms ("squashes") the stimulus F_m into an analog response $a_m \in [0, 1]$ is assumed to be monotonic (thus a sigmoid). (Note: in some formulations, the range of analog activity is chosen as $[-1, 1]$ instead of $[0, 1]$.) Neurons in the input layer serve only as data registers: their activities a_k are set externally and hence their response characteristics are irrelevant. In many problems, one may just as well take the output units to be linear.

When a given activity pattern is imposed on the input layer, the system computes an output according to the following dynamical rules:

(i) All units within a layer update their states in parallel.
(ii) Successive layers are updated sequentially, starting at the input layer and proceeding forward until the output layer is reached.

The artificial neural system we have described may be taught by example, using a supervised learning routine designed to reduce the errors made by the network when it is repeatedly exposed to examples from a set of training patterns (see Fig. 4). Generally, the input patterns of the training set are presented to the system in random order, errors in the responses of the output neurons are observed, and the weights (including biases) $V_{mm'}$ that parametrize the "education" of the network are adjusted step by step so as to decrease the value of a positive semi-definite *cost function* or *objective function* $C(\{V_{mm'}\})$. The cost function is chosen to provide a useful overall measure of learning error, and vanishes for perfect response to the training patterns. Various techniques are available for minimizing the function $C(\{V_{mm'}\})$ in weight space. One is just gradient-descent minimization; as we shall see, this is the basis of the standard backpropagation algorithm (Rumelhart et al 1986b,c, Le Cun 1985, Parker 1986). Another procedure (widely regarded as superior) is the conjugate-gradient method (Luenberger 1984). Both gradient-descent and conjugate-gradient methods have the drawback that they can get stuck in local minima.

This scenario depicted in Fig. 4 is sometimes called "learning with a teacher," since the system gets a "grade" that tells it how badly it is doing. (In reinforcement learning –"learning with a critic"– it would only be told that its response is right or wrong (Hertz et al. 1991).) Typically, some hundreds, even thousands of passes through the training set are needed before

Three Layer Feedforward Network

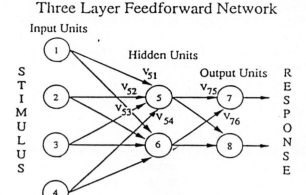

Generic Neuron

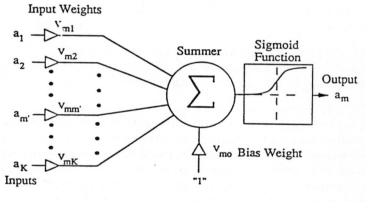

$$a_m = g(\sum_{m'=o}^{K} V_{mm'} a_{m'}) \quad \text{where} \quad g(y) = \frac{1}{1 + e^{-y}}$$

Fig. 3. Three-layer feedforward neural network and analog neuron.

the cost function is effectively reduced to its asymptotic value. For real-world applications, this value will rarely be zero, i.e. perfect learning is not to be expected. If, as is customary, the initial values of the weights are chosen by random sampling from a delimited uniform distribution, and the order of presentation of the training patterns is random, different implementations of the learning procedure will yield different trained networks, which amount to different *models* of the training data. Obviously, the real point of the whole

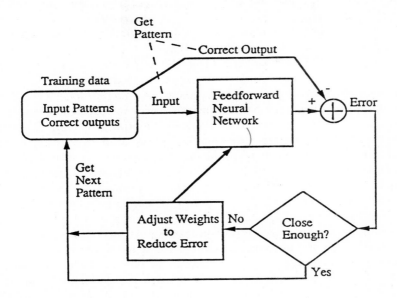

Fig. 4. Generic training scheme for a feedforward network.

exercise lies in the expected *generalization ability* of such networks. Under favorable circumstances, they may be able to capture the essential aspects of the relevant input-output map and generate correct responses for novel input patterns not contained in the training set.

Any mapping I → O can be modeled by this approach, provided only that the exemplars I^μ of I and the corresponding instances O^μ of O can be expressed, respectively, in terms of the activities a_k and a_i of sets of input neurons k and output neurons i. However, most applications fit into one of two basic problem types. In the first – the standard *classification problem* – the output pattern serves to assign a given input pattern to one of a discrete number of mutually exclusive classes. (In Sect. 7, this is called a "1 of M" problem.) It will be useful to distinguish two subtypes of this kind of problem, namely (i) *detection problems*, in which the network is asked to indicate the presence or absence of a *single* attribute and (ii) *sorting problems*, in which the net is asked to assign the input pattern to one of *several* categories or "pigeonholes." In either case, the nature of the input pattern can be quite general: it can represent the presence or absence of each member of a certain set of qualitative attributes and thus have Boolean character; or it can consist of a set of integers and/or a set of real numbers from continua, which might represent the presence of certain attributes to various degrees or simply the values of real input variables. We may append the trivial observation that a sorting problem with only two classes can be treated as a detection problem in which membership in the second class is inferred from absence of the attribute determining membership in the first class. A more general kind of classification problem than embraced by (i) and (ii) leaves open the possibility that an input pattern may belong to *more* than one class. This case surfaces briefly in Sect. 7, but is otherwise not considered in these notes.

In the second broad problem type, known as *function approximation* or (commonly) *regression*, the output pattern serves to represent a set of dependent variables that may take values from a continuum or from a discrete spectrum. The input pattern may represent input variables of the same general nature as for classification problems. Admittedly, the second problem type can be construed to include the first, but the intended distinctions may be important to rational algorithmic design.

Indeed, successful modeling of a given problem domain by a feedforward neural net is often crucially dependent on proper input and output coding schemes and appropriate pre- or post-processing. At this point we will simply list a few of the more obvious possibilities for representing input data and targeted output response in terms of neuronal activities. The first two specifications will be specialized to input patterns; with straightforward rephrasing they will apply to *targeted* output patterns. (It must be kept in mind that the output neurons are necessarily of analog type in our development.)

- o *Unary, or "grandmother" coding.* Each neuron in the input layer stands for a particular feature that may occur in the input patterns, being clamped fully "on" or fully "off," depending on whether that feature is present or absent. For example, the input patterns may be five-letter English words, which we may encode by 5 groups of 26 neurons, with one dedicated neuron for each possible letter in each position within such a word. This is a highly localized form of coding. The name "grandmother coding" comes from the idea in brain theory (now almost universally discounted) that there might be a neuron in your brain which turns on when and only when you see your grandmother – the ultimate in local memory storage.

- o *Binary coding* as a string of bits. The members of a row of input units are individually clamped fully "on" or fully "off" so as to provide a binary representation of exemplars of a discrete set of input patterns. Specifically, the patterns in question might correspond simply to integers, or they might be two-dimensional images composed of light and dark pixels. This type of coding is more general and generally more distributed than unary coding, in that the presence of a given feature (e.g. the integer 17) may activate more than one neuron.

- o *Analog coding* of one or more continuous input or output variables. The value of each such variable is represented by the activity of a single dedicated analog input or output neuron. Consider the case of one input variable $X \in [X_1, X_2]$ and one output variable $Y \in [Y_1, Y_2]$. An input value for X may be encoded in the activity of the sole input neuron a_k as $(X - X_1)/(X_2 - X_1)$, and a target value of Y may be encoded by a similar squashing transformation. The output computed by the network may be decoded from the activity a_k of the sole output neuron as $Y_1 + (Y_2 - Y_1)a_i$, i.e., by a scaling and shifting transformation. The extension to several analog input variables and several analog output variables is obvious.

The above coding/decoding specifications are not complete in cases where the target pattern is a bit string, as in unary or binary representation of the desired output. Some overt postprocessing recipe needs to be imposed to interpret the *actual* activities of the output neurons, since the squashing function $g(F_i)$ will generally deliver a value of a_i lying *between* 0 and 1. For detection problems, one can use a single analog output neuron to signal the presence of the relevant attribute according to the criterion $a_i \geq a_c$, with $a_i < a_c$ indicative of its absence. The choice to be made for the dividing line a_c will depend on the nature of the problem at hand, and especially on the seriousness of false negatives relative to false positives. For sorting problems, each of the various possibilities, or categories, is normally represented by a dedicated output neuron (unary coding) and the assignment made by the network would most simply be that category whose output neuron has the largest activity ("winner takes all").

In sorting problems, it is possible to design the network (including the choice of output-layer squashing functions) in such a way that the activities of the output neurons can be directly interpreted as estimated *probabilities* that the input pattern belongs to their respective designated categories (Stolorz et al. 1991, Gernoth and Clark, 1995a, Clark et al. 1999). This possibility is addressed in Sect. 7 and exemplified in Chapter 3. Similarly, one may arrange for the single analog output neuron used in a detection problem to provide a direct readout of the inferred probability that the input pattern has the attribute being sought.

The analog coding scheme can be refined by the introduction, in the input or output layer, of a *group* of analog neurons to represent a given real variable. Each neuron of such a group is responsible for a disjoint portion of the range of the variable. When applied to an output variable, this "real-number" coding strategy must be accompanied by an explicit postprocessing rule for readout of the value computed for the variable. Relevant details may be found in Gazula et al. (1992).

The notation $^c(I + H_1 + H_2 + \ldots H_L + O)_{c'}[P]$ permits an easy identification of particular choices of architecture and coding. In this expression, I and O are the numbers of input and output neurons and H_l is the number of neurons in the lth hidden layer; c and c' denote the input and *target*-output coding schemes, respectively; and P is the number of adaptive weights (including biases). The biases of the input neurons do not contribute to P, since these units function only as data registers. Among other possibilities, the labels c

and c' may assume the values u for unary, b for binary, a for analog, and r for "real-number."

5.2   The Credit Assignment Problem

Now for a brief retrospection.

The Elementary Perceptron (Rosenblatt 1962), the most primitive nonlinear version of a layered, feedforward net, consists of an input layer of units that simply record each input pattern as a bit string, plus an output layer of binary, McCulloch-Pitts threshold neurons. The input units are joined directly to the output neurons by a system of modifiable connections. Often the output layer contains only a single neuron, which renders a simple yes-no, 1-0 detection judgment, and we can focus on this case without serious loss of generality. It was established in the 1960s (Minsky and Papert 1969) that such systems have rather limited capabilities for the modeling of input-output maps. They work by finding a hyperplane $\sum_k V_{ik} S_k + V_{io} = 0$ in the multidimensional space of input states $\{S_k\}$, which separates those inputs corresponding to "yes" from those corresponding to "no." Finding such a hyperplane means finding proper weights V_{ik} for the connections from the input units to the output neuron, and a proper bias V_{io} for the output neuron. Rosenblatt's Perceptron Learning Rule is guaranteed to converge to a solution for these parameters in a finite number of pattern presentations, *assuming* that the Elementary Perceptron is capable of solving the problem at hand. However, one quickly encounters problems in which the "yes" and "no" cases *cannot* be separated by a linear manifold. Consequently it is not possible to find values of the couplings V_{ik} and bias V_{io} that allow the Elementary Perceptron to solve these problems. The XOR (exclusive-OR) problem, and its generalization, the parity problem (to determine whether the number of "1" entries in a bit string is even or odd) are the most famous examples. The demonstrable inadequacy of the Elementary Perceptron was a major factor in the demise of neural networks as a major branch of research in the infancy of the field of artificial intelligence (Cowan and Sharp 1988).

Why not try more layers? In fact, it turns out that the *three-layer* architecture suffices to form arbitrarily complex decision regions. These may be convex, concave, multiply connected, or disconnected, along with the simple hyperplane-bounded regions of the Elementary Perceptron. Moreover, Hornik et al. (1989) have shown that any Borel-measurable function from one finite-dimensional space to another can be approximated to any desired degree of accuracy by some three-layer feedforward net built from neuron-like units with arbitrary activation (squashing) functions. For other results of this kind establishing universal computational properties of multilayer perceptrons, see Denker et al. (1987), Cybenko (1989), and Funahashi (1989). (We hasten to interject that these are just existence theorems and tell us nothing about how many hidden neurons are needed and how to determine the weight/bias parameters. In practice, it can be advantageous to use *more* than three layers, as attested by some of the examples studied in Sect. 6.)

The vastly greater versatility of multilayer systems was already appreciated in the 1960s. But it was also recognized that the introduction of intermediate layers of hidden units gives rise to the *credit assignment problem*. It was easy enough to see how to modify the connections of two-layer devices (of Elementary Perceptron or Adaline types) to improve their performance in tasks within their abilities (Rosenblatt 1962, Widrow and Hoff 1960, Widrow 1962), since the faulty connections directly affect the output. But in the case of a feedforward network with hidden units, or in a recurrent network, it is generally far from obvious how to identify the connections that are contributing to incorrect [correct] behavior and determine their degree of responsibility, so that they can be punished [rewarded] accordingly. Another way of stating the problem is to observe that the hidden units must be turned into feature detectors which effect an appropriate mapping between configurations of input and output units. The difficulty is to figure out what features they should detect. Several workable solutions of the credit assignment problem were offered in the 1980s, including, most notably, the *Boltzmann machine learning algorithm* (Ackley et al. 1985, Hinton and Sejnowski 1986) and *backpropagation of error signals* (Rumelhart et al. 1986b,c). The former applies to symmetrically coupled recurrent nets, and the latter to layered feedforward architectures. For further discussion of perceptrons and the credit assignment problem, the reader is referred to the historical survey by Cowan and Sharp (1988) and the mathematical analysis by Minsky and Papert (1969).

5.3   Backpropagation Learning Algorithm

Backpropagation (or some variant of it) is by far the most widely practiced method for dealing with the credit assignment problem and training neural nets to do useful things. In a departure from traditional perceptrons, but in accord with the general scenario for supervised learning set forth at the beginning of this section, the networks to be trained by backpropagation are made up of analog neuronal units with smooth, continuous output signals, since the procedure requires derivatives of these activities to be calculated. To see how the scheme works, let us consider a three-layer net composed of input, intermediate (hidden), and output layers, with their respective units indexed by k, j, and i, in our usual notation. The goal is to adjust the couplings V_{jk} (from input neurons to hidden neurons) and V_{ij} (from hidden neurons to output neurons), as well as the biases V_{jo} and V_{io}, to achieve a prescribed input-output mapping,

$$I^{(\mu)} \to O^{(\mu)} = T^{(\mu)}, \quad \forall\, \mu. \tag{34}$$

In words: ideally, the input stimulus $I^{(\mu)}$ should lead to an output response $O^{(\mu)}$ that coincides with the desired target $T^{(\mu)}$, for every pattern μ in the set of training patterns. To be more concrete, the mapping might associate parities with bit strings, connectivities with figures, authors with prose, etc. – or, to cite a scientific example, stability or instability with a given nuclear species.

As indicated earlier, the activity of a generic neuronal element m is assumed to obey

$$a_m = g(F_m) \,, \tag{35}$$

where, to be definite, the logistic choice $g_L(u) = [1 + \exp(-u)]^{-1}$ may be adopted for the sigmoidal response function g. The argument F_m appearing in (35) is either an external stimulus that sets the activity of the neuron at some prescribed value in $[0,1]$ (if m is a neuron k in the input layer); or it has the standard form $\sum_{m'} V_{mm'} a_{m'}$ (for neurons $m = j, i$ in hidden and output layers). For pattern μ, we identify $\{a_k^{(\mu)}\}$ with $\mathrm{I}^{(\mu)}$, $\{a_i^{(\mu)}\}$ with $\mathrm{O}^{(\mu)}$, and $\{t_i^{(\mu)}\}$ with $\mathrm{T}^{(\mu)}$. The notation should, by now, be obvious: the $a_k{}^{(\mu)}$ are the input-neuron activities corresponding to pattern μ, while the $a_i{}^{(\mu)}$ and $t_i^{(\mu)}$ are respectively the *actual* and *target* activities of the output neurons.

Backpropagation is a supervised learning procedure in which the cost function

$$C[V] = \frac{1}{2} \sum_{i,\mu} [a_i^{(\mu)} - t_i^{(\mu)}]^2 \,, \tag{36}$$

which measures the deviation of the actual output activities from their correct values, is minimized by *gradient descent*. The process is an iterative one; at each step, the connection strengths are improved incrementally by altering them in a way that is guaranteed to decrease $C[V]$ (or as a limiting case, leave it unchanged). This behavior is assured if the change $\Delta V_{mm'}$ of $V_{mm'}$ (where $mm' = ij$ or jk) is taken to have the form

$$\Delta V_{mm'} = -\eta \frac{\partial C}{\partial V_{mm'}} \quad (\eta > 0) \,, \tag{37}$$

where the constant η assumes the role of a learning-rate parameter. (We observe that

$$\Delta C = \sum_{mm'} \frac{\partial C}{\partial V_{mm'}} \Delta V_{mm'} \tag{38}$$

then becomes the negative of a sum of squares.) The required derivatives can be computed from

$$a_i^{(\mu)} = g\left(\sum_{j=o}^{H} V_{ij} a_j^{(\mu)} \right) = g\left(\sum_{j=o}^{H} V_{ij} g\left(\sum_{k=o}^{I} V_{jk} a_k^{(\mu)} \right) \right) , \tag{39}$$

the biases V_{io}, V_{jo} being considered as connection weights from an auxiliary neuron o with perpetual activity $a_o = 1$. Working backward from the output units to the hidden layer, it is found that

$$\Delta V_{ij} = \eta \sum_{\mu} [t_i^{(\mu)} - g(F_i^{(\mu)})] g'(F_i^{(\mu)}) a_j^{(\mu)} \equiv \eta \sum_{\mu} \delta_i^{(\mu)} a_j^{(\mu)}, \qquad (40)$$

with $F_i^{(\mu)} = \sum_{j=o}^{H} V_{ij} a_j^{(\mu)}$. The correction is thus governed by the *error signal* $\delta_i^{(\mu)}$, implicitly defined in (40). This part of the problem was solved nearly four decades ago by Widrow and Hoff (1960), dealing with two-layer nets. The "hard" part of the three-layer problem is to correct the weights of the connections from the input neurons to the hidden neurons, since it is usually not clear what kind of couplings the hidden neurons should develop with their inputs to produce the required behavior of the output units. While this problem appears difficult conceptually, backpropagation offers a starkly mechanical approach to its solution: knowing the error signals $\delta_i^{(\mu)}$, one just uses the chain rule of partial differentiation to calculate the derivative of $a_i^{(\mu)}$ of (39), and hence the derivative of C of (40), with respect to the couplings V_{jk}. The result is

$$\Delta V_{jk} = \eta \sum_{\mu,i} \delta_i^{(\mu)} V_{ij} g'(F_j^{(\mu)}) a_k^{(\mu)} \equiv \eta \sum_{\mu} \delta_j^{(\mu)} a_k^{(\mu)}, \qquad (41)$$

with $F_j^{(\mu)} = \sum_k V_{jk} a_k^{(\mu)}$. Eqs. (40) and (41) constitute the "generalized delta rule" (Rumelhart et al. 1986b), so named because they have basically the same form as the delta rule (16). The important new feature is that the error signals $\delta_j^{(\mu)}$ entering (41) are obtained by a recursive computation in which the error signals $\delta_i^{(\mu)}$ for the *next following* layer (reckoned in the direction from input to output) are first determined.

Extension of the algorithm to more layers is straightforward, weight modifications being implemented using a series of error corrections $\delta_i^{(\mu)}$, $\delta_{j_1}^{(\mu)}$, $\delta_{j_2}^{(\mu)}, \dots$. The δ values for the couplings to the Lth layer are determined from those for the couplings to the $(L + 1)$th layer in just the same manner as was done above for $L = 2$. Thus one can say that the error correction – "the medicine" – is propagated *backward* from the output to the preceding layers, the "doses" to the individual connections being administered in a manner that assures steady improvement of the performance of the system as reflected in a stepwise decrease of the cost function C toward a local minimum. For each training item μ presented to the network, the dose $\Delta V_{mm'}^{(\mu)}$ received by a given synapse is proportional to the output $a_{m'}$ of the presynaptic unit m' and the error $\delta_m^{(\mu)}$ attributed to the postsynaptic unit m.

One pass through the training set is called an "epoch." According to the above derivation, updates of the weights are to be made *after each epoch* (note sums over μ in (40) and (41)). This is called *batch updating*. It is often advantageous instead to make weight corrections *after each pattern* (thus removing the μ summation). This strategy (called *on-line* updating) will cause some deviation from true gradient descent in C, but this can actually help the system avoid local minima. (In any case, true gradient descent would require a vanishingly small learning rate η.)

Another useful variation on "pure" backpropagation involves the introduction of a momentum term (Rumelhart et al. 1986b). Let $\Delta V_{mm'}(n)$ be the correction that was determined for the weight of connection mm' upon the nth pattern presentation. The original on-line formula for the correction to this weight at the $(n+1)$th presentation is modified by the addition of a term proportional to $\Delta V_{mm'}(n)$:

$$\Delta V_{mm'}(n+1) = \eta \delta_m^{(\nu)} a_{m'} + \alpha \Delta V_{mm'}(n) \,. \tag{42}$$

(We suppose here that pattern ν happens to be chosen for the $(n+1)$th presentation.) The constant of proportionality α is called the momentum parameter. This refinement allows past weight changes to influence the current direction of corrections in weight space, with the possibility of avoiding large oscillations in cases where the structure of the cost surface in weight space includes deep ravines with steep walls and gently sloping floors.

Backpropagation is not without its limitations and pitfalls. Usually, there is no guarantee that the system will respond satisfactorily when shown novel stimuli, i.e. that the trained system is able to generalize from what it has learned. Moreover, since the method is based on gradient descent, there is the possibility of getting stuck in a local minimum of C, which may be far above the global minimum corresponding to the "best" internal representation of the regularities of the stimulus-response ensemble. Perhaps most important, it is not clear that the backpropagation algorithm will scale economically when confronted with large-scale, real-life problems. In many interesting cases, the empirical results have been so encouraging that there is a prevalent tendency to discount the seriousness of these concerns. On the other hand, failures are not likely to be reported in great numbers.

The theoretical foundations of learning and generalization in perceptron-like systems are still very incomplete. The considerable technological impact of these systems warrants a continued theoretical effort similar in spirit to that which has proven so successful in the analysis of content-addressable memory models. Such an effort, exploiting advanced methods of statistical physics and probability theory, is well under way, as documented in the lengthy review article by Watkin et al. (1993). In addition, we may point to significant early contributions to the statistical mechanics of perceptron learning by Sompolinsky et al. (1990), Barkai et al. (1990), Levin et al. (1990), Krogh and Hertz (1991), Kanter (1992), and Seung et al. (1992). One of the primary goals of fundamental research on neural networks should be the invention of learning algorithms, applicable to multilayer systems or some other architecture, for which a convergence theorem analogous to that for the Elementary Perceptron can be proven, or for which at least the probability of success within a given training time can be estimated (Domany 1988). For a given learning algorithm – say backpropagation – it is of great practical importance to have good theoretical bounds on the number of training examples needed to achieve useful generalization and to have criteria and results for the optimal numbers of layers and hidden neurons.

There is no tangible evidence that the backpropagation process underlies any form of learning in biological nerve nets. Backpropagation of error signals and corresponding synaptic modifications would appear to require (i) antidromic transmission of information, backward along axon fibers, and (ii) the existence of structures that compare the output of each neuron of the learning assembly with signals from a teacher (itself an unknown element with no obvious biological counterpart). Be that as it may: backpropagation was not introduced in an attempt to imitate the workings of the brain, but rather to solve a fundamental computational problem.

As is the case with many useful inventions, it is surprising that the backpropagation solution of the credit assignment problem was not found much earlier. The key ingredients are easily at hand: (i) choice of a suitable objective function, (ii) use of analog neurons in the hidden and output layers, and (iii) the chain rule of partial differentiation. The chain rule terminates because the number of layers is finite and there are no recurrent connections. In fact, backpropagation was actually discovered by Werbos (1974), a decade before the widely publicized work of Rumelhart et al. (1986b,c). However, a discovery must also be timely to receive notice, and, with the rapid development of computers, the time was ripe in 1986. It is also worth mentioning here that Pineda (1987,1989) and Almeida (1987,1988) have extended the backpropagation approach to recurrent nets. A good summary of this work appears in Hertz et al. (1991).

Let us now turn to successful applications of the backpropagation algorithm and closely related techniques for supervised learning. Accounts of performance on a number of standard test problems, including the XOR and parity problems, the encoding problem, symmetry problems, addition, negation, and the "T-C" problem, are given in Rumelhart et al. (1986b). Among some of the more interesting applications of real-world character we may list:

- Extraction of family-tree relationships (Hinton 1986)
- Discrimination between noisy speech patterns (Plaut et al. 1986)
- NETtalk – learning to read English text aloud (Sejnowski and Rosenberg 1987)
- Classification of sonar signals returned from undersea targets (Gorman and Sejnowski 1988)
- Inference of an object's shape from its shading (Lehky and Sejnowski 1988)
- Recognition of handwritten zip codes (Le Cun et al. 1990a)
- Navigation of a car (Pomerleau 1989)
- Medical diagnostics including analysis of clinical images
- Processing of employment and loan applications
- Financial forecasting – stock market, exchange rates, etc.
- Discrimination, classification, and other data-processing tasks in experimental science

- Statistical modeling of scientific databases

A list many pages long could be generated, in the wake of the neural-network "gold rush" that took place in the late 1980s and early 1990s, across a broad sweep of problem areas from robotics to weather forecasting to speech recognition. Much of this work is poorly documented or proprietary, and it is difficult to judge its quality. Still, it is probably fair to say that a substantial fraction of neural-network applications, based mainly on backpropagation, yield promising results rather quickly, with rather little effort. In nearly all cases, there exist highly developed conventional methods that show superior performance. Nevertheless, the flexibility and power of neural-network approaches must be acknowledged, especially when it is considered that the more established techniques that provide the existing performance benchmarks have been expertly tailored, over many years, to narrow problem domains.

6 Multilayer Perceptrons: Scientific Applications

To the physicist, the most estimable applications of artificial neural networks are those that contribute directly to the advancement of science. Such efforts have progressed beyond the stage of mere curiosities, having scored successes (not always without qualification!) in a number of fields. Those involving multilayer networks or perceptrons include:

- o Recovery of atmospheric phase distortion from stellar images in astronomy based on adaptive optics (Angel et al. 1990, Sandler et al. 1991, 1994a,b, Lloyd-Hart et al. 1991, Stahl and Sandler 1995)
- o Star/galaxy discrimination in sky-survey photographic plates (Odewahn et al. 1992)
- ▷ Classification of atomic energy levels (Peterson 1990,1991,1998)
- ▷ Modeling of internal energy flow and potential energy surfaces in molecular physics and physical chemistry (Sumpter et al. 1992, Blank et al. 1995)
- ▷ Learning and prediction of nuclear systematics (Clark and Gazula, 1991, Gazula et al. 1992, Clark et al. 1992,1999, Gernoth et al. 1993, Gernoth and Clark 1995a,b, Mavrommatis et al. 1998)
- o Impact-parameter determination in heavy-ion collisions (Bass et al. 1996)
- o Triggering, event selection, event classification, and other applications in experimental high-energy physics (Denby and Linn 1990, Lönnblad et al. 1990,1991b,1992, Bortolotto et al. 1991,1992, Becks et al. 1993, Stimpfl-Abele and Yeps 1993, Babbage and Thompson 1993, Peterson et al. 1994)
- o Classification of low-resolution mass spectra of unknown compounds in organic chemistry (Curry and Rumelhart 1990)
- o Identification of ^{1}H-NMR spectra of complex oligosaccharides in carbohydrate chemistry (Thomsen and Meyer 1989, Meyer et al. 1991)

▷ Prediction of protein structure based on sequence information (Qian and Sejnowski 1988, Bohr et al. 1988,1990,1992,1993a, Holley and Karplus 1989, Andreassen et al. 1990, Kneller et al. 1990, Wilcox et al. 1990, Stolorz et al. 1991, Wade et al. 1992, Rost and Sander 1992,1993a-c,1994, Rost et al. 1993,1994, Bohr and Brunak 1994, Rezko et al. 1995, Bohr 1998)

▷ Analysis of nucleic-acid sequences (Brunak et al. 1990a,b,1991)

In the above list, applications marked with a circle (○) are concerned with the interpretation of individual *experimental* observations and the identification of interesting objects and events, exploiting the pattern-recognition capabilities of neural nets. Items labeled with an arrowhead (▷) are more *theoretical* in nature, as they require the network, under training with a subset of the existing database for a given property of a class of physical or chemical systems, to create a statistical model of that property, for subsequent use in prediction. Examples of both types of applications are singled out below for closer study.

To see why scientists find this work appealing or provocative, or both, let us examine once more the scenario in which a layered feedforward network is trained to associate members of a set of input patterns with appropriate output patterns. The inputs and their corresponding outputs may be considered as the "environment" of the artificial neural system. We have seen that in all but the simplest classification tasks soluble by Elementary Perceptrons, successful neural networks must have the ability to form internal representations of their environments, through the receptive field patterns that the hidden neurons develop during the supervised learning process. "Receptive field" is a term borrowed from the physiology of visual systems; here it is taken to mean the set of weights of the incoming connections of a neuronal unit. These weights determine the stimulus patterns or, more abstractly, the "features" to which the neuron responds most strongly. If the classification or regression problem assigned to the network is to be performed reliably with limited numbers of units and interconnections, a network must discover economical rules for describing the correlations between input and output patterns. Suppose a network has indeed developed working rules that give a reasonably faithful representation of a set of experienced associations (the training set), to the extent that the response of the network to the input patterns of the training set is nearly always correct or close to the target. The broader applicability of these rules, or the generalization ability of the system, may then be tested by exposing the network to novel patterns (the test set) and observing its response. A high percentage of correct (or nearly correct) responses to unfamiliar stimuli indicates that the system is not simply using its free parameters (weights/biases) to make an input-output lookup table, but is actually capturing the most important rules underlying the associations it has experienced.

Now, re-read the foregoing paragraph, with the following substitutions: *data* for *training-set associations*; *"independent" physical variable* for *input pattern*; *dependent physical variable* for *output pattern*; *model* or *theory* for *rules*; and *prediction* for *response to unfamiliar stimuli*. It may then be recognized that, in essence, what the neural net is doing in the above scenario has a strong resemblance to what a scientist does when he or she is doing science. Of course, what has been described above is just the familiar process of statistical inference, clothed in "connectionist" (Rumelhart et al. 1986a) language. The network learns the statistics of the problem in the training phase; in actual operation on new inputs, it performs the required inference within the limits set by its acquired knowledge and by its architecture. Given these limitations, it is hardly likely that current or conceived neural networks will replace human scientists, who have at their disposal a superb inference machine containing 10^{11} multiplex neurons in a network built and honed over 10^9 years of evolution and decades of on-line learning. The fancy architecture and priors of the human machine simply can't be matched by the lowly neural nets of today's technology. But it is reasonable to think that they can help with the tedious heavy lifting of statistical analysis, when we are confronted with complex problems involving many degrees of freedom.

The use of multilayer feedforward networks for classification and regression tasks in science will have obvious practical merit if trustworthy results can be produced in important physical contexts. The approach is most likely to be successful in cases where a large database exists, furnishing some thousands of training patterns for the creation of an accurate representation of the relevant associations. Since feedforward nets are much better at interpolation than extrapolation, their predictive faculty is most likely to be useful for filling gaps in the database, rather than projection far beyond the range of current data.

Beyond such utilitarian considerations, more intriguing possibilities and issues may arise, especially as more advanced architectures and learning rules are developed. In building a workable internal model, the scientific neural network may uncover rules or regularities we already know, but it is also possible that it may reveal aspects of nature that might otherwise elude the human scientist. Unfortunately, neural nets represent what they have learned in ways that are alien to us, and this has opened neural-network modeling to much criticism, notwithstanding the fact that other nonlinear nonparametric procedures are just as opaque. The "knowledge" gained by a neural network is buried in the connections and the patterns of weights, and in practice it may be difficult or impossible to derive a set of clearly-expressed rules from the matrix of neuron-neuron interactions. One is confronted with a new and challenging class of ill-posed inverse problems whose analysis has scarcely begun. Some interesting approaches to rule extraction are described by Denker et al. (1987), Sanger (1989), and McMillan et al. (1991).

While lacking practicable general methods for rule extraction, we can still obtain helpful glimpses into the innards of the neural-network machine. One strategy that has been fruitful in a number of problems is analysis of the receptive fields of the hidden neurons (Rumelhart et al. 1986b, Sejnowski and Rosenberg 1987, Gorman and Sejnowski 1988, Lehky and Sejnowski 1988). Receptive-field studies can inform us on what the features of the data environment have been singled out by the trained network as especially significant. Under training, hidden neurons may develop that turn on strongly if and only if certain features are present in the input pattern − e.g., a magic neutron or proton number in the case of a network designed to detect nuclear stability. Insights can sometimes be gained from a cluster analysis:

(i)   Expose the network to a selection of input patterns.
(ii) Record the hidden-neuron activities produced by each of these inputs.
(iii) Group together the pairs of hidden-neuron responses that are most alike.
(iv) For each such pair, examine the corresponding inputs to see what they have in common.
(v)   Repeat this process for pairs of pairs, pairs of pairs of pairs, etc.

For the NETtalk system (Sejnowski and Rosenberg 1987), which translates written text to phonemes, this kind of analysis has revealed similarities in hidden-neuron responses to p's and b's, and, working up the hierarchy of pairs of pairs, pairs of pairs of pairs, etc., to all vowels as a group and to all consonants as a group. We can also expect to find clues to underlying invariances and symmetries of a given problem in the pattern of weights that emerges in the learning process.

Another useful probe is sensitivity analysis, with the aim of determining which input variables are really important and which can be eliminated. In one approach, networks can be constructed and trained with all proposed input variables in play, and then with each such variable omitted in turn (by the deletion of corresponding input-layer neurons). Comparative performance on the test set should reveal redundant or irrelevant variables, if present. (The fact that some input variables may not actually be independent is the reason why quotes were attached to this adjective four paragraphs backstream.) Alternatively, a network may be trained using all of the proposed input variables, and tested on runs through the test set in which artificial noise has been injected into each of the input-variable channels in turn. Those variables for which the predictive performance of the net is most seriously affected by noise are evidently the most important. Variables whose corruption has little impact on performance may be eliminated.

Concerning actual rule extraction, it is worth noting that in some classification problems (e.g. the stability/instability discrimination task discussed in Sect. 6.3) one can use an automatic logic minimizer to generate a Boolean logic network counterpart of a favored neural-network model. We outline the procedure for the case of a single hidden layer and a single binary output

(Gernoth et al. 1993). (In the case of an analog output, a postprocessing rule can be applied to enforce a binary decision. See Sect. 5.1.)

(i) Choose a suitable activity threshold a_t (e.g. 0.5).
(ii) Expose the trained network to each pattern of the training set and replace the resulting hidden-unit states a_h by binary values $\Theta(a_h - a_t)$.
(iii) For consistency and economy, delete those cases in which the same set of binary hidden-unit states corresponds to different binary output states, and eliminate any duplicate cases of binary-hidden–output associations.
(iv) Binary patterns not appearing in the laundered set of hidden-unit activities are assumed to produce "don't care" outputs.
(v) The resulting set of pairs of binary hidden-unit patterns and binary outputs is then fed to a logic minimizer (widely available as part of an integrated-circuit design package).

The logic minimizer returns a description of a Boolean logic network that yields the correct binary output when a given pattern of binary hidden-unit states is presented as input. For this strategy to be useful, the resulting logic network must be insensitive to the choice of activity threshold a_t (i.e., in normal operation, most of the hidden units must show activities near zero or near saturation). The treatment may be generalized to allow for multilevel units (as well as multiple hidden layers and multiple multilevel outputs) with concomitant increase of complexity of the logical system that simulates the neural net.

From these varied considerations, it would seem that the neural-network black box is, after all, not totally opaque, and one can often learn quite a lot about the problem under study by attempts to look inside. In addition, we should point out that the "black box" objection of critics of neural networks carries less weight when they are used as tools for image analysis and data processing in experimental physics than it does for the more theoretically motivated applications.

We are now prepared to discuss selected applications of layered feedforward networks to scientific problems.

6.1   Star-Galaxy Discrimination

A prototypical example of the utility of neural networks in classification and object identification is provided by the development of an automatic star/galaxy discriminator for processing images generated by the University of Minnesota Automatic Plate Scanner (APS) (Odewahn et al. 1992). The goal of the larger scientific project is to catalog images contained on the 936 plate pairs of the first-epoch Palomar Sky Survey. The APS is well suited to digitizing the data gathered in such surveys. The images are parametrized in terms of a number of densitometric measures, including diameter D, ellipticity, average transmission, central transmission, "jitter," various gradients, etc. – 14 parameters in all, in the work to be described. The task of the classifier system is to determine, on the basis of these parameters, whether each spot on a sky-survey plate corresponds to a star or to a galaxy. Odewahn et al. tried (two-layer) Elementary Perceptrons (Rosenblatt 1962, Minsky and Papert 1969) as well as multilayer systems trained by backpropagation.

Analog units with logistic squashing functions were used in all layers of the backpropagation networks. The 14 input parameters corresponding to a given image were preprocessed before using them to set the activities of the 14 dedicated analog input units, the values of these parameters being scaled and shifted to fall within the interval [0,3]. (This was found to accelerate the learning process, which started from random weights.) The favored multilayer architecture consisted of 14 input neurons, 13 neurons in one hidden layer, and 2 output neurons. One output neuron was assigned to "star" identification, with a target activity of 1 [0] in case data for a star [galaxy] image was presented at the input. The other output neuron was dedicated to "galaxy" identification in the corresponding manner. The output unit with the larger activity determined the star/galaxy decision ("winner-takes-all" criterion). In the notation introduced at the end of Sect. 5.1, the architecture/coding scheme chosen for the multilayer backpropagation networks was $^a(14 + 13 + 2)_u$.

The training sets consisted of images from two Palomar Sky Survey plates containing a total of 2082 star and 2665 galaxy images (identified by human discriminators). A third such plate, containing 2380 star and 936 galaxy images, provided images for testing. All three plates belong to the survey field containing the Coma cluster of galaxies. Separate training sets were formed from "small-diameter" ($73\mu m \leq D \leq 137\mu m$) and "large-diameter" ($146 \leq D \leq 330\mu m$) images. The small-diameter subset contained 1050 stars and 1116 galaxies; the large-diameter subset, 719 stars and 1584 galaxies. Separate networks of each type (Elementary Perceptron and multilayer) were constructed for each of these training sets. The trained networks were labeled SP, LP, S1, and L1, with "S" for small-diameter, "L" for large-diameter, "P" for Elementary Perceptron, and "1" for backpropagation net. The SP and LP classifiers were defined by the hyperplane yielding the smallest number of errors after 50,000 epochs of training on the respective image samples. (In either case, no improvement was seen after 28,000 epochs.) The backpropagation networks were trained until learning appeared to saturate. The learning rate η was taken as 0.01 and the momentum parameter α as 0.9. The favored large-diameter and small-diameter backpropagation classifiers, S1 and L1, were attained after 925 and 6540 passes through their respective training sets. The greater number of training epochs required in the latter case is symptomatic of greater difficulty in learning to distinguish stars from galaxies for smaller images.

All four classifiers were able to learn their training sets with satisfactory or high accuracy. In the large-diameter case, network L1 showed nearly perfect performance, errors being made on only 2 galaxies and 3 stars; whereas LP misclassified 9 stars and 9 galaxies for an overall error rate of 0.83%. In the small-diameter case, the error rate for S1 was 1.39% overall (with a breakdown of 2.36% for stars, 0.95% for galaxies); for SP it was 6.04% overall (with a breakdown of 9.74% for stars and 4.36% for galaxies). It is not surprising that the learning performance of the backpropagation networks was superior, in view of the greater parametric resources provided by the hidden layer. While the classification accuracy of both S1 and SP was noticeably worse than that of their large-image counterparts, the falloff in performance was more serious for the Elementary Perceptron classifier. These studies indicate rather convincingly that the parameter space corresponding to the large-parameter images is nearly linearly separable – implying that the classification problem for this set of images is a simple one.

In the parameter domain of the smaller images, the boundary between stars and galaxies is considerably more complex, and the greater discriminatory powers of the three-layer backpropagation network are needed. However, this advantage may become moot if the small-diameter training set is significantly corrupted with errors.

In order to assess generalization ability, each image from the test sample was processed with the classifiers appropriate to its diameter class. The results were pooled to obtain overall predictive scores for examples novel to the trained networks. Jointly, the backpropagation networks L1 and S1 were able to classify both star and galaxy images in the test set with success rates exceeding 90% down to the faint O-plate B-band magnitude of 19.5. (The reader is reminded that magnitude is measured on a logarithmic scale, with higher magnitudes being dimmer. A sixth-magnitude star is barely visible to the naked eye.) Combined performance of the L1 and S1 systems was better than 90% for galaxy images down to magnitude 20, while the success rate for stars oscillated in the range 92-100% for magnitudes from 14 down to 19.5. Surprisingly, the Elementary Perceptron classifiers LP and SP jointly scored comparable success rates on the test sample. The expected superiority of the multilayer networks was not realized. The falloff in performance of the S1 network at the faint end (most notably for galaxies) was attributed to one or more of several causes, including: (i) paucity of information in the small-D sets, (ii) inability of the backpropagation algorithm to cope with noisy data, and (iii) incorrect human classification of training/test images.

It is to be expected that galaxy images, especially if faint, will be harder for a network to deal with, since they are necessarily more complex and variable (Odewahn et al. 1992). Therefore a large galactic training sample is needed to identify them reliably. The images for stars, ordinarily being round and compact, are simpler and more consistent. Thus one should be able to get by with a small stellar training sample, although a problem can arise at large image sizes, where diffraction spikes will become prominent and confuse the classifier.

To gauge the relative importance of the 14 input parameters, Odewahn et al. subjected the large- and small-image backpropagation networks (L1 and S1) to a sensitivity analysis, following the noise-injection procedure described above. For both classifiers, the average transmission emerged as the dominant parameter. For L1, gradient parameters were also identified as good discriminators. In the case of S1, gradient parameters were less important, the central transmission and an image-area parameter being more significant for classification.

The results obtained by the neural-network classifiers were deemed to be satisfactory, especially in demonstrating that the catalog being compiled by the Minnesota group will be "as deep as" (or deeper than) the commonly used Lick galaxy survey. On the other hand, the simplicity of the star/galaxy classification problem – compared with others that will be considered in this volume – suggests that other more conventional classification algorithms (Duda and Hart 1969) would be equally successful.

Neural networks have also proven their worth in more challenging problems in the field of astronomy, most notably in adaptive optics, where they are used for wavefront sensing, reconstruction, and prediction. This important area of research is reviewed in Chapter 2.

6.2   Fast Interpolators for Potential-Energy Surfaces

Blank et al. (1995) have developed feedforward neural networks that model global properties of potential-energy surfaces in chemical systems on the basis of information from a limited set of configurations. This work serves to exemplify the application of multilayer perceptrons to regression tasks (also called function approximation).

The study carried out by Blank et al. has a compelling practical motivation. Computer-intensive simulations of chemical systems based on molecular dynamics and Monte Carlo algorithms call for values of the potential energy at arbitrary configurations. The required potential-energy surfaces may be determined from experimental data or from electronic-structure calculations based on band-structure techniques or density functional theory. However, due both to experimental limitations and to the difficulty and expense of "ab initio" calculations, accurate results for the potential energy are only available at a restricted number of points. Therefore a fast and accurate interpolator is needed. It is natural to try multilayer perceptrons trained on the available data. If successful, such interpolators could be of great value not only in physical chemistry, but also in condensed-matter physics, materials science, and molecular biology (cf. Bohr 1998).

To furnish initial demonstrations of feasibility, Blank et al. made several fits to error-free data derived from an empirical model of CO adsorbed on a Ni(111) surface. Three-layer feedforward nets with analog input and output coding were trained with an adaptive, global, extended Kalman filter (Blank and Brown 1994) instead of backpropagation. The value of the energy was delivered by a linear output neuron, while the hidden units had the usual logistic squashing functions. The data points used for training were spaced uniformly on grids in two or three dimensions, depending on the trial problem considered. Predictive accuracy of interpolators was measured by the mean absolute deviation (MAD) of model predictions from test data. The first problem involved two degrees of freedom: the lateral position x of the center of mass along a line between two stable sites and the angle θ of the molecular axis relative to the surface normal. Surface atoms were fixed in their equilibrium positions. Working with a 42-point training set on a 6×7 grid in (x, θ), the optimal number of hidden neurons was determined to be 6. In this case the MAD was 0.019 kcal/mole (0.3% of the total potential range). Increasing the size of the training set to 81 points, on a 9×9 grid, a network model of type $^a(2+6+1)_a[25]$ achieved a MAD of 0.010 kcal/mole. This accuracy is quite sufficient, considering that the predictive error corresponds to the ambient thermal energy at about 5 K. To compare with a more familiar interpolation procedure, Blank et al. made simple cubic spline fits of the same sets of training data, with resulting MAD values of 0.072 kcal/mole (42-point set) and 0.014 kcal/mole (81-point set). This comparison, and others to follow, will illuminate a discussion in Sect. 8.

In another trial problem, a third degree of freedom was added, namely the z coordinate of the center of mass, measured normal to the surface. Since the energy depends strongly on z, this problem is significantly more difficult than the first. Training sets of 343 points ($7 \times 7 \times 7$ grid in (x, θ, z)), 441 points ($7 \times 7 \times 9$ grid), and 637 points ($7 \times 7 \times 13$ grid) were adopted and networks of type $^a(3+H+1)_a$, with H ranging from 7 to 15, were studied. By increasing the density of points along the z coordinate, more hidden nodes may be included in the network without overfitting. (For a discussion of overfitting by neural networks, see Sect. 6.3.) A network with $H = 15$ and $P = 76$ weights, developed using the larger training set, gave a predictive MAD below 0.02 kcal/mole with a maximum error of 0.14 kcal/mole, these results being obtained on a test set of 2000 points. The absolute error is double the smallest value reached in the problem with two degrees of freedom; on the other hand, the error relative to the total potential-energy range of 17 kcal/mole is somewhat smaller. A cubic spline fit to the larger training set yields a MAD of 0.007 kcal/mole, thus showing better accuracy than the best neural-network model of Blank et al. However, larger networks may be able to reach lower MAD values.

An important consideration in comparing the merits of neural-network models and spline fits is the much greater parametric efficiency of neural-network function approximation. The spline procedure forces a local fit, implying as many parameters as data points – up to 637 in the problem with three degrees of freedom, as opposed to 76 for the best available neural net. In the network representation of the data, each weight contains information about many data points. As pointed out by Blank et al., this redundancy can make the neural-network model less sensitive to noise in the training data than the spline fit.

The neural-net approach has another intrinsic advantage over standard polynomial spline fitting. In the case of nonuniform spacing of the data, the neural-network procedure would remain the same, whereas a more complicated spline procedure would be needed, such as multivariate adaptive regression splines (MARS) (Friedman 1991).

The efficiency and accuracy of the neural-net interpolator were exhibited in quantum transition-state calculations for surface diffusion of $CO/Ni(111)$ based on a Monte Carlo path-integral algorithm. Potential values can be calculated from the trained network model ten times faster than from the original analytical expression for the empirical potential. If millions of potential evaluations are needed in a calculation, this speedup compensates many-fold for the time expended in training the network.

Blank et al. went on to consider a more demanding test in which the interaction between an H_2 molecule and a $Si(k100)2 \times 1$ surface was described with twelve degrees of freedom. The training set consisted of potential energies calculated by local density functional theory for 750 configurations. "Optimal" results were found with a network of type $^a(12 + 8 + 1)_a[113]$, which gave a MAD of 1.7 kcal/mole on the training set and 2.1 kcal/mole on a test set containing 617 data points, also generated by density functional theory. The predictive error in this case was 1.5% of the full range of the potential energy. The higher error level in this problem was attributed to the fact that the energy depends on coordinates not present in the model. Effectively, restriction to 12 degrees of freedom introduces noise into the training and test data.

Again a comparison with traditional statistical methods was sought, but the nonuniform spacing of the data and the presence of noise limited the choice of methods. Attempts to implement the MARS algorithm for more than two degrees of freedom were not successful.

6.3   Statistical Modeling of Nuclear Properties

The essential message of this chapter is that multilayer feedforward networks of analog neurons can be trained to approximate a physical mapping from a set of "independent" (or input) variables X to a set of dependent (output) variables Y. When applied to the complex of data and mappings that constitute the body of empirical knowledge of a class of physical systems, this possibility opens the way to a novel adaptive approach to global modeling that complements more conventional theoretical approaches, in both practical and conceptual senses. In precision and range, such neural-network statistical models may in some cases rival or surpass traditional phenomenological and microscopic treatments. However, as statistical models, they will generally contain more adjustable parameters than traditional models, and they may be expected to falter if they are required to extrapolate to examples that are too different from those in the training set.

The mature field of nuclear physics, supported by an enormous collection of data reflecting the fundamental principles of quantum mechanics and the behavior of strong, electromagnetic, and weak forces at the femtometer scale of distance, provides a wealth of opportunities for testing and exploiting this more theoretical use of neural-network techniques (Clark and Gazula 1991, Gazula et al. 1992, Clark and Gernoth 1992, Clark et al. 1992, Gernoth et al. 1993). There is currently a strong incentive for the development of global models of nuclear properties, driven by the manufacture of many new nuclear species at radioactive-beam facilities and by the needs of complex reaction-network calculations involved in models of nucleosynthesis.

For our purposes, the properties of a given nuclear species, or nuclide, are determined entirely by its proton (Z) and neutron (N) numbers, or by its neutron number and mass number $A = Z + N$. Nuclides are conveniently represented in a nuclidic chart, a plot of Z versus N containing a point or square for each known nuclide. The stable nuclides form a jagged line on this chart, with Z increasing more slowly than N. This line is taken to define the bottom of the so-called valley of beta stability, in the sense that increasing stability implies stronger binding and lower total energy.

The National Nuclear Data Center at Brookhaven maintains a comprehensive archive of the available experimental information on a wide range of nuclear properties, for all known nuclides. These properties include (i) the mass, spin, parity, electromagnetic moments, rms radius, various cross sections, and other observables of the nuclidic ground state; (ii) level schemes and data on excited states; and (iii) decay modes, branching probabilities, lifetimes, and decay chains of unstable states. In contrast to elaborate and time-consuming quantum-mechanical calculations tailored to specific nuclides and specific states, *global* nuclear models are constructed to describe selected nuclear properties for a substantial class of nuclides, or over a substantial region of the nuclidic chart.

In principle, one should be able to calculate all properties of any given nuclide from fundamental theory. However, such a program, requiring application of QCD in the nonperturbative regime, is currently intractable. Most nuclear theory is in fact practiced at a "second-principles" or "third-principles" level based on hadrons or nucleons and their effective interactions, rather than quarks and gluons. This situation leaves room for innovative theoretical approaches and fresh modes of description that may educe new insights from the existing database.

Over the last few years, neural-network methods have been used to create global models of a number of prominent nuclear properties:

(a) Nuclear stability (Clark and Gazula 1991, Gazula et al. 1992, Clark et al. 1992,1999, Gernoth et al. 1993, Gernoth and Clark 1995a)

(b) Atomic masses (Gazula et al. 1992, Clark et al. 1992, Gernoth et al. 1993, Clark and Gernoth 1995, Gernoth and Clark 1995b, Athanassopoulos et al. 1998)

(c) Neutron separation energies (Gazula et al. 1992, Clark et al. 1992)

(d) Ground-state spins and parities (Clark et al. 1992, Gernoth et al. 1993)

(e) Branching probabilities for different decay channels (Clark et al. 1994, Gernoth and Clark 1995a)

(f) Halflives for β^- decay (Mavrommatis et al. 1998)

The current status of applications (b) and (e) is summarized in Chapter 3. Here we shall concentrate on (a) and (d), making only brief comments on the other entries.

Stability/Instability Discrimination. The first concerted application of neural networks to nuclear physics addressed the simplest and most basic of questions: is the ground state of a given nuclide stable or unstable? Feedforward networks were taught to discriminate between stable and unstable nuclides. It will be worthwhile to examine this effort in some detail. In addition to its pedagogical value, the nuclear stability/instability discrimination problem has provided a testbed for supervised learning algorithms. (See Sect. 5 of Chapter 3, Gernoth and Clark (1995a), and Clark et al. (1999).) In addition, it will be illuminating to compare the quality of the results with those obtained in the star/galaxy discrimination problem and with those obtained in a similar classification problem in protein chemistry.

There is clearly some arbitrariness in the criterion for stability. On the one hand, no nuclei may be stable on a time scale of 10^{50} years; and on the other, it seems reasonable to include long-lived naturally occurring radioactive isotopes in the stable class. In the initial work (Clark and Gazula 1991, Gazula et al. 1992, Clark et al. 1992, Gernoth et al. 1993), the "stable" nuclides were taken as those identified as such on the General Electric Chart of the Nuclides available in 1990. (Later work (e.g. Clark et al. 1999) has adopted the convention used in the Brookhaven Database. The differences are immaterial for present purposes.) The full data base inferred from the GE chart consisted of 2226 nuclides, with unstables outnumbering stables by about 9 to 1. A training set consisting of 1909 patterns was formed by deleting, at random, approximately 15% of the unstable examples and a like percentage of the stables. The set of 317 deleted nuclides served as a test set for evaluating predictive performance. Another training set and complementary test set (with respectively 1689 and 537 nuclides) were formed in the same way, based on 25% random deletion.

The choice of binary input coding is strongly suggested by the importance of pairing and shell effects in nuclear binding (Bohr and Mottelson 1969), these effects being associated with the integral (quantal!) nature of Z and N. Accordingly, the first 8 of 16 input units were clamped "on" ($a_k = 1$) or "off" ($a_k = 0$) so as to express Z as a binary number; the remaining 8 input units were used to encode N in the same manner. The hidden and output layers were composed of analog neurons, with logistic squashing functions.

We note that the stability/instability discrimination problem may be treated either as a detection or a sorting task, the attribute in the former option being "stability" and the pigeonholes in the latter being "stable" and "unstable." (We further note that the star/galaxy discrimination problem can also be treated either way.) Both options were considered, and unary coding of outputs was adopted in both cases.

Stability detection was carried out by networks of type $^b(16 + H + 1)_u[P]$. Since the single output neuron was assumed to have a logistic squashing function, its activity $a_i \in [0, 1]$ could be interpreted as the network's estimate of the probability that the input nuclide (Z, N) is stable (Stolorz et al. 1991). A "best-guess" decision was made using the simple postprocessing rule that $a_i \geq 0.5$ means "stable" and $a_i < 0.5$, by default, means unstable. Execution of the sorting task was based on models of type $^b(16 + H + 2)_u[P]$ (as well as some four-layer nets). Two output neurons (again having logistic squashing functions) represented the "stable" and "unstable" categories, and a "winner-takes-all" rule was imposed to crystallize the network's decision between these mutually exclusive choices.

Separate sets of training runs employed (i) ordinary backpropagation (involving gradient descent) and (ii) the conjugate-gradient method (Luenberger 1984) to search for a minimum of the cost function (36) in weight space. The backpropagation runs were carried out for detection networks, $^b(16 + H + 1)_u[P]$, and included the choices $H = 0$, 10, 19, and 24 for the number of hidden units. The momentum parameter α was set at 0.9, and various learning rates were tried. Weights, initialized randomly, were updated after each pattern presentation. The conjugate-gradient runs were performed for sorting networks, mainly of type $^b(16 + H + 2)_u[P]$, with $H = 10$, 15, and 20. For these experiments, weights were necessarily updated after each epoch.

Summarizing the results for the backpropagation networks, learning performance was very poor without a hidden layer ($H = 0$), improved for $H = 5$, and appeared to be saturating around $H = 19 - 24$. For $H = 19$ (implying $P = 343$ weight parameters), training on the *full* database yielded an overall discriminatory accuracy of 94% in a representative experiment, with a score of 75% in the identification of stables and 96% in the identification of unstables. It is useful to introduce the terms *efficiency* and *impurity*, to denote, respectively, the percentage of examples with a given attribute (here, stability) which are *correctly* identified ("true positives") and the percentage of input patterns without the attribute which are *incorrectly* identified by the network ("false positives"). Thus, the net in the cited example learned with 75% efficiency at a 4% impurity level. In a predictive run after training a network on the 1909-pattern training set, responses to the corresponding 317-pattern test set showed 69% efficiency and 6% impurity. For a network trained on the 1689-pattern training set, the performance figures on the corresponding 537-pattern test set were 63% efficiency and 7% impurity.

The conjugate-gradient experiments demonstrated improved learning, especially for the stable nuclides, and somewhat better predictive performance. Since the stability and instability decisions are complementary, one may still use efficiency and impurity as performance measures, where again these terms refer to the classification of nuclides as stable. A network of type $^b(16 + H + 2)_u[P]$ with $H = 10$ and $P = 192$ weights learned the larger training set (1909 patterns) with an efficiency of 100% and an impurity of 4.6% (meaning that the net decided correctly for 95.9% of the patterns). Testing prediction on the remaining 317 patterns, this net scored 77.1% in efficiency and 7.5% in impurity. The networks with larger numbers of hidden units did slightly better in learning the same training set. A net with $H = 20$ ($P = 382$) learned perfectly, while one with $H = 15$ ($P = 287$) failed to learn only a single pattern. On the other hand, predictive performance deteriorated, with the efficiency dropping to the 66% level but impurity remaining about the same.

This last comparison provides an example of the phenomenon of *over-training* (also called *overlearning* or *overfitting*). A network with more adjustable parameters has the capacity to learn the training data more accurately. However, as the training continues, the more amply endowed network has a tendency to learn the fine details of the training set at the expense of generalization ability. In the limit of a very large number of weights, a network may evolve into a "dumb" lookup table that identifies all the training patterns perfectly but performs at the chance level on novel patterns. In neural networks, as elsewhere in statistics, science, and life, we run up against Occam's razor: other things being equal, simpler is usually better.

An extreme case of overtraining was encountered in a backpropagation experiment based on a highly localized unary input coding scheme. A dedicated "on-off" input neuron was installed for each value of Z from 1 to 126 and for each value of N from 1 to 184. In a three-layer network of type $^u(310 + 5 + 1)_u$ having 1561 parameters, learning was close to perfect, with an overall accuracy of 98% for the full database. By contrast, prediction was extremely unreliable. In one test, stability was predicted for only 9 out of 62 novel stable examples.

Although raw percentages of correct assignments are commonly used to judge the quality of a discriminator, such performance measures can be quite misleading, especially if one output category occurs far more frequently than the other(s). This is the case in the problem being considered, since any of the data sets employed contains approximately 9 unstable nuclides for every stable example. A "lazy" network that assigns instability to *all* input patterns would score at the 89% correctness level overall. It is true that a zero or low efficiency measure would tell us if this were happening. The relatively high efficiencies achieved by the nets described above indicate that they are using their resources to form economical internal representations of what they have been taught.

The quality of a discriminator can be assessed more systematically in terms of the Mathews correlation coefficient. For a given input pattern, the task of the system is to decide whether or not a designated attribute is present. Continuing the nuclear theme, we take "stability" to be that attribute, but it could just as well be "star-ness" in the star/galaxy discrimination problem, or presence of an α-helix in the protein secondary-structure problem to be discussed later. The Mathews coefficient

$$C = \frac{p\bar{p} - q\bar{q}}{\sqrt{(p + q)(\bar{p} + q)(p + \bar{q})(\bar{p} + \bar{q})}}\,,\tag{43}$$

is constructed to eliminate the effect of bias from first-order frequencies. In this formula, p is the number of stable input nuclides correctly identified as stable, $\bar{p}$ the number of unstable input nuclides correctly identified as unstable, q the number of stable input nuclides incorrectly identified as unstable, and $\bar{q}$ the number of unstable input nuclides incorrectly identified as stable. The value of the coefficient C ranges between -1 and $+1$, taking its maximum or ideal value of $+1$ when all patterns are correctly identified and its minimum value of -1 when all identifications are incorrect. In the case of a lazy network model that assigns all patterns to one class (e.g. unstable), C vanishes, indicating trivial performance. Evidently, $p + q$ is just the total number of stables in the data set considered and $\bar{p} + \bar{q}$ is the total number of unstables, so the Mathews coefficient is in effect a function only of the number p of correctly identified stables and the number $\bar{p}$ of correctly identified unstables. By symmetry, the Mathews coefficient for stability equals that computed for the complementary attribute of instability.

This more incisive performance measure is also useful in sorting problems. If several outcomes are possible, a Mathews coefficient can be defined for each: for a given input pattern, the classification made by the network tells us whether or not a preselected outcome is judged to be present. This situation arises in the assignment of spin to nuclear ground states (see below). It also arises in protein secondary-structure classification when the possibilities are opened (say) to α-helix, β-sheet, and random coil. However, in sorting problems with more than two categories, the Mathews coefficients for the alternative attributes will in general all be different. It will still be true, of course, that the coefficient for the presence of the "α-helix" property will be the same as that for the (lumped!) "non-α-helix" property.

The Mathews coefficient is to be evaluated independently for the training and test sets. It is illuminating to quote some specific values obtained in the stability/instability discrimination problem. Recalling the specific backpropagation results described earlier, the experiment in which a network of type $^{b}(16 + 19 + 1)_{u}[343]$ was trained on the full database led to a Mathews coefficient $C_{\text{learn}} = 0.69$. The predictive runs on the smaller and larger test sets (after training on the corresponding 1909-pattern and 1689-pattern training sets) yielded, respectively, $C_{\text{pred}} = 0.58$ and 0.51. The experiment with local coding gave coefficients $C_{\text{learn}} = 0.92$ and $C_{\text{pred}} = 0.12$, indicating excellent "memorization" but abysmal generalization.

Turning to the nets trained with the conjugate-gradient algorithm, the example of type $^b(16 + 10 + 2)_u[192]$ had coefficients $C_{\text{learn}} = 0.84$ and $C_{\text{pred}} = 0.61$. Not mentioned previously is a four-layer network of type $^b(16 + 9 + 6 + 2)_u[227]$, which was also trained on the 1909-pattern training set (Gernoth et al. 1993). This network attained predictive accuracies of 83% and 94%, respectively, for the stables and unstables in the corresponding test set. The associated Mathews coefficients were $C_{\text{learn}} = 0.87$ and $C_{\text{test}} = 0.68$, the latter representing the highest predictive significance yet achieved for any stability/instability discriminator. Either of these conjugate-gradient nets showed better performance in both learning and prediction than the best of the available backpropagation models of the stability/instability dichotomy, with fewer weight parameters. (It should be mentioned, however, that the result $C_{\text{test}} = 0.68$ has been matched by a network designed to generate branching probabilities among different nuclear decay modes, or signal the stability of a given input nuclide (Gernoth and Clark 1995a). In fact, the network in question reaches $C_{\text{learn}} = 0.93$ and may therefore be judged superior to the best of the conjugate-gradient discriminators. On the other hand, as explained in Chapter 3, the decay/stability models have a broader purpose and are correspondingly more elaborate in architecture and training regimen.)

At this point, we may compare with the Mathews coefficients obtained in the star/galaxy discrimination problem. In this case, the reliance on raw success rates is not so problematic, since the frequencies of stars and galaxies in the relevant samples are roughly equal. Table 3 of Odewahn et al. gives the data needed for evaluation of C_{learn} for the models they investigated. In the case of the backpropagation networks (S1 and L1) trained respectively on small-diameter and large-diameter images, the results are 0.97 and essentially unity (in that order). To obtain a meaningful Mathews coefficient corresponding to joint use of the two backpropagation nets for prediction, we consider the magnitude binning interval 19.0–20.0, where the problem is clearly nontrivial. The data in Table 9 of the cited paper then yield $C_{\text{pred}} = 0.88$, attesting to the high reliability of this discriminator system as well as the relative simplicity of the problem.

Indeed, as has already been remarked, the star/galaxy discrimination problem is linearly separable to a rather good approximation (at least for the larger-diameter images). Consequently, the Elementary Perceptron classifiers are not substantially inferior to the three-layer backpropagation networks, in either learning or prediction. The situation is quite different for the nuclear stability/instability problem, for which hidden neurons are essential (Clark and Gazula 1991). Two-layer networks taught by backpropagation learn little more than the higher frequency of unstable nuclides. Considering the comparative difficulty of the nuclear problem, a good representation of the database, accompanied by significant predictive power (with $C_{\text{pred}} = 0.5 - 0.7$), has been achieved by fairly simple networks. However, a yes-no, black-and-white formulation of the nuclear stability problem is unlikely to admit significantly better results, since the networks are given no information about degrees of stability or instability. This remark leads into the far more sophisticated regression studies of nuclear binding in terms of the atomic mass table, developed at length in Chapter 3.

As a modest refutation of the alleged inscrutability of neural-network models, an issue raised earlier in this section, we may point to some remarkable features of the $^b(16 + H + 1)_u$ stability-detection networks trained by backpropagation (Clark and Gazula 1991, Gazula et al. 1992). These features are presumably shared by the sorting networks trained by the conjugate-gradient algorithm. Examination of the learning dynamics of the (3-layer) backpropagation nets, and inspection of the receptive field patterns $\{V_{jk}\}$ of their hidden neurons j after the completion of training, indicate clearly that these systems are able to capture the importance, for stability, of pairing and shell structure. They quickly learn to make distinctions between even-Z–even-N, even-Z–odd-N, odd-Z–even-N, and odd-Z–odd-N nuclei, based on the least significant input bits of Z and N. Moreover, certain hidden neurons develop into detectors for magic numbers (e.g. Z or $N = 2, 8, 20, 28, 50, 82$, and $N = 126$) and tend to excite the output neuron, which codes for stability. Other hidden neurons map out the unstable boundary regions of the valley of beta stability and inhibit activity of the output neuron for nuclidic inputs in those regions. Of course, most of the receptive field patterns are highly complex and difficult to interpret. Nevertheless, there is ample evidence that some of the networks have developed physically sensible internal models of the stability data on which they have been trained.

Spin-Parity Assignment. Another – less academic – problem in nuclear physics that reveals the strengths and weaknesses of statistical modeling with neural networks is learning and prediction of the spin and parity J^π of nuclear ground states (Clark et al. 1992, Gernoth et al. 1993). This sorting problem was studied to see if neural nets can adequately internalize and express certain relations obeyed by angular momenta in quantum mechanics, as well as regularities of the available data that are commonly described within the shell and collective models of nuclear structure (Bohr and Mottelson 1969). Ideally, one would like to design networks that can predict J and π with some confidence for "new" nuclides for which no measurements or empirical assignments are available.

Appropriate data sets for training and testing were formed from entries in the 1990 Brookhaven table of nuclides, which listed J^π assignments for 1889 nuclides. This database was broken down into data subsets corresponding to even-Z–even-N (EE), even-Z–odd-N (EO), odd-Z–even-N (OE), and odd-Z–odd-N (OO) nuclides, with respective populations 575, 437, 442, and 435.

Corresponding training sets were created by random deletion of nuclides from these subsets with probability 0.1, leaving complementary test sets containing 52, 51, 41, and 43 nuclides.

In most of the experiments to be described, the input variables Z and N were each encoded by 8 binary "on-off" neurons as in the network models constructed for the stability/instability problem. The obvious choice of target-output coding, adopted in the first round of experiments, is a unary representation in which one neuron is associated with each quantized angular momentum value up to some reasonable cutoff, and an additional output neuron detects the presence of even parity. Setting the cutoff at 15/2, the output layer then consists of 17 neurons: 15 to code for the spin values $J = 0, 1/2, 1, 3/2, \ldots, 7$, one to code for $J > 15/2$, and one for (even) parity.

Standard backpropagation (updated on-line, with a momentum term) was used to train three-layer networks, i.e., models of the type $^b(16 + H + 17)_u[P]$. Rather poor performance in learning and prediction was found when a network with $H = 20$ and $P = 697$ was trained on the full database of 1889 examples or on the training set of approximately 90% of these patterns. Greatly improved results were obtained when separate networks were developed for the four classes of nuclides (EE, EO, OE, and OO), using the corresponding training and test sets. Such networks (again with $H = 20$ and $P = 697$) learned the quantum mechanical restrictions to integral spin values for EE and OO nuclides and to half-odd-integral spins for odd-A nuclei; these rules were *never* violated by the mature nets, either on the training set or the test sample.

A further round of experiments was performed in which the output options were restricted to those permitted by the quantum-mechanical rules of angular-momentum addition, thus cutting down the number of output neurons by nearly one-half. Specifically, networks of type $^b(16 + 26 + 9)_u[685]$ were considered for odd-A nuclei (EO and OE classes); and of type $^b(16 + 25 + 10)_u[685]$ for even-A nuclei (EE and OO classes). The last spin output neuron was responsible for the cases $J \geq 8$ and $J \geq 15/2$ for the even and odd-A nuclei, respectively. (Another useful coding scheme maintains both integral and half-odd integral options, again with one unit for parity but also using one unit to signal half-odd-integral character for the spin; a further 10 output units serve to represent the integral part int(J) of the spin, 0 up to 8, with a "grab bag" unit to indicate int(J) ≥ 9. In passing, we remark that, based on this coding scheme, good results have been achieved with a *single net* trained on data sets containing all four classes of nuclides.)

In compiling success rates for the first two rounds of experiments, the angular momentum computed by the given network for an input nuclide (Z, N) was decided by a "winner-takes-all" competition between the output neurons dedicated to spin, and the parity assignment was taken as even [odd] when the parity neuron had activity $a_\pi \geq 0.5$ [$a_\pi < 0.5$]. It is tempting to interpret the actual activities of the output neurons as relative probabilities attributed by the net to the corresponding outcomes. This interpretation (indeed, in terms of absolute probabilities) can be made rigorous if the logistic squashing functions assumed for the output neurons are replaced by "soft-max" functions, as described in Chapter 3 or in Stolorz et al. (1991).

Mature networks developed for the EE class showed perfect performance on both the training and test sets belonging to that class. With sufficient training, the simple empirical fact that all even-even nuclei have spin-0, even-parity ground states was trivially grasped by these systems, since they never saw any counterexamples. At the opposite extreme, performance was generally rather poor for OO nuclides. Human theorists also have trouble with this class, for which the angular-momentum coupling rules are clearly the most complicated. Accordingly, attention focuses on performance levels for odd-A nuclei, which admit the most salient comparisons.

Percentages of successful assignments of spins and parities by networks of type $^b(16 + 20 + 17)_u[697]$ were as follows. For the EO class, the learning scores were respectively 93% and 97% for spin and parity, based on the 386-nuclide EO training set; whereas the predictive scores were 49% for spin and 78% for parity, based on the 51-nuclide EO test set. Somewhat better performance was obtained for the OE network, based on the 401-nuclide OE training set and the 41-nuclide OE test set: respectively 92% and 76% in learning and predicting spins, 94% and 73% in learning and predicting parities. Further studies involved modified EO and OE test sets of similar size but including 5 "simple shell-model nuclei." (A simple shell-model nucleus is an odd-A nucleus in which the proton or neutron subsystem containing the odd nucleon forms a closed shell plus (or minus) one nucleon.) The J^π values for these nuclides were predicted with significantly higher reliability than for other examples. Collectively, the success rates of all networks of type $^b(16 + 20 + 17)_u[697]$ in predicting the spins [parities] of the nuclides in their corresponding odd-A test sets averaged just above 60% [just above 80%].

The network models with output architectures restricted by the laws of angular-momentum addition showed learning scores similar to those found for the "17-output" nets (slightly better for parities, slightly worse for spins). Prediction rates were slightly worse for the odd-A classes, dropping to around 50% for the few networks that were trained. The main effect of the restrictive output coding appeared to be a sharpening in the output decision, with the "winning" spin neuron showing considerably greater dominance than in the experiments with nets that have output neurons for both integral and half-odd-integral spin values.

In the spin/parity experiments described to this point, no attempt was made to optimize network architecture; in particular, the number of neurons in the single hidden layer was chosen rather arbitrarily. Therefore a more systematic study was carried out, narrowing consideration to spin assignment for the odd-Z–even-N class, while retaining the three-layer architecture and the same input and output coding schemes (except for omission of the parity neuron). The OE training and test sets used were the same as in the initial study. However, networks were trained with the conjugate-gradient algorithm rather than with backpropagation. The best predictive performance was obtained for $H = 20$ hidden units and $P = 508$, with a score of 73.2%. Learning was nearly perfect, at 99.5%.

Results for conjugate-gradient nets with other numbers of hidden units are displayed in Table 1. We observe that the predictive score of 73.2% can be matched by a network with 15 hidden units and 125 fewer weight parameters. However, in the case shown, performance on the training set was significantly lower than for the net with 20 hidden units. The network with 25 hidden neurons showed clear signs of overtraining – learning was perfect, but predictive power was down sharply to 58.5%.

Table 1. Accuracy of performance of three-layer feedforward nets in the assignment of ground-state spins of odd-Z–even-N nuclides. Binary coding of Z and N is employed, and target outputs are represented by saturated activity of one of 8 output units associated with half-odd-integral spin values running from $\frac{1}{2}$ to $\frac{15}{2}$ (higher spins being lumped into the $\frac{15}{2}$ category). Networks with different numbers H of hidden neurons have been trained by the conjugate-gradient procedure. Learning refers to the training set of 401 nuclei, prediction to the test set of 41. The classification decision is imposed by a winner-takes-all criterion.

$I + H + O$	Learning (%)	Prediction (%)
$16 + 10 + 8$	84.5	58.5
$16 + 13 + 8$	89.2	65.9
$16 + 15 + 8$	93.5	73.2
$16 + 18 + 8$	93.7	70.7
$16 + 20 + 8$	99.5	73.2
$16 + 25 + 8$	100	58.5

An expected limitation of these models became apparent when the favored network with 20 hidden units was asked to decide on the spins of EO nuclei, a class it had not seen at all. Predictive performance dropped to a level of only 24%. Based on this result, the two classes of nuclides (OE and EO) should not be considered "homologous," since information on the $(N, Z) \to J$ mapping for the odd-even class does not give predictive leverage when the system is presented with even-odd input nuclides.

An additional set of experiments involving more elaborate coding and learning strategies yielded network models that were remarkably successful in the prediction of spin and parity of odd-A nuclei (Gernoth et al. 1993). An automated real-number coding scheme (Gazula et al. 1992) was used to create multi-unit analog representations of Z and N at the input interface and of J^π at the output interface. In these experiments, the J^π assignment for given (Z, N) was taken as that of the ground state *or* that of the longest-lived isomeric state, whichever has the greater halflife. A sophisticated backpropagation training scheme was implemented in which units are successively added to a single hidden layer (as well as to input and output layers) to achieve nearly optimal performance. Training involved repeated presentation of 417 even-Z–odd-N nuclides, and the mature networks were tested on 22 fresh examples in this nuclear class. For a case with architecture $16 + 16 + 15$, the training set was learned with 95.3% accuracy, and the mature net correctly predicted the J^π categories of better than 83% of the nuclides in the EO test set. The latter score compares very favorably with the 62% predictive accuracy reached for spins of odd-A nuclei by global nuclear structure calculations employing the macroscopic-microscopic approach (Möller and Nix 1990). On the other hand, when the network was tested on odd-Z–even-N nuclei, the predictive performance fell to around 10%. A net of similar type trained on OE data again scored at better than 80% on test nuclides from the same class, but only at the 20% level for EO examples.

The foregoing presentation has slipped back into the habit of giving raw percentages as a measure of classifier success. What about the gold standard of quality measures, the Mathews correlation coefficients? When tested on nuclides from the same class ("homologous nuclei"), the various neural-network spin or spin/parity models that have been developed for even-odd and odd-even nuclei generally show high predictive Mathews coefficients for the various spin possibilities, with values that are typically well above 0.5 even for nets which predict spins only at the 50% level overall. Predictive Mathews coefficients for the two parity choices are uniformly higher than for individual spin values.

Other Nuclear Properties. The two nuclear applications of multilayer perceptrons that have been discussed in detail – stability/instability discrimination and spin-parity assignment – involve classification problems. Other nuclear applications provide good examples of the use of such networks for nonlinear nonparametric regression:

o With the current surge in laboratory creation of new nuclei far from beta stability, *atomic-mass* predictors based on neural networks (Gernoth et al. 1993, Clark and Gernoth 1995, Gernoth and Clark 1995b) have attracted special attention due to their potential for supporting and complementing calculations with conventional theoretical methods. The cited work suggests that neural-network global models of mass systematics can compete with or surpass traditional models in the vicinity of the valley of stability, the figures of merit being rms errors in learning (fitting) and prediction (interpolation or extrapolation). However, this same work indicates a problem with extrapolation to new (Z, N) regions away from the "known" nuclei. Recent work (Athanassopoulos et al. 1998), which implements significant improvements in input coding and in the training algorithm, shows that the problem is not nearly so serious as had been thought, and that quite useful extrapolation capability ("extrapability") can in fact be achieved (see Chapter 3). The new coding scheme (two neurons for analog representation of Z and N; two parity units for unary representation of even or odd character of Z and N) increases parametric efficiency, while the new training scheme makes it easier to avoid local minima of the cost function.

o With Z and N as inputs, three-layer feedforward networks have been used to model *neutron separation energies* S_n of even-Z–odd-N nuclei. Real-number coding was employed for N and Z, as well as for the output S_n. Networks of types $^r(10 + 10 + 9)_r[209]$, $^r(18 + 18 + 18)_r[684]$, and $^r(18 + 38 + 18)_r[1424]$ were able to fit training data and generalize to test nuclei with average absolute relative errors of about 1% in learning and 2% in prediction (Gazula et al. 1992, Clark et al. 1992). A network of the second type made an impressive prediction for a whole line of data corresponding to $N - Z = 19$ (cf. Bohr and Mottelson 1969).

o Statistical modeling of *branching probabilities for different modes of decay* of nuclear ground states has turned out to be an extraordinarily advantageous and successful application of neural-network techniques (Clark et al. 1994, Gernoth and Clark 1995a). Multilayer perceptrons were trained on data from the Brookhaven archive, using a modified backpropagation algorithm based on the relative entropy or Kullback-Leibler distance (Kullback 1959) as cost function. Inputs Z and N were once again encoded in binary, while an array of output neurons provided an analog representation of the probabilities of decay into different modes (α decay, β^- decay, electron-capture, etc.), with an extra output neuron to allow for the exclusive possibility of stability. To permit the output of the network to be interpreted as a probability distribution, the output neurons were given normalized soft-max squashing functions. The procedure is described in more detail in Chapter 3. The best models resulting from this work are noteworthy in showing performance at mean error levels for branching probabilities that are conservatively measured at 5% on the training set and 15% on the test sample, with total close matches in learning and prediction at levels near 90% and 75%, respectively. (In a "close match," the output activities elicited by the input pattern must be within 5% of their targets.) Such performance figures could not easily be achieved by traditional quantum-theoretic approaches. Estimation of branching probabilities requires the evaluation of the partial widths (or partial decay rates) for all the decay modes involved. Calculation of these quantities generally entails direct consideration of energetics, selection rules, preformation rates, and barrier penetration probabilities. In practice, theoretical decay rates may differ from experiment by one or more orders of magnitude. Accordingly, this is a problem in which neural-network statistical modeling can more than hold its own.

○ In ongoing efforts (Mavrommatis et al. 1998), feedforward neural networks are being used to model the *systematics of β^- decay*. In design as well as intent, this work is patterned after well-known investigations within conventional nuclear theory (Staudt et al. 1990). Consideration is restricted to nuclides whose ground states decay 100% via the β^- mode. In the first study, training and test sets consisting, respectively, of 575 and 191 nuclides were drawn from the Brookhaven database. For the most part, the treatment was rather standard. At the input interface, Z and N were encoded in binary; and at the output, a single analog neuron was employed to represent the logarithm of the halflife. Architectures with one and two hidden layers were considered, and training was by the familiar ("vanilla") on-line backpropagation algorithm with a momentum term. The overall performance of the best networks, assessed in terms of the measures introduced by Staudt et al., approached that of the conventional models constructed by these authors. For example, the experimental halflives of almost all nuclides in the data sets with halflives less than 10^6 sec were reproduced within a factor 10. (Larger deviations occurred for 1 of the training-set nuclides and for 17.2% of the test-set examples.) Improvements upon this performance have been gained by adding another input neuron that codes for the Q-value of the decay, thus providing the network with information on the difference between the initial and final nuclear mass-energies. Further improvements are to be expected upon adoption of a more economical input-coding scheme and a more advanced training procedure.

6.4   Prediction of Protein Structure

The protein folding problem (Richards, 1991) is among the most urgent, fundamental, and difficult problems in all of science. The creation of a giant new high-tech industry based on "designer proteins" awaits its solution. Neural-network ideas and techniques have provided the impetus for a number of highly innovative contributions in this problem area, not least the development of associative-memory models of protein folding. (For citations, see the last paragraph of Sect. 4, attention being drawn especially to Sasai and Wolynes (1990) and Bohr and Wolynes (1992).) In keeping with the central theme of the current section, the discussion will focus instead on the application of layered feedforward nets to prediction of secondary and tertiary protein structure from the primary amino-acid sequence (Qian and Sejnowski 1988, Bohr et al. 1988,1990,1993a, Holley and Karplus 1989, Kneller et al. 1990, Wilcox et al. 1990, Stolorz et al. 1991, Rost and Sander 1992,1993a-c,1994, Rost et al. 1993,1994, Bohr and Brunak 1994, Bohr 1998).

It is most instructive to introduce this important effort by reviewing the original work of Bohr et al. (1988). In this work, backpropagation was used to teach three-layer networks to associate primary and secondary structure by classifying amino-acid residues into one of two categories for each of three types of secondary feature: α-helix (or not), β-sheet (or not), and random coil (or not). This approach to the prediction of protein secondary structure offered a novel alternative to traditional statistical approaches (e.g. Chou and Fasman 1978) and to computer-intensive *ab initio* calculations taking account of interactions of residues with one another and with the environment (e.g. Paine and Scheraga, 1987).

The problem has much in common with that of pronouncing written English text. A single character of text does not correspond uniquely to one particular phoneme (compare the "a" in "fat" with that in "meat"); rather, the correct pronunciation depends on context, i.e. on correlations with other nearby characters. Likewise, the residue-to-secondary-unit mapping depends, in a highly nontrivial way, on preceding and successive residues in the series. Accordingly, the design of nets for prediction of secondary protein structure was patterned after that of NETtalk (Sejnowski and Rosenberg 1987), the layered system taught by backpropagation to convert written to spoken English. In the favored arrangement, an input layer views a range of 51 successive residues, the center residue being the target and the 25 on either side furnishing its context. Each of the 51 places in the "window" may be occupied by any of the 20 possible individual amino acids. A unary or "grandmother" input representation was adopted: Each residue was encoded as 19 zeros and 1 one (the latter appearing in a different position for each different amino acid). With the window size of 51, this coding prescription implies an input layer of 1020 neurons. The hidden layer contained 40 units. The output layer consisted of 2 neurons, the sum of their activities being unity, so that in any particular case one or the other was dominant, yielding an exclusive indication of whether or not the target residue participates in the secondary feature being investigated. The total number of weight parameters was 40,922. (In the shorthand notation we have been using, the network type chosen is $^u(1020 + 40 + 2)_u[40,922]$.) Separate networks were constructed to predict participation (or not) in each of the three secondary forms: α-helix, β-sheet, and random coil. (Thus the problem of predicting secondary structure was subdivided into three classification problems of the simple detection type.) Training and testing was based on a sample of 56 proteins, selected from the Brookhaven Protein Data Bank, for which the amino-acid-to-secondary-structure mapping has been determined by X-ray diffraction. A net was trained on a subset of n proteins from the sample; the remaining $56 - n$ were reserved to test predictive performance. For the network that learns to detect α-helix structure, the percentage of correctly predicted associations approached 73% (with a Mathews correlation coefficient $C_{\text{pred}}(\alpha) = 0.38$) as the training set was enlarged stepwise to include the full sample. A detailed comparison with the results of the Chou-Fasman analysis in the case of the membrane-bound rhodopsin protein gave apparent

evidence of a superior discriminatory ability of the neural net in the assignment of α-helix structure to segments of the protein molecule, in that the
network indicated seven (rather than six) α-helices. The correlation between
actual and target responses of the trained network declined when the window
size was taken smaller than 51, but did not increase appreciably for larger
sizes.

We may pause for a moment and reflect on the quality of predictive performance realized by neural networks in the dual-option classification problems
that have been discussed. The three problems in question are discrimination
of: (i) stars from galaxies; (ii) stable from unstable nuclides; and (iii) residue
involvement in an α-helix rather than another protein backbone structure.
(All three are of course equivalent to detection problems, in the terminology
of Sect. 5.1.) In the first of these problems, Mathews coefficients for star or
galaxy detection well above 0.8 were obtained; and in the second, the highest
Mathews coefficients found for stability (or instability) were just below 0.7.
Since the level of refinement of the treatments of these two problems was
similar, it is plausible to suppose that their difficulty [(i) easier than (ii)] is
aptly reflected in the predictive significance achieved, as measured by the
quoted Mathews coefficients.

By the same token, the protein secondary structure problem (iii) appears
to be much more difficult than either of the other two. Indeed, a number of
groups have expended considerable effort to push $C_{\mathrm{pred}}(\alpha)$ significantly higher
than 0.4, without success. (See Stolorz et al. (1991) for a particularly incisive
study.) The "ground rules" of this game required adherence to the spirit
of the scheme described above, in which the network receives information
only about the primary structure, i.e., the sequence of amino-acid residues
in the protein. A very important consideration was and is "fair play" in
the choice of training and test sets, with degrees of homology being a key
issue (Rost et al. 1993). More recently, $C_{\mathrm{pred}}(\alpha)$ values in excess of 0.5 *have*
been achieved with improved and properly sanitized training and test sets,
although at the expense of supplying the network with additional information
(e.g. of evolutionary nature), through modifications of the training procedure,
inputs, or architecture relative to the standard scheme. For penetrating views
on these and other matters, see Chapters 5 and 6 and Bohr (1998).

Three groups (Qian and Sejnowski 1988, Bohr et al. 1988, Holley and
Karplus 1989) were quick to exploit the transparent analogy of the secondary-
structure problem with that of mapping written text to spoken phonemes and
its "solution" by NETtalk. However, it was Bohr et al. (1990) who followed
with a pathbreaking step toward prediction of tertiary or folded structure by
designing a feedforward network capable of generating inter-residue distance
information as well as secondary-structure assignments. The architecture introduced by these authors contains three unary-coded output nodes, one for
each of the three nominal secondary structures, plus a set of 30 output units
whose activities provide estimates for elements of a binary distance matrix.

These elements constrain the three-dimensional, folded structure of the protein by specifying which of the M α-carbon atoms of the backbone preceding the central amino acid of a W-residue window are within a given threshold distance d_o of this target residue. The published results refer to $M = 30$, $W = 61$, $d_o = 0.8$ nm. In recording the decisions of the output neurons representing distance-matrix entries, an activity near 1 [near 0] was interpreted as a "yes" ["no"] answer to the question: is the corresponding residue within distance d_o? With 300 units in a single hidden layer, the enlarged network was of type $^u(1220 + 300 + 33)_u[376, 233]$. (The number of weight parameters has reached the staggering figure of 376,233.) Using backpropagation, the system was trained on a set of proteins that are functionally, but not structurally, homologous. The distance constraints predicted from the primary sequence of a test protein were employed in a steepest-descent routine that folds the protein backbone until a maximal number of such constraints is satisfied. Comparing with tertiary structures derived from X-ray data, good results were obtained for the trypsin ITRM (223 residues long, and 74% homologous to the starting configuration, 4PTP). The steepest-descent procedure was tested on the trypsin inhibitor 6PTI (56 residues long). Using an experimentally derived binary distance matrix, an excellent fit of the tertiary form of 6PTI was obtained starting from a random initial configuration. This work must be viewed as an impressive start on a problem of awesome difficulty.

Protein-structure studies based on neural networks have proliferated in the decade that has elapsed since the pioneering efforts sketched above. For reports on more recent developments and overviews of the status of this new field of computational research, the reader is directed to Chapters 5 and 6, to the cited articles by Rost and coworkers, to the conference collection edited by Bohr and Brunak (1994), and to the monograph of Bohr (1998).

7 Neural Networks Approximate Bayesian Inference

The status of neural-network models within Bayesian estimation theory is a subject rich with fundamental and practical ramifications. The structural relation between layered feedforward nets and Bayes' famous rule for the assembly of posterior distributions is already present, in seminal form, in Minsky and Papert's pivotal treatise on perceptrons (Minsky and Papert 1969). This connection is also close to the surface in the classic book of Duda and Hart (1973). Shifting the perspective from structure to implementation, more recent works (Ruck et al. 1990, Wan 1990, Richard and Lippmann 1991) have shown that neural-network classifiers estimate Bayesian posterior probabilities. More specifically, it has been demonstrated, under modest assumptions, that perceptron-based approaches to classification yield architecture-limited approximations to the posterior probabilities of Bayes optimal classifiers.

There are many other facets of the Bayes $\leftrightarrow$ neural-network connection. Anderson and Abrahams (1987) were the first to point out that the original

Hopfield model can be viewed as implementing Bayesian inference on analog quantities in terms of probability density functions. More generally, they showed that Bayesian statistics can provide a formal framework for designing associative neural networks. In a similar spirit, Lansner and Ekeberg (1989) introduced a one-layer, feedback-coupled model composed of analog neurons and implemented a learning algorithm derived from Bayes' rule. In this model, designed to perform a content-addressable memory task, learning is a process of collecting statistics and recall is a statistical inference process. Buntine and Weigend (1991) and MacKay (1992a,b) have elaborated upon Bayesian formulations of backpropagation. The comprehensive review of progress in learning theory and generalization published by Watkin et al. (1993) adopts Bayesian probability theory as a unifying perspective.

These and other developments have greatly advanced our understanding of artificial neural networks as new tools for statistical analysis and statistical inference (Smith 1993, Cherkassky et al. 1994, Haykin 1999). On the other side of the coin, recent developments in computational neuroscience, pioneered by Charles Anderson and collaborators (Anderson 1994,1996, Anderson and Van Essen 1994, Eliasmith and Anderson 1998, Hakimian et al. 1998) signal a major thrust toward Bayesian probabilistic theories of population-coded information processing in the brain (see, for example, Girosi and Poggio (1995), Lewicki and Sejnowski (1996), Zemel and Hinton (1995), Zemel and Dayan (1997), and Zemel et al. (1997,1998)).

The purpose of this section is to make the statistical foundation of artificial neural networks both more concrete and more transparent by introducing a very flexible type of two-layer perceptron which has the capacity to represent all higher-order correlations between input variables and match the probabilities generated by a Bayes classifier. In the context of the proofs given by Ruck et al. (1990), Wan (1990), and Richard and Lippmann (1991), this will remove architecture as an obstacle to achievement of Bayes-optimal performance and place the onus of imperfection entirely on the training data and the training process.

We shall first determine what structural properties a Bayes classifier should have in general. It will then be rather easy to see how we can match those properties with a suitably designed two-layer perceptron.

An important set of problems in pattern recognition involves the classification of patterns represented by vectors $x = (x_1, ..., x_K)$ of finite length K. The classifier system is to assign each such pattern to one of a finite number L of distinct categories λ. It is well known (Duda and Hart 1973) that Bayes' rule provides the basis of an optimal strategy for dealing with this decision problem, which was called a "sorting problem" in Sects. 5 and 6. According to *Bayes' rule of inference*, the posterior probability that the correct category is λ when the pattern is known to be x is given by

$$p(\lambda|x) = \frac{p(x|\lambda)P(\lambda)}{p(x)} . \tag{44}$$

In this expression, $p(x|\lambda)$ is the class- (or state-) conditional probability that the pattern is x when the category is known to be λ (the "likelihood" of the data x) and $P(\lambda)$ is the prior probability of finding λ. The denominator $p(x) = \sum_\nu p(x|\nu)P(\nu)$ appearing in rule (44) is just a normalization constant that ensures satisfaction of $\sum_\lambda p(\lambda|x) = 1$, as required for a probability distribution over exhaustive outcomes λ. The optimal decision strategy is to select that category μ among $\lambda = 1, ..., L$ for which $p(\lambda|x)$ assumes the largest value. This recipe is optimal in the sense that it minimizes the probability of misclassification.

In the simplest situation, the input variables x_1, x_2,..., x_K are independent of one another, which implies that the class-conditional probability $p(x|\lambda)$ must be just a product of independent factors $\rho_i(x_i|\lambda)$, one for each variable x_i, $i = 1, ..., K$. In general, however, correlations between these variables of all orders up to K may be present in the vectors of the data ensemble. This general situation can be represented by the product decomposition

$$p(x|\lambda) = \prod_i \rho_i(x_i|\lambda) \prod_{i<j} \rho_{ij}(x_i x_j|\lambda) \prod_{i<j<k} \rho_{ijk}(x_i x_j x_k|\lambda) \cdots \rho_{1...K}(x_1...x_K|\lambda),$$

$$(45)$$

where the indices $i, j, k, ...$ have the range 1 to K subject to the indicated restrictions, and commas between arguments x_i, x_j, x_k, ... are omitted for brevity. The first product represents the contribution to the class-conditional probability from the x_i regarded as independent variables; the second, the modification produced by pairwise correlations of the x_i; the third, the modification due to triplet correlations; and so on up to the effects of K-wise correlations. The factors $\rho_{...}$ are positive semi-definite and together preserve $0 \le p(x|\lambda) \le 1$. The order chosen for the arguments $x_{k_1}, ..., x_{k_m}$ is irrelevant, but the $K!/m!(K-m)!$ factors $\rho_{k_1 \cdots k_m}(x_{k_1} \ldots x_{k_m}|\lambda)$ arising for a given order m may be different functions of their arguments.

The decomposition (45) may be regarded as an identity. Whether or not it is a useful representation depends on the complexity, or rather the simplicity, of the decision tree of the problem being solved. It will be useful if $p(x|\lambda)$ can be accurately approximated by the nontrivial inclusion of a relatively small number of factors in the general product, with the other factors set to their trivial values of unity. Keeping only the first $\prod$ factor in this sense, i.e., assuming the input variables to be independent of one another, we obtain what is called a *naive Bayes classifier*. Discussions of the architectural connection between neural nets and the Bayes rule normally stop at this level (Minsky and Papert 1969).

It is a somewhat amusing fact that the product decomposition of $p(x|\lambda)$ has a conspicuous analog in quantum many-body physics. The ground-state wave function $\Psi(x_1 x_2 \ldots x_K)$ of a system of indistinguishable bosons has an exact Feenberg product representation (Clark 1979)

$$\Psi(x_1 x_2 \ldots x_K) = \prod_i \phi(x_i) \prod_{i<j} f_2(x_i x_j) \prod_{i<j<k} f_3(x_i x_j x_k) \cdots f_K(x_1 \ldots x_K)$$

$$(46)$$

in terms of one-body, two-body, three-body, ... factors reflecting the influence of a mean field and the existence of a hierarchy of interparticle correlations. The structural analogy with (45) is illuminating in spite of the fact that the wave function for the Bose system is symmetrical in all particle coordinates, whereas the inputs to the classifier system are distinguishable from one another.

Guided by this analogy, it is natural to imitate variational approaches in many-body physics (Navarro and Polls 1998) and investigate successive approximants $M = 1, 2, \ldots$ in which higher factors $\rho_{k_1 \ldots k_m}(x_{k_1} \ldots x_{k_m} | \lambda)$, with $m > M$, are set equal to unity. The decomposition (45) is more general than the product representation used by Duda and Hart to generate the Chow expansion (Chow 1968). Curiously, the Duda-Hart expansion has a structure analogous to that of the wave function used to model fermion pairing in superconductors (Schrieffer 1964). Another decomposition (Bahadur 1961) is the counterpart of a perturbation expansion of the many-body wave function. In fact, the connection between perceptrons and Bayes' theorem developed by Stolorz et al. (1991) is made through the Bahadur expansion.

We now resume to the main thread of our argument. Having exposed through expression (45) the essential structural requirements for a Bayes classifier, we consider a two-layer feedforward neural network consisting of K input neurons (with labels i, j, k, etc.) whose activities register the components of a given pattern vector x, and L output neurons λ whose activities y_λ are to be interpreted as probabilities. To guarantee that the outputs y_λ are positive and sum to unity, we suppose that they are determined by a "soft-max" squashing function (Stolorz et al. 1991)

$$y_\lambda(x) = \frac{e^{u_\lambda(x)}}{\sum_\nu e^{u_\nu(x)}}, \qquad (47)$$

where u_λ is the net stimulus to unit λ from the units of the input layer. The nature of the interactions between input and output neurons is still to be specified.

Next comes the crucial step in the constructive proof. To establish an explicit formal connection between Bayes' rule and the operation of the two-layer probabilistic neural network, the Bayes posterior probability for each class λ is identified with the activity y_λ of the output neuron corresponding to that class:

$$p(\lambda|x) = \frac{p(x|\lambda)P(\lambda)}{\sum_\nu p(x|\nu)P(\nu)} = y_\lambda(x) \equiv \frac{e^{u_\lambda(x)}}{\sum_\nu e^{u_\nu(x)}}. \qquad (48)$$

We make the additional obvious identification

$$u_\lambda(x) = \ln\left[p(x|\lambda)P(\lambda)\right] = \ln p(x|\lambda) + \ln P(\lambda) \qquad (49)$$

and evaluate $\ln p(x|\lambda)$ using the product decomposition (45):

$$\ln p(x|\lambda) = \sum_i \ln \rho(x_i|\lambda) + \sum_{i<j} \ln \rho(x_i x_j|\lambda) + \sum_{i<j<k} \ln \rho(x_i x_j x_k|\lambda) +$$
$$+ \cdots + \ln \rho(x_1 ... x_K|\lambda) . \tag{50}$$

Simplifying the notation, subscripts have been removed from the various ρ functions, which must now be identified by the labels on their arguments. It may be helpful to emphasize that i, j, k, etc. denote *distinguishable* input neurons.

To reduce expression (50) and translate (49) into a more familiar form involving neuronal activities and connection weights, we restrict considerations to the case that the input patterns are bit strings, i.e. $x_k \in \{0,1\}$, with $k = 1, ..., K$. It is convenient to write $1 - x_k = \bar{x}_k$ and to introduce an abbreviated notation made transparent by the examples

$$\rho(x_1 = 1, x_2 = 0|\lambda) = \rho_{1\bar{2},\lambda} , \quad \rho(x_1 = 1, x_2 = 1, x_3 = 0, x_4 = 1|\lambda) = \rho_{12\bar{3}4,\lambda} . \tag{51}$$

A sequence of identities beginning with

$$\rho(x_1 x_2|\lambda) = \rho_{12,\lambda}^{x_1 x_2} \rho_{1\bar{2},\lambda}^{x_1 \bar{x}_2} \rho_{\bar{1}2,\lambda}^{\bar{x}_1 x_2} \rho_{\bar{1}\bar{2},\lambda}^{\bar{x}_1 \bar{x}_2} ,$$

$$\rho(x_1 x_2 x_3|\lambda) = \rho_{123,\lambda}^{x_1 x_2 x_3} \rho_{12\bar{3},\lambda}^{x_1 x_2 \bar{x}_3} \rho_{1\bar{2}3,\lambda}^{x_1 \bar{x}_2 x_3} \rho_{\bar{1}23,\lambda}^{\bar{x}_1 x_2 x_3} \rho_{1\bar{2}\bar{3},\lambda}^{x_1 \bar{x}_2 \bar{x}_3} \rho_{\bar{1}2\bar{3},\lambda}^{\bar{x}_1 x_2 \bar{x}_3} \rho_{\bar{1}\bar{2}3,\lambda}^{\bar{x}_1 \bar{x}_2 x_3} \rho_{\bar{1}\bar{2}\bar{3},\lambda}^{\bar{x}_1 \bar{x}_2 \bar{x}_3} , \tag{52}$$

and continuing with $\rho(x_1 x_2 x_3 x_4|\lambda)$, etc., then follows from the assumption that the inputs x_k can only take on the values 0 and 1.

After introducing the identities of type (52) into (50), routine manipulations lead to an expression for the stimulus (49) of the form

$$u_\lambda(x) = V_{\lambda,0} + \sum_i V_{\lambda,i} x_i + \sum_{i<j} V_{\lambda,ij} x_i x_j + \sum_{i<j<k} V_{\lambda,ijk} x_i x_j x_k +$$
$$+ \cdots + V_{\lambda,12\cdot\cdot K} x_1 \cdots x_K . \tag{53}$$

Explicit formulas for the connection weights $V_{\lambda,k_1...k_m}$ and bias $V_{\lambda,0}$, in terms of the conditionals $\rho(x_{k_1} \ldots x_{k_m}|\lambda)$ and the prior $P(\lambda)$, may be found in Clark et al. (1994,1999). The familiar architecture of the Elementary Perceptron, involving only pairwise couplings from input to output neurons, is recognizable in (53) upon suppression of all terms nonlinear in the input variables.

From this demonstration we may conclude that, in general, our two-layer probabilistic network with soft-max output functions (47) can match Bayes' rule if and only if forward couplings from the input neurons to the output neurons of all orders up to K are permitted. Such a system will be called a *higher-order probabilistic perceptron* (HOPP).

It is interesting that this demonstration shares an important property with the probability density functional or PDF approach to neural information processing introduced by Anderson (1996). Upon invoking Bayesian principles of statistical inference, this approach provides a framework for the design of neural-network models that can carry out prescribed information-processing tasks. The resulting models come complete not only with architecture, but also with connection weights, which are determined by conditional probabilities associated with the assigned task and by basis functions that are related to prior probabilities.

Returning to artificial neural networks and classification, architecture is only part of the story – as has already been intimated earlier in this section. In real-world problems one hardly ever has direct access to the conditionals and priors needed to evaluate the weights and biases in expression (53). They must be learned from examples.

It is straightforward to formulate a procedure for training HOPPs based on incremental gradient-descent minimization of a squared-error cost function (*à la* backpropagation, *sans* hidden layers). Alternatively, a relative-entropy cost function may be used. The development of such training algorithms is sketched in Chapter 3. Hopefully, these procedures will move the weight parameters ever closer to those needed to match Bayes performance, but it is not at all obvious in what sense this is being accomplished and what obstacles remain. It is here that the cited results of Ruck et al., Wan, and Richard and Lippmann come into the picture.

As already indicated, these authors have established rather strong theorems relating perceptron training procedures to the estimation of conditional probabilities, and indeed to the estimation of Bayesian posterior probabilities. The essential results have been stated most succinctly by Richard and Lippmann (1991). It is assumed that there is one output for each pattern class. Two special cases are examined. The first is the standard one in which the desired (i.e., target) outputs are "1 of M" (or "1 of L" in our notation), meaning that one output should be unity, and all others zero. In the other case, the target outputs are binary, i.e., unity or zero, without restriction on the number that can simultaneously take the value 1. Inputs may be continuous or binary. Paraphrasing:

⋄ When network parameters (i.e., the weights) are chosen to minimize a squared-error cost function, outputs estimate the conditional expectations of the desired outputs so as to minimize the mean-squared estimation error.

⋄ For a 1 of M problem, when network parameters are chosen to minimize a squared-error cost function, the outputs provide direct estimates of the posterior probabilities of the Bayes classifier so as to minimize the mean-squared estimation error.

⋄ Finally, when the desired outputs are binary but not necessarily 1 of M and the network parameters are chosen to minimize a squared-error cost function, the outputs estimate the conditional probabilities that the desired outputs are unity, given the input.

For completeness, we append the remark that if the target outputs are the Bayes posterior probabilities themselves (supplied somehow!), these quantities are tautologically estimated in the same minimum mean square-squared error sense if network parameters are chosen to minimize a squared-error cost function.

Parallel results hold when network parameters are chosen to minimize a cross-entropy cost function or the Kullback-Leibler relative entropy (Hampshire and Perlmutter 1990a,b, Richard and Lippmann 1991, Clark et al. 1999). The stated results apply, in particular, for networks with sigmoidal squashing functions trained by the backpropagation algorithm, for radial basis function networks (Moody and Darken 1989, Specht 1990, Niranjan and Fallside 1990, Poggio and Girosi 1990a,b, Hertz et al. 1991, Haykin 1999), and certainly for HOPP networks trained by the procedures introduced in Sect. 5 of Chapter 3 or in Clark et al. (1999).

According to Richard and Lippmann, the implications and practical benefits for pattern recognition are that network outputs can be used as Bayesian probabilities in simple classification tasks and can be treated as probabilities when making higher-level decisions. As always, there is a catch or two (or three). Trained networks will estimate Bayesian probabilities or conditionals poorly in cases of (a) inadequate or improper network architecture, size, or connectivity; or (b) limited and/or biased training data; or (c) inadequate search procedures for finding the global minimum of the chosen cost function.

It might appear that "catch" (a) is removed by adopting HOPP architecture with its higher-order connectivity. However, in structures allowing for all possible higher-order connections, totaling $L \sum_{m=0}^{K} \binom{K}{m}$ for K input units and L output units, there is the obvious danger that the incipient combinatoric explosion of weight parameters will overwhelm available computational resources and preclude tractable application when the size of the system grows beyond modest numbers of input units. Some means must be found for trimming the connection tree down to its indispensable branches, i.e., for optimizing the HOPP architecture.

We may remark that if many higher-order correlations between input variables are present in the problem at hand, the difficulty is intrinsic and must also be confronted when seeking to solve the problem using more traditional multilayer, pairwise-coupled feedforward networks. Accurate modeling will necessitate large numbers of hidden units, optimally arranged in one or multiple hidden layers. The HOPP scheme has the virtue of making all higher-order correlations explicit from the outset.

Although the number of input variables may be large in a real-world problem, it will often be true that only a few higher-order correlations are important and that most of the weight parameters in the general HOPP architecture are not needed. A good knowledge of the problem domain may allow one to identify and eliminate these irrelevant parameters. Yet in general it is desirable to have available some systematic procedure for skeletonization of HOPP networks that will – with some acceptable tradeoff of accuracy – quench the combinatoric explosion and reduce parametric costs to manageable proportions. Such a scheme is offered and tested in Chapter 3, where the story of HOPP networks is resumed.

By this point, the operation of the two-layer HOPP system as an inference machine should be apparent:

⋆ In the training phase, the HOPP learns the statistics of the problem, determining its weight parameters in terms of conditionals and priors.

⋆ In the computational phase, it effectively applies Bayes rule of inference.

Moreover, the work of a variety of authors, including Anderson and Abrahams, Lansner and Ekeberg, and Richard and Lippmann, has informed us that this same statistical description has broad validity for neural-network models that perform pattern-recognition tasks.

8   Doing Science With Neural Nets: Pride and Prejudice

It is appropriate to close this introductory chapter with a collection of observations and opinions on the utility of artificial neural networks as tools for modeling and data analysis in science. The remarks below will focus primarily on network models designed for pattern recognition and function approximation. At issue are the strengths and weaknesses of neural-network techniques compared to established approaches.

A number of researchers have addressed this issue, either within the context of scientific deployments or more generally. Zupan and Gasteiger (1991,1993) have evaluated and compared a wide range of applications in chemistry. A recent summary of neural computational methods and applications, "with high-energy physics taking central stage," has been given by Horn (1997). Contrasting with the generally positive tone of these and other commentators, Duch and Diercksen (1994) have leveled sharp criticisms at what they consider inappropriate or indiscriminate use of backpropagation networks in physics and chemistry and at gratuitous interpretations of the results obtained. In a provocative exercise, Ripley (1994) carried out a number of comparative studies within a broader context, pitting multilayer perceptrons against state-of-the-art methods from mathematical statistics, notably MARS (Friedman 1991) and projection pursuit (Friedman and Stuetzle 1981). Ripley's analysis of the results cast serious doubt on the claims being made for the superiority of neural-network techniques in classification and function approximation. On the other hand, the critical conclusions drawn from this work lose some of their impact when considered in the light of Friedman's comprehensive discussion of predictive learning and function approximation. This balanced overview (Friedman 1994) places neural networks and more traditional statistical approaches within a common framework, providing guidelines for objective comparisons.

When neural networks re-emerged on the scene in the mid-80s as a new and glamorous computational paradigm, the initial reaction in some sectors of the scientific community was perhaps too enthusiastic and not sufficiently critical. There was a tendency on the part of practitioners to oversell the powers of neural-network or "connectionist" solutions relative to conventional techniques – where conventional techniques can include both traditional theory-rich modeling and established statistical methods. The last five years have seen a correction phase, as some of the practical limitations of neural-network approaches have become apparent, and as scientists have become better acquainted with the wide array of advanced statistical tools that are currently available.

A number of the objections and warnings expressed during the corrective phase are well taken; others are less justified. It is certainly inappropriate to use neural networks to carry out arithmetic or symbolic manipulation tasks, which are much more efficiently performed by sequential computer programs. The potential strengths of neural networks, which are inherently parallel structures, lie rather in pattern recognition or more broadly in statistical inference. Like the brain, they might be expected to excel in situations where a high error rate is tolerable and the problem to be solved is mathematically ill-posed. As we have seen, neural networks offer novel approaches to solution of constrained-optimization problems, although they will rarely yield the best solutions.

In fact, the vast majority of applications, in science or otherwise, do play to the potential strengths of connectionist systems and primarily involve regression or classification by layered feedforward networks (i.e., perceptrons) that have received supervised training. Although backpropagation is the industry standard, more sophisticated training algorithms – e.g. conjugate-gradient and genetic algorithms (Hertz et al. 1991) – have come into wide use. Self-organizing feature maps (Haykin 1999), also referred to as topology-conserving maps (Ritter and Schulten 1988) and Kohonen nets (Kohonen 1989,1997), afford an increasingly popular alternative to perceptrons in some problems of pattern recognition and classification (see, e.g., Chapter 7).

In presenting the results of such applications, one should be careful not to overinterpret what a network has actually accomplished. Reiterating an example cited by Duch and Diercksen, one should not be deluded that ordinary feedforward network models "understand" or even "embody" the Schrödinger equation just because they are able to learn – by example – the correlations between the parameters of some textbook Hamiltonian and its lowest energy eigenvalue. Neural networks should not be expected to do the formal or conceptual work of human scientists.

Once it is realized that neural-network algorithms are just another set of statistical tools, their mystique largely evaporates and the temptation to anthropomorphize their function is greatly reduced. Attention has already been called to the irony of this reaction (albeit a welcome one), noting that biological neural networks in the brain can also be viewed as statistical inference machines.

Moving on more directly to questions of technical utility, it will be convenient for brevity of expression to absorb "classification" into the broader term of "function approximation" and lump it with "regression," noting with Friedman (1994) that categorical variables (our "attributes") can be mapped into real-valued variables.

It is surely a wasteful exercise to invoke the heavy machinery of neural networks to fit a simple, relatively smooth function of a small number of variables. This is most obvious in situations where physical grounds dictate or motivate a simple global form for the function, whose parameters can be easily determined by a least-squares minimization. More generally, neural-network modeling may be an inefficient choice in low-dimensional problems when the familiar method of linear expansion in a set of basis functions (including, as special cases, Fourier expansion, spline interpolation, and polynomial fits) would give an accurate representation (Duch and Diercksen 1994).

Still, the pendulum should not swing too far in the direction of downgrading neural-network approaches. While many low-dimensional problems might best be attacked by familiar linear methods such as polynomial fitting, the situation is not always so clear-cut. To give one example, the nuclear problems considered in Sect. 6.3 involve only the proton and neutron numbers Z and N as input variables, but the physical properties that are being modeled are generally *not* simple, smooth functions of these variables. The integral character of Z and N is essential to the description of quantum phenomena such as shell structure and pairing. This important aspect of nuclear physics has been successfully incorporated in the global models based on feedforward networks. Another example may be found in the work of Blank et al. (1995), who trained multilayer neural networks for use as fast interpolators of interaction potentials in chemical systems. In this work, detailed comparisons were made with cubic-spline fits in problems with two and with three degrees of freedom. Comparable predictive accuracy was achieved with the two regression techniques (neural-net and cubic-spline). On the other hand, the neural-network approach has distinct advantages, including (i) greater parametric economy and attendant robustness of performance in the presence of noisy data and (ii) the ability to handle data sets with nonuniform spacing between points.

Turning to problems where the input vector is of high dimension and/or traditional linear regression methods prove inadequate, multilayer neural networks may have much to offer, *along with* other sophisticated methods of statistical analysis that have come on the scene in the last two or three decades. Neurally inspired approximation systems occupy a valid place in the taxonomy of nonparametric estimation (Friedman 1994). Moreover, it is now generally accepted that in some problem domains, multilayer perceptrons can compete effectively with other, more established nonparametric regression procedures such as projection pursuit and recursive partitioning techniques (e.g. CART and MARS) (Cherkassky et al. 1994, Smith 1993). For example, it has been shown that neural-network function approximation is superior to the MARS procedure in some cases with noisy data (De Veaux et al. 1993).

In passing judgment on neural networks, it should be understood that all approaches to nonlinear nonparametric regression must contend with the "curse of dimensionality" (Bellman 1961, Friedman 1994). In a worst case scenario, the number of sample points needed to estimate a function increases exponentially with the problem dimensionality. Thus, in high dimensions, all feasible training samples are pitifully sparse, and all sample points are effectively "outliers." Thus, the dilemma of bad extrapability encountered in big problems is hardly unique to backpropagation networks. This commonality has the advantage that statistical methods already developed for dealing with poor generalization performance (notably, regularization techniques) can be transplanted to neural-network modeling – a "misery-loves-company" effect.

Any universal method for function approximation (whether based on multilayer perceptrons or some more conventional approach) has its own distinctive strengths and weaknesses; hence its performance relative to other universal methods may be very sensitive to the choice of problem. Another consideration that must be kept in mind when evaluating a critical numerical comparison of methods is that the author of the comparison will usually be implementing the different methods with different levels of skill (Friedman 1994, Cherkassky et al. 1994). Even if, in fair competition, a neural-network treatment proves not to be the optimal choice for speed and accuracy in a given application, it may still be preferred for practical reasons of convenience, such as readily available and easily adaptable software.

It is also important not to define artificial neural networks too narrowly when weighing their practical value. Feedforward nets with logistic-sigmoid units trained by "vanilla" backpropagation are the most popular and visible realizations, but in truth they are only the beginning of the story. The power and flexibility of neural-network approaches can be substantially enhanced by such elaborations on the usual perceptron theme as:

(a) Adaptive architectures, which allow a network to grow or shrink to fit the problem (Mézard and Nadal 1989, Fahlman and Lebiere 1990, Hinton 1986, Le Cun et al. 1990b, Gernoth et al. 1993, Gernoth and Clark 1995a), prominent examples being the Fahlman-Lebiere cascade-correlation algorithm and various weight-decay and pruning recipes.

(b) Recurrent connectivity, prominently represented by Boltzmann machines (Ackley et al. 1995, Hinton and Sejnowski 1986), recurrent backpropagation (Pineda 1987,1989, Almeida 1987,1988), Elman nets (Elman 1990), and real-time recurrent learning (Williams and Zipser 1989a,b).

(c) Combination of supervised and unsupervised learning schemes, as in counterpropagation networks (Hecht-Nielsen 1987,1988, Huang and Lippmann 1988) and hybrid approaches falling under the term "radial basis functions" (Moody and Darken 1989, Specht 1990, Niranjan and Fallside 1990, Poggio and Girosi 1990a,b, Hertz et al. 1991).

It should be no surprise that all three of these themes are present in the work described in subsequent chapters of this book, for example: (a) in Chapters 3 and 5, (b) in Chapter 4, and (c) in Chapters 4 and 9. The reference to radial basis functions under (c) serves as a reminder that one is by no means restricted to the choice of logistic sigmoids for the neuronal transfer or squashing functions. Certainly, the transfer function should be adjusted to the specific needs of the problem at hand.

The idea of committee machines (Nilsson 1965, Haykin 1999) opens another attractive option for enhancing the performance of connectionist systems. The outputs of several different predictors are combined according to a linear or nonlinear rule (e.g. ensemble averaging (Horn 1997)) to produce an overall output that, at least in some statistical sense, is guaranteed to be more reliable.

A related issue is cross-validation (leave out one example, train on the remainder, test on the one left out; repeat for all examples; repeat for subsets of examples to the extent feasible). This standard statistical exercise (Friedman 1994, Haykin 1999) has not usually been practiced in the great wave of neural-network applications, a fact which has understandably led statisticians to be suspicious of the whole enterprise (witness the Ripley (1994) article). The computational expense and tedium of cross-validation is an obstacle that is now being relieved by faster computers and better software packages. As a useful alternative, one may work with three data sets, called here the training set, the test set, and the validation set. The test set (used to monitor generalization performance) is allowed to influence modeling strategies, but the validation set is left untouched and preferably hidden throughout the whole model-building process. Indeed, the validation set may consist of new experimental results unknown at the time of the theoretical work. Scientific applications of this kind (in nuclear physics and protein chemistry) are described in Chapters 3 and 6.

One very attractive feature of neural networks is that they allow us to attack complex problems – or simple ones, for that matter – without any prior understanding (Horn 1997). This feature has played a large part in their popularity. Because of its not-infrequent abuse, this feature is also largely responsible for the unpopularity of neural networks in some quarters. (Other nonparametric methods may of course be tarred with the same brush.) At any rate, it is often possible to use neural nets to obtain a reasonable "first draft" of the solution of a problem without any great effort, though in many instances it is very hard to go much beyond this first draft with cookbook neural-network techniques. When some knowledge of the problem domain is at hand, it would be foolish not to use it in the design of the network, including architecture, coding, and neuronal response. There is no point in asking the network to learn regularities we already know about (except possibly as a test its abilities). To do so occupies resources the network can put to better use and, in increasing the complexity of the network, tends to work against good generalization.

In a complicated problem there will typically be many variables (prospective inputs) that might influence the quantity that is to be estimated or the decision that is to be made. It then becomes very important to optimize network design with respect to the choice of input variables. Principal component analysis and independent component analysis (Hertz et al. 1991, Haykin 1999) offer natural criteria for making this choice (Horn 1997). In particular, the first few principal components of the covariance matrix may be considered to define the most relevant variables. Restricting the input variables to this set economizes in network structure, favoring better predictive performance. In fact, the preprocessing step of principal component analysis can be implemented with a neural network (Oja 1989, Sanger 1989). Such a network can then be used as a "front end" for a perceptron classifier or interpolator. Self-organizing maps may be regarded as a nonlinear generalization of principal component analysis, and indeed the network algorithms of Kohonen offer another powerful avenue to dimensionality reduction. Thus one is led rather smoothly to the notion of cooperative neural networks or hybrid network structures (cf. Chapters 4 and 9) – and then to the idea of more general hybrids containing neural networks linked with information processors of other kinds. It does not take much vision to forecast that this is the wave of the future, since the concept is already reality in a number of industrial applications.

As remarked earlier in the instance of backpropagation, neural-network algorithms have a distinct competitive advantage in the relative simplicity of their implementation, and the consequent reliability of associated software. Excellent software packages – fast, flexible, and user-friendly – are offered in considerable variety on the commercial market. On the academic circuit, two that stand out are (i) JETNET (Peterson et al. 1994), a versatile fortran program available via anonymous ftp through the Lund web site www.thep.lu.se/tf2/, and (ii) the Stuttgart Neural Network Simulator, which may be downloaded from www.informatik.uni-stuttgart.de/ipvr/bv/projekte/ snss/snss.html. Useful information and advice on neural-network software, theory, applications, and activities may be found in the internet newsgroup comp.ai.neural-nets.

In conclusion, as a methodology for classification or function approximation in scientific problems, computational analysis based on neural networks is expected to prove most valuable in applications for which (i) the data set is large and complex, (ii) there is as yet no coherent theory of the underlying phenomenon, or quantitative theoretical explication is impractical, and (iii) no simple global model (involving simple parametrized functions) exists (cf. Duch and Diercksen 1994). Under these circumstances, multilayer feedforward nets, when taught by example, may be capable of recognizing higher-order correlations in the input data and making appropriate decisions or computations based on their presence. In this sense, a successful network model could then be considered to embody fundamental features of the underlying phenomenon. Neural-network systems that arguably fulfill these specifications include those developed for discrimination between star and galaxy images in optical sky survey plates (Odewahn et al. 1992); for the prediction of the branching probabilities for nuclear decay modes (Gernoth and Clark 1995a); and for the mapping of donor and acceptor sites in human genes (Brunak et al. 1991). On similar grounds, experimental high-energy physics has been considered a natural domain for application of neural-network hardware and software to data acquisition and analysis (Denby 1988), not least because of the prospects for fast on-line implementation in environments with ultrahigh event rates.

Acknowledgments

Research support from the U.S. National Science Foundation, under Grant Nos. PHY96-02727 and IBN96-34314, is gratefully acknowledged. Some of the results surveyed in Sect. 6.3 derive from collaborations with H. Bohr, K. A. Gernoth, E. Mavrommatis, J. S. Prater, and M. L. Ristig, as well as A. Athanassopoulos, S. Dittmar, S. Gazula, and J. Hasenbein. Special thanks are extended to M. Barber, T. Lindenau, M. L. Ristig, and M. Samiullah for valuable discussions and for editorial assistance in the preparation of this chapter.

References

Ackley, D. H., Hinton, G. E., Sejnowski, T. J. (1985): A Learning Algorithm for Boltzmann Machines. Cognitive Science **9**, 147–169.

Alkon, D. L. (1984): Calcium-Mediated Reduction of Ionic currents: A Biophysical Memory Trace. Science **226**, 1037–1045.

Alkon, D. L. (1988): *Memory Traces in the Brain* (Cambridge University Press, Cambridge, England).

Alkon, D. L., Blackwell, K. T., Barbour, G. S., Rigler, A. K., Vogl, T. P. (1990): Pattern Recognition by an Artificial Network Derived from Biologic Neuronal Systems. Biol. Cybern. **62**, 363–376.

Almeida, L. B. (1987): A Learning Rule for Asynchronous Perceptrons with Feedback in a Combinatorial Environment. *IEEE First International Conference on Neural Networks*, San Diego, 1987, Vol. II, edited by Caudill, M., Butler, C. (IEEE, New York), 609–618.

Almeida, L. B. (1988): Backpropagation in Perceptrons with Feedback. *Neural Computers* (Neuss 1987), edited by Eckmiller, R., von der Malsburg, Ch. (Springer-Verlag, Berlin), 199–208.

Amit, D. J., Gutfreund, H., Sompolinsky, H. (1985a): Spin-Glass Models of Neural Networks. Phys. Rev. A **32**, 1007–1018.

Amit, D. J., Gutfreund, H., Sompolinsky, H. (1985b): Storing Infinite Numbers of Patterns in a Spin-Glass Model of Neural Networks. Phys. Rev. Lett. **55**, 1530–1533.

Amit, D. J., Gutfreund, H., Sompolinsky, H. (1987a): Statistical Mechanics of Neural Networks Near Saturation. Ann. Phys. (NY) **173**, 30–67.

Amit, D. J., Gutfreund, H., Sompolinsky, H. (1987b): Information Storage in Neural Networks with Low Levels of Activity, Phys. Rev. A **35**, 2293–2303.

Amit, D. J. (1989): *Modeling Brain Function: The World of Attractor Neural Networks* (Cambridge University Press, Cambridge, England).

Anderson, C. H. (1987): *Proceedings of the IEEE First International Conference on Neural Networks*, San Diego, June 1987, edited by Caudill, M., Butler, C. (IEEE, New York), 105-112.

Anderson, C. H. (1994): Basic Elements of Biological Computational Systems. Int. J. Mod. Phys. C **5**, 135-137.

Anderson, C. H., Van Essen, D. C. (1994): Neurobiological Computational Systems. *Computational Intelligence Imitating Life*, edited by Zurada, J. M. et al. (IEEE Press, New York), 213–223.

Anderson, C. H. (1996): Unifying Perspectives on Neuronal Codes and Processing. *Condensed Matter Theories*, Vol. 6, edited by Ludeña, E., Vashishta, P., Bishop, R. F. (Nova Science Publishers, Commack, NY), 365–373.

Anderson, J. A. (1970): Two Models for Memory Organization Using Interacting Traces. Math. Biosci. **8**, 137–160.

Andreassen, H., Bohr, H., Bohr, J., Brunak, S., Bugge, T., Cotterill, R. M. J., Jacobsen, C., Kusk, P., Lautrup, B., Petersen, S. B., Saermark, T., Ulrich, K. (1990): Analysis of the Secondary Structure of the Human Immunodeficiency Virus (HIV) Proteins p17, gp120, and gp41 by Computer Modeling Based on Neural Network Methods. Journal of Acquired Immune Deficiency Syndromes (AIDS) **3**, 615–622 .

Angel, J. R. P., Wizinowich, P., Lloyd-Hart, M., Sandler, D. (1990): Adaptive Optics for Array Telescopes Using Neural Network Techniques. Nature **348**, 221–224.

Anninos, P. A., Beek, B., Csermely, T. J., Harth, E. M., Pertile, G. (1970): Dynamics of Neural Structures. J. Theoret. Biol. **26**, 121–148.

Anninos, P., (1972): Cyclic Modes in Artificial Neural Nets: Kybernetik **11**, 5–14.

Arbib, M. A. (1995): *The Handbook of Brain Theory and Neural Networks* (MIT Press, Boston).

Athanassopoulos, S., Mavrommatis, E., Gernoth, K. A., Clark, J. W. (1998): To be published.

Babbage, W. S., Thompson, L. F. (1993): The Use of Neural Networks in $\gamma - \pi^0$ Discrimination. Nucl. Instrum. Methods **A330**, 482–486.

Bahadur, R. R. (1961): A Representation of the Joint Distribution of Responses to n Dichotomous Items. *Studies in Item Analysis and Prediction*, edited by Solomon, H. (Stanford University Press, Stanford, CA) 158–168.

Barkai, E., Hansel, D., Kanter, I. (1990): Statistical Mechanics of a Multilayered Neural Network. Phys. Rev. Lett. **65**, 2312–2315.

Bass, S. A., Bischoff, A., Maruhn, J. A., Stöcker, H., Greiner, W. (1996): Neural Networks for Impact Parameter Determination. Phys. Rev. C **53**, 2358-2363.

Becks, K. H., Block, F., Drees, J., Langefeld, P., Seidel, F. (1993): B-quark Tagging Using Neural Networks and Multivariate Statistical Methods – A Comparison of Both Techniques. Nucl. Instrum. Methods **A329**, 501–517.

Bellman, R. E. (1961): *Adaptive Control Processes*. (Princeton University Press, Princeton, NJ).

Blank, T. B., Brown, S. D. (1994): Adaptive, Global, Extended Kalman Filters for Training Feedforward Neural Networks. J. Chemometrics **8**, 391-407.

Blank, T. B., Brown, S. D., Calhoun, A. W., Dorn, D. J. (1995): Neural Network Models of Potential Energy Surfaces. J. Chem. Phys. **103**, 4129–4137.

Bliss, T. V. P., Collingridge, G. L. (1993): A Synaptic Model of Memory: Long-Term Potentiation in the Hippocampus. Nature **361**, 31–39.

Bohr, A., Mottelson, B. R. (1969): *Nuclear Structure*, Vol. I (New York: W. A. Benjamin).

Bohr, H., Bohr, J., Brunak, S., Cotterill, R. M. J., Lautrup, B., Nøskov, L., Olsen, O. H., Petersen, S. B. (1988): Protein Secondary Structure and Homology by Neural Networks: the α-Helices in Rhodopsin. FEBS Letters **241**, 223–228.

Bohr, H., Bohr, J., Brunak, S., Cotterill, R. M. J., Fredholm, H., Lautrup, B., Petersen, S. B. (1990): A Novel Approach to Prediction of the 3–Dimensional Structures of Protein Backbones by Neural Networks. FEBS Letters **261**, 43–46.

Bohr, H., Wolynes, P. G. (1992): Initial Events of Protein Folding from an Information-Processing Viewpoint. Phys. Rev. A **46**, 5242–5248.

Bohr, H., Goldstein, R. A., Wolynes, P. G. (1992): Predicting Surface Structures of Proteins by Neural Networks. AMSE Periodicals, Modeling, Measurements, and Control C **31**, 35–58.

Bohr, J., Bohr, H., Brunak, S., Cotterill, R. M. J., Lautrup, B., Fredholm, H., Petersen, S. B. (1993a): Protein Structure from Distance Inequalities. J. Molec. Biol. **231**, 861–869.

Bohr, H., Irwin, J., Mochizuki, K., Wolynes, P. G. (1993b): Classification and Prediction of Protein Side-Chains by Neural Network Techniques. Int. J. Neural Syst. (Supplementary Issue), 177-182.

Bohr, H., Brunak, S. (1994): *Protein Structure by Distance Analysis* (IOS Press, Amsterdam).

Bohr, H. G. (1998): *Neural Network Prediction of Protein Structures*. (Polyteknisk Forlag, Lyngby, Denmark).

Bonhoefter, T., Staiger, V., Aertsen, A. (1989): Synaptic Plasticity in Rat Hippocampal Slice Cultures: Local "Hebbian" Conjunction of Pre- and Postsynaptic Stimulation Leads to Distributed Synaptic Enhancement. Proc. Nat. Acad. Sci. USA **86**, 8113–8117.

Bortolotto, C., de Angelis, A., Lanceri, L. (1991): Tagging the Decays of the Z^0 Boson into b Quark Pairs with a Neural Network Classifier. Nucl. Instr. Meth. **A306**, 459-466.

Bortolotto, C., de Angelis, A., de Groot, N., Seixas, J. (1992): Neural Networks in Experimental High Energy Physics. Int. J. Mod. Phys. **C3**, 733-771.

Bounds, D. G. (1987): New Optimization Methods from Physics and Biology. Nature **329**, 215–219.

Bressloff, P. C., Taylor, J. G. (1990): Random Iterative Networks. Phys. Rev. A **41**, 1126–1137.

Bressloff, P. C. (1991): Stochastic Dynamics of Time-Summating Binary Neural Networks. Phys. Rev. A **44**, 4005–4016.

Bressloff, P. C. (1992): Analysis of Quantal Synaptic Noise in Neural Networks Using Iterated Function Systems. Phys. Rev. A **45**, 7549-7559.

Brown, T. H., Chapman, P. F., Kairiss, E. W., Keenan, C. L. (1988): Long-Term Synaptic Potentiation, Science **242**, 724–728.

Bruce, A. D., Gardner, E. J., Wallace, D. J. (1987): Dynamics and Statistical Mechanics of the Hopfield Model. J. Phys. A: Math. Gen. **20**, 2909–2934.

Bruck, J., Goodman, J. W. (1988): On the Power of Neural Networks for Solving Hard Problems. *Neural Information Processing Systems*, edited by Anderson, D. Z. (American Institute of Physics, New York).

Brunak, S., Engelbrecht, J., Knudsen, S. (1990a): Cleaning Up Gene Databases, Nature **343**, 123.

Brunak, S., Engelbrecht, J., Knudsen, S. (1990b): Neural Network Detects Errors in the Assignment of MRNA Splice Sites. Nucleic Acids Res. **18**, 4797-4801.

Brunak, S., Engelbrecht, J., Knudsen, S. (1991): Prediction of Human mRNA Donor and Acceptor Sites from the DNA Sequence. J. Molec. Biol. **220**, 49–65.

Bryngelson, J. D., Wolynes, P. G. (1987): Spin Glasses and the Statistical Mechanics of Protein Folding. Proc. Nat. Acad. Sci. USA **84**, 7524–7528.

Bryngelson, J. D., Hopfield, J. J., Southardi, S. N., Jr. (1990): A protein structure predictor based on an energy model with learned parameters, Tetrahedron Comp. Meth. **3**, 129–141.

Buntine, W. L., Weigend, A. S. (1991): Bayesian Backpropagation. Complex Systems **5**, 603-643.

Caianiello, E. R. (1961): Outline of a Theory of Thought Processes and Thinking Machines. J. Theor. Biol. **2**, 204–235.

Cherkassky, V., Friedman, J. H., Wechsler, W., editors (1994): *From Statistics to Neural Networks. Theory and Pattern Recognition Applications* (Springer Verlag, Berlin).

Chou, P. Y., Fasman, G. D. (1978): Empirical Predictions of Protein Conformation. Ann. Rev. Biochem. **47**, 251–276.

Chow, C. K., Liu, C. N. (1968): Approximating Discrete Probability Distributions with Dependence Trees. IEEE Trans. Information Theory, **IT-14**, 462–467.

Clark, J. W. (1979): Update on the Crisis in Nuclear-Matter Theory: A Summary of the Trieste Conference. Nucl. Phys. **A328**, 587–595.

Clark, J. W., Rafelski, J., Winston, J. V. (1985): Brain Without Mind: Computer Simulation of Neural Networks with Modifiable Neuronal Interactions. Physics Reports **123(4)**, 215–273.

Clark, J. W. (1988): Statistical Mechanics of Neural Networks. Physics Reports **158**, 9–157.

Clark, J. W. (1991): Neural Network Modelling. Phys. Med. Biol. **36**, 1259–1317.

Clark, J. W., Gazula, S. (1991): Artificial Neural Networks That Learn Many-Body Physics. *Condensed Matter Theories*, Vol. 6, edited by Fantoni, S., Rosati, S. (Plenum, New York), 1–24.

Clark, J. W., Gazula, S., Gernoth, K. A., Hasenbein, J., Prater, J. S., Bohr, H. (1992): Collective Computation of Many-Body Properties by Neural Networks. *Recent Progress in Many-Body Theories*, Vol. 3, edited by Ainsworth, T. L.,

Campbell, C. E., Clements, B. E., Krotscheck, E. (Plenum Press, New York), 371–386.

Clark, J. W., Gernoth, K. A. (1992): Teaching Neural Networks to Do Science. *Structure: From Physics to General Systems*, Vol. 2, edited by Marinaro, M., Scarpetta, G. (World Scientific, Singapore), 64–77.

Clark, J. W., Gernoth, K. A., Ristig, M. L. (1994): Connectionist Many-Body Phenomenology. *Condensed Matter Theories*, Vol. 9, edited by Clark, J. W., Shoaib, K. A., Sadiq, A. (Nova Science Publishers, Commack, NY), 519–537.

Clark, J. W., Gernoth, K. A. (1995): Statistical Modeling of Nuclear Masses Using Neural Network Algorithms. *Condensed Matter Theories*, Vol. 10, edited by Casas, M., de Llano, M., Navarro, J., Polls, A. (Nova Science Publishers, Commack, NY), 317–333.

Clark, J. W., Gernoth, K. A., Dittmar, S., Ristig, M. L. (1999): Higher-Order Probabilistic Perceptrons as Bayesian Inference Engines. Phys. Rev. E, to be published.

Cohen, M. A., Grossberg, S. (1983): Absolute Stability of Global Pattern Formation and Parallel Memory Storage by Competitive Neural Networks. IEEE Trans. Syst. Man Cybern. **SMC-13**, 815–825.

Cooper, L. N. (1973): A Possible Organization of Animal Memory and Learning. *Proceedings of the Nobel Symposium on Collective Properties of Physical Systems*, edited by Lundquist, B., Lundquist, S. (Academic Press, New York), 252–264.

Cover, T. M. (1965): Geometrical and Statistical Properties of Systems of Linear Inequalities with Applications in Pattern Recognition. IEEE Trans. Electron. Comput. **EC-14**, 326–334.

Cowan, J. D. (1967): *A Mathematical Theory of Central Nervous Activity*, Ph.D. Thesis, University of London.

Cowan, J. D. (1970): A Statistical Mechanics of Nervous Activity. *Lectures on Mathematics in the Life Sciences*, Vol. 2, edited by Gerstenhaber, M. (American Mathematics Society, Providence, R.I.), 1–57.

Cowan, J. D., Sharp, D. H. (1988): Neural Nets. Quarterly Reviews of Biophysics **21**, 365–427.

Cowan, J. D. (1990): Discussion: McCulloch-Pitts and Related Neural Nets from 1943 to 1989. Bull. Math. Biol. **52**, 73–97.

Cragg, B. G., Temperley, H. N. V. (1954): The Organisation of Neurones: A Cooperative Analogy. EEG Clin. Neurophysiol. **6**, 85–92.

Curry, B., Rumelhart, D. E. (1990): MSnet: A Neural Network which Classifies Mass Spectra. Tetrahedron Comp. Meth. **3**, 213–237.

Cybenko, G. (1989): Approximation by Superpositions of a Sigmoidal Function. Mathematics of Control, Signals, and Systems **2**, 303–314.

Dayan, P., Willshaw, D. J. (1991): Optimising Synaptic Learning Rules in Linear Associative Memories. Biol. Cybern. **65**, 253–265.

Denby, B. (1988): Neural Networks and Cellular Automata in Experimental High Energy Physics. Comput. Phys. Commun. **49**, 429–448.

Denby, B., Linn, S. L. (1990): Spatial Pattern Recognition in a High Energy Particle Detector Using a Neural Network Algorithm. Comput. Phys. Commun. **56**, 293–297.

Denker, J. S. (1986): Neural Network Models of Learning and Adaptation. Physica **D22**, 216–232.

Denker, J., Schwartz, D., Wittner, B., Solla, S., Hopfield, J., Howard R., Jackel, L. (1987): Automatic Learning, Rule Extraction and Generalization. Complex Systems **1**, 877–922.

De Veaux, R., Psichogios, D. C., Ungar, L. H. (1993): Comput. Chem. Eng. **17**, 819.

Domany, E. (1988): Neural Networks: A Biased Overview. J. Stat. Phys. **51**, 743–775.

Duch, W., Diercksen, G. H. F. (1994): Neural Networks as Tools to Solve Problems in Physics and Chemistry. Comput. Phys. Commun. **82**, 91-103.

Duda, R. O., Hart, P. E. (1973): *Pattern Classification and Scene Analysis* (Wiley, New York).

Eccles, J. C. (1957): *The Physiology of Nerve Cells* (Johns Hopkins University Press, Baltimore).

Eccles, J. C. (1964): *The Physiology of Synapses* (Springer-Verlag, Berlin).

Eliasmith, C., Anderson, C. H. (1998): Developing and Applying a Toolkit for a General Neurocomputational Framework. *Neurocomputing 1998*, in press.

Elman, J. L. (1990): Finding Structure in Time. Cognitive Science **14**, 179–211.

Fahlman, S. E., Lebiere, C. (1990): The Cascade-Correlation Learning Architecture. *Advances in Neural Information Processing Systems*, Vol. 2, edited by Touretzky, D. S. (Morgan Kaufmann, San Mateo, CA), 524–532.

Farley, B. G., Clark, W. A. (1954): Simulation of Self-Organizing Systems by Digital Computer. I.R.E. Transactions on Information Theory **4**, 76–84.

Friedman, J. H., Stuetzle, W. (1981): Projection Pursuit Regression. J. Amer. Statis. Assoc. **76**, 817–823.

Friedman, J. H. (1991): Multivariate Adaptive Regression Splines (with Discussion and Rejoinder). Ann. Statist. **19**, 1–141.

Friedman, J. H. (1994): An Overview of Predictive Learning and Function Approximation. *From Statistics to Neural Networks. Theory and Pattern Recognition Applications*, edited by Cherkassky, V., Friedman, J. H., Wechsler, W. (Springer-Verlag, Berlin), 1–61.

Friedrichs, M., Wolynes, P. G. (1989): Toward Protein Tertiary Structure Recognition by Means of Associative Memory Hamiltonians. Science **246**, 371–373.

Friedrichs, M. S., and Wolynes, P. G. (1990): Molecular Dynamics of Associative Memory Hamiltonians for Protein Tertiary Structure Recognition. Tetrahedron Comp. Meth. **3**, 175–190.

Funahashi, K., (1989): On the Approximate Realization of Continuous Mappings by Neural Networks. Neural Networks **2**, 183–192.

Gardner, E. (1988): The Space of Interactions in Neural Network Models. J. Phys. A: Math. Gen. **21**, 257–270.

Gasteiger, J., Li, X., Rudolph, Ch., Sadowski, J., Zupan, J. (1994): Representation of Molecular Electrostatic Potentials by Topological Feature Maps. J. Am. Chem. Soc. **116**, 4608–4620.

Gasteiger J. (1998): Neural Networks in Drug Design. Survey talk presented at the 194th Heraeus Seminar, "Scientific Applications of Neural Nets," Physikzentrum, Bad Honnef, May 13–15.

Gazula, S., Clark, J. W., Bohr, H. (1992): Learning and Prediction of Nuclear Stability by Neural Networks. Nucl. Phys. A **540**, 1–26.

Gernoth, K. A., Clark, J. W., Prater, J. S., Bohr, H. (1993): Neural Network Models of Nuclear Systematics. Phys. Lett. B **300**, 1–7.

Gernoth, K. A., Clark, J. W. (1995a): Neural Networks that Learn to Predict Probabilities: Global Models of Nuclear Stability and Decay. Neural Networks **8**, 291–311.

Gernoth, K. A., Clark, J. W. (1995b): A Modified Backpropagation Algorithm for Training Neural Networks on Data with Error Bars. Comput. Phys. Commun. **88**, 1–22.

Girosi, F., Poggio, T. (1995): Regularization Theory and Neural Network Architectures. Neural Computation **7**, 219–269.

Glauber, R. J. (1963): Time-Dependent Statistics of the Ising Model. J. Math. Phys. **4**, 294-307.

Gorman, R. P., Sejnowski, T. J. (1988): Analysis of Hidden Units in a Layered Network Trained to Classify Sonar Targets. Neural Networks **1**, 75–89.

Grondin, R. O., Porod, W., Loeffler, C. M., Ferry, D. K. (1983): Synchronous and Asynchronous Systems of Threshold Elements. Biol. Cybern. **49**, 1–7.

Grossberg, S. (1976): Adaptive Pattern Classification and Universal Decoding: Part I. Parallel Development and Coding of Neural Feature Detectors. Biol. Cybern. **23**, 121–134.

Gustafsson, B., Wigström, H., Abraham, W. S., Huang, Y.-Y. (1987): Long-Term Potentiation in the Hippocampus Using Depolarizing Current Pulses as the Conditioning Stimulus to Single Volley Synaptic Potentials. J. Neurosci. **7**, 774–780.

Gyulassy, M., Harlander, M. (1991): Elastic Tracking and Neural Network Algorithms for Complex Pattern Recognition. Comput. Phys. Commun., **66**, 31-46.

Hakimian, S., Anderson, C. H., Thach, T. (1998): A PDF Model of Populations of Purkinje Cells: Non-linear Interactions and High Variability. *Neurocomputing 1998*, in press.

Hampshire, J. B. II, Perlmutter, B. A. (1990a): A Novel Objective Function for Improved Phoneme Recognition Using Time-Delay Neural Networks. IEEE Trans. Neural Networks **1**, 216-228.

Hampshire, J. B. II, Perlmutter, B. A. (1990b): Equivalence Proofs for Multilayer Perceptron Classifiers and the Bayesian Discriminant Function. *Proceedings of the 1990 Connectionist Models Summer School*, edited by Touretzky, D., Elman, J. Sejnowski, T., Hinton, G. (Morgan Kaufmann, SanMateo, CA).

Harth, E. M., Csermely, T. J., Beek, B., Lindsay, R. D. (1970): Brain Functions and Neural Dynamics. J. Theoret. Biol. **26**, 93–120.

Haykin, S. (1999): *Neural Networks: A Comprehensive Foundation*, Second Edition (Prentice Hall, Upper Saddle River, NJ).

Hebb, D. O. (1949): *The Organization of Behavior: A Neuropsychological Theory* (Wiley, New York).

Hecht-Nielsen, R. (1987): Counterpropagation Networks. Applied Optics **26**, 4979–4984.

Hecht-Nielsen, R. (1988): Applications of Counterpropagation Networks. Neural Networks **1**, 131–139.

Hertz, J., Krogh, A., Palmer, R. G. (1991): *Introduction to the Theory of Neural Computation* (Addison–Wesley, Redwood City, California).

Herz, A., Sulzer, B., Kühn, R., van Hemmen, J. L. (1989): Hebbian Learning Reconsidered: Representation of Static and Dynamic Objects in Associative Neural Nets. Biol. Cybern. **60**, 457–467.

Hinton, G. E. (1986): Learning Distributed Representations of Concepts. *Proceedings of the Eighth Annual Conference of the Cognitive Science Society*, Amherst 1986 (Lawrence Erlbaum, Hillsdale, NJ), 1–12.

Hinton, G. E., Sejnowski, T. J. (1986): Learning and Relearning in Boltzmann Machines. *Parallel Distributed Processing: Explorations in the Microstructure of Cognition*, Vol. 1, edited by Rumelhart, D. E., McClelland, J. L, and the PDP Research Group (MIT Press, Cambridge, MA), 282–317.

Holley, L., Karplus, M. (1989): Protein Secondary Structure Prediction with a Neural Network. Proc. Nat. Acad. Sci. USA **86**, 152–156.

Holzgrabe, U., Wagener, M., Gasteiger, J. (1996): Comparison of Structurally Different Allosteric Modulators of Muscarinic Receptors by Self-Organizing Neural Networks. J. Mol. Graphics **14**, 185-193, color plates on pages 217–221.

Hopfield, J. J., (1982): Neural Networks and Physical Systems with Emergent Collective Computational Abilities. Proc. Nat. Acad. Sci. USA **79**, 2554–2558.

Hopfield, J. J. (1984): Neurons with Graded Response Have Collective Computational Properties Like Those of Two-State Neurons. Proc. Nat. Acad. Sci. USA **81**, 3088–3092.

Hopfield, J. J., Tank, D. W. (1985): "Neural" Computation of Decisions in Optimization Problems. Biol. Cybern. **52**, 141–152.

Hopfield, J. J., Tank, D. W. (1986): Computing with Neural Circuits: A Model. Science **233**, 625–633.

Horn, D. (1997): Neural Computation Methods and Applications: Summary Talk of the AI Session. Nucl. Instr. Methods **A389**, 381-387.

Hornik, A. K., Stinchcombe, A. M., White, A. H. (1989): Multilayer Feedforward Networks Are Universal Approximators. Neural Networks **2**, 359–366.

Huang, W. Y., Lippmann, R. P. (1988): Neural Net and Traditional Classifiers. *Neural Information Processing Systems* (Denver 1987), edited by Anderson, D. Z. (American Institute of Physics, New York), 387–396.

Johnston, D., Wu, S. M. (1995): *Foundations of Cellular Neurophysiology* (MIT Press, Cambridge, MA).

Kandel, E. R., Hawkins, R. D. (1992): The Biological Basis of Learning and Individuality. Scientific American **267**(3), 79-86.

Kanter, L. (1992): Information Theory of a Multilayer Neural Network with Discrete Weights, Europhys. Lett. **17**, 181–186.

Keeler, J. D. (1986): Comparison Between Sparsely Distributed Memory and Hopfield-Type Neural-Network Models. Institute for Nonlinear Sciences, University of California at San Diego, preprint.

Kelso, S. R., Ganong, A. H., Brown, T. H. (1986): Hebbian Synapses in Hippocampus. Proc. Nat. Acad. Sci. USA **83**, 5326–5330.

Kirkpatrick, S., Sherrington, D. (1978): Infinite-Ranged Models of Spin Glasses. Phys. Rev. B **17**, 4384–4403.

Kirkpatrick, S., C. D. Gelatt, C. D., Jr., Vecchi, M. P. (1983): Optimization by Simulated Annealing, Science **220**, 671–680.

Kneller, D. G., Cohen, F. E., Langridge, R. (1990): Improvements in Protein Secondary Structure Prediction by an Enhanced Neural Network. J. Mol. Biol. **214**, 171–182.

Knight, B. W. (1972): Dynamics of Encoding in a Population of Neurons. J. Gen. Physiol. **59**, 734-766.

Kohonen, T. (1989): *Self-Organization and Associative Memory*, Third Edition (Springer-Verlag, Berlin).

Kohonen, T. (1997): *Self-Organizing Maps*, Second Edition (Springer-Verlag, Berlin).

Krogh, A., Hertz, J. A. (1991): Dynamics of Generalization in Linear Perceptrons. *Advances in Neural Information Processing Systems*, Vol. 3, edited by Touretzky, D. S., Lippmann, R. (Morgan Kaufmann, San Mateo, CA).

Kullback, S. (1959): *Information Theory and Statistics* (Wiley, New York).

Kürten, K. E., Clark, J. W. (1986): Chaos in Neural Systems. Phys. Lett. **114A**, 413–418.

Kürten, K. E. (1988a): Critical Phenomena in Model Neural Networks. Phys. Lett. **A129**, 157–160.

Kürten, K. E. (1988b): Phase Transitions in Quasirandom Neural Networks. *Proceedings of the IEEE First Annual International Conference on Neural Networks*, Vol. 2, (IEEE, New York), 197–204.

Kürten, K. E. (1988c): Transition to Chaos in Asymmetric Neural Networks. *Condensed Matter Theories*, Vol. 3, edited by Arponen, J. S., Bishop, R. F., Manninen, M. (Plenum, New York), 333–338.

Kürten, K. E. (1988d): Self-Organization in Model Neural Networks with Activity–Dependent Synaptic Interactions. *Proceedings of the Ninth European Meeting on Cybernetics and Systems Research Vienna, 1988*, edited by Trappl, R. (Kluwer Academic Publishers, Amsterdam), 495–500.

Kürten, K. E. (1988e): "Training" Quasirandom Neural Networks. *Chaos and Complexity*, Torino, 1987, edited by Levi, R., Ruffo, S., Ciliberti, S., Buiatti, M. (World Scientific, Singapore).

Kürten, K. E. (1989): Dynamical Phase Transitions in Short-Ranged and Long–Ranged Neural Network Models. J. Phys. (France) **50**, 2313–2323.

Lansner, A., Ekeberg, Ö. (1989): A One-Layer Feedback Artificial Neural Network with a Bayesian Learning Rule. Int. J. Neural Systems, **1**, 77-87.

Le Cun, Y. (1985): A Learning Scheme for Asymmetric Threshold Networks. Proc. Cognitiva **85**, 599–604.

Le Cun, Y., Boser, B., Denker, J. S., Henderson, D., Howard, R. E., Hubbard, W., Jackel, L. D. (1990a): Handwritten Digit Recognition with a Backpropagation Network. *Neural Information Processing Systems*, Vol. 2, edited by Touretzky, D. S. (Morgan Kaufmann, San Mateo, CA), 396–404.

Le Cun, Y., Denker, J. S., Solla, S. A. (1990b): Optimal Brain Damage. *Neural Information Processing Systems*, Vol. 2, edited by Touretzky, D. S. (Morgan Kaufmann, San Mateo, CA), 598–605.

Lehky, S. R., Sejnowski, T. J. (1988): Network Model of Shape from Shading: Neural Function Arises from Both Receptive and Projective Fields. Nature **333**, 452–454.

Levin, E., Tishby, N., Solla, S. A. (1990): A Statistical Approach to Learning and Generalization in Layered Neural Networks. Proc. IEEE **78**, 1568–1574.

Lewicki, M. S., Sejnowski, T. J. (1996): Bayesian Unsupervised Learning of Higher Order Structure. *Advances in Neural Information Processing Systems 9: Proceedings of the 1996 Conference.* edited by Mozer, M. C., Jordan, M. I., Thomas, P. (MIT Press, Cambridge).

Li, X., Gasteiger, J., Zupan, J. (1993): On the Topology Distortion in Self-Organizing Feature Maps. Biol. Cybern. **70**, 189–198.

Lin, S., Kernighan, B. W. (1973): An Effective Heuristic Algorithm for the Traveling Salesman Problem. Operations Research **21**, 498–516.

Lippmann, R. P. (1987): An Introduction to Computing with Neural Nets, IEEE ASSP Mag. **4**(2), 4–22.

Little, W. A. (1974): The Existence of Persistent States in the Brain. Math. Biosci. **19**, 101–120.

Littlewort, G. C., Clark, J. W., Rafelski, J. (1988). Transition to Cycling in Neural Networks. *Computer Simulation in Brain Science*, edited by Cotterill, R. M. J. (Cambridge University Press, Cambridge, England), 345–356.

Lloyd-Hart, M., Wizinowich, P., McLeod, B., Wittman, D., Colucci, D., Dekany, R., McCarthy, D., Angel, J. R. P., Sandler, D. (1992): First Results of an On-Line Adaptive Optics System with Atmospheric Wavefront Sensing by an Artificial Neural Network. Ap. J. **390**, L41–L44.

Lönnblad, L., Peterson, C., Rögnvaldsson, T. (1990): Finding Gluon Jets with a Neural Trigger *Phys. Rev. Lett.* **65**, 1321-1324.

Lönnblad, L., Peterson, C., Pi, H., Rögnvaldsson, T. (1991a): Self-Organizing Networks for Extracting Jet Features. Comput. Phys. Commun. **67**, 193–209.

Lönnblad, L., Peterson, C., Rögnvaldsson, T. (1991b): Using Neural Networks to Identify Jets. Nucl. Phys. **B349**, 675-702.

Lönnblad, L., Peterson, C., Rögnvaldsson, T. (1992): Pattern Recognition in High Energy Physics with Artificial Neural Networks – JETNET 2.0. Comput. Phys. Commun. **70**, 167–182.

Luenberger, D. G. (1984): *Linear and Nonlinear Programming*, Second Edition (Addison-Wesley, Reading, MA).

MacKay, D. J. C. (1992a): Bayesian Interpolation. Neural Computation **4**, 415-447.

MacKay, D. J. C. (1992b): Bayesian Framework for Backpropagation Networks. Neural Computation **4**, 448-472.

Malinow, R., Miller, J. P. (1986): Postsynaptic Hyperpolarization During Conditioning Reversibly Blocks Induction of Long-Term Potentiation. Nature **320**, 529–530.

Mavrommatis, E., Dakos, A., Gernoth, K. A., Clark, J. W. (1998): Calculations of Nuclear Half-Lives with Neural Nets. *Condensed Matter Theories*, Vol. 13, edited by da Providência, J., Malik, F. B. (Nova Science Publishers, Commack, NY).

McCulloch, W. S., Pitts, W. (1943): A Logical Calculus of the Ideas Immanent in Nervous Activity. Bull. Math. Biophys. **5**, 115–137.

McMillan, C., Mozer, M. C., Smolensky, P. (1991): The Connectionist Scientist Game. *Proceedings of the Thirteenth Annual Conference of the Cognitive Science Society* (Erlbaum, Hillsdale, NJ).

Metropolis, N., Rosenbluth, A. W., Rosenbluth, M. N., Teller, A. H., Teller, E. (1953): Equation of State Calculations for Fast Computing Machines. J. Chem. Phys. **6**, 1087–1092.

Meyer, B., Hansen, T., Nute, D., Albersheim, P., Darville, A., York, W., Sellers, J. (1991): Identification of the ^{1}H-NMR Spectra of Complex Oligosaccharides with Artificial Neural Networks. Science **251**, 542–544.

Mézard, M., Nadal, J.-P. (1989): Learning in Feedforward Layered Networks: The Tiling Algorithm. J. Phys. A: Math. Gen. **22**, 2191–2204.

Minsky, H., Papert, S. (1969): *Perceptrons* (MIT Press, Cambridge, MA).

Möller, P., Nix, J. R. (1990): Global Nuclear-Structure Calculations. Nucl. Phys. **A520**, 369c-376c.

Moody, J., Darken, C. (1989): Fast Learning in Networks of Locally-Tuned Processing Units. Neural Computation **1**, 281–294.

Müller, B., Reinhardt, J. (1990): *Neural Networks–An Introduction* (Springer-Verlag, Berlin).

Navarro, J., Polls, A., editors (1998): *Microscopic Quantum Many-Body Theories and Their Applications*, LNP 510 (Springer-Verlag, Berlin).

Nilsson, N. J. (1965): *Learning Machines Foundations of Trainable Pattern-Classifying Systems* (McGraw-Hill, New York).

Niranjan, M., Fallside, F. (1990): Neural Networks and Radial Basis Functions in Classifying Static Speech Patterns. Computer Speech and Language **4**, 275–289.

Odewahn, S. C., Stockwell, E. B., Pennington, R. L., Humphreys, R. M., Zumach, W. A. (1992): Automated Star/Galaxy Discrimination with Neural Networks. Ap. J. **103**, 318–331.

Ohlsson, M., Peterson, C., Yuille, A. (1992): Track Finding with Deformable Templates – The Elastic Arms Approach. Comput. Phys. Commun. **71**, 77-98.

Oja, E. (1989): Neural Networks, Principal Components, and Subspaces. Int. J. Neural Syst. **1**, 61–68.

Paine, G. H., Scheraga, H. A. (1987): Prediction of the Native Conformation of a Polypeptide by a Statistical-Mechanical Procedure. III. Probable and Average Configurations of Enkephalin. Biopolymers **26**, 1125–1162.

Palm, G. (1982): *Neural Assemblies: An Alternative Approach to Artificial Intelligence* (Springer-Verlag, Berlin).

Parker, D. B. (1986): A Comparison of Algorithms for Neuron-like Cells. *Neural Networks for Computing*, AIP Conference Proceedings, Vol. 151, edited by Denker, J. S. (American Institute of Physics, New York), 327–332.

Peretto, P. (1984): Collective Properties of Neural Networks: A Statistical Physics Approach. Biol. Cybern. **50**, 51–62.

Peretto, P. (1988): On Learning Rules and Memory Storage Abilities of Asymmetrical Neural Networks. J. de Physique (France) **49**, 711–726.

Peretto, P. (1992): *An Introduction to the Modeling of Neural Networks* (Cambridge University Press, Cambridge, England).

Personnaz, L., Guyon, I., Dreyfus, G. (1986): Collective Computational Properties of Neural Networks. New Learning Mechanisms. Phys. Rev. A **34**, 4217–4228.

Peterson, C., Anderson, J. R. (1987): A Mean Field Theory Learning Algorithm for Neural Networks. Complex Systems **1**, 995–1019.

Peterson, C., Anderson, J. R. (1988): Neural Networks and NP-Complete Optimization Problems – A Performance Study on the Graph Partition Problem. Complex Systems **2**, 59-89.

Peterson, C. (1989): Track Finding with Neural Networks. Nucl. Instr. Methods **A279**, 537–545.

Peterson, C., Söderberg, B. (1989): A New Method for Mapping Optimization Problems onto Neural Networks. Int. J. Neural Syst. **1**, 3-22.

Peterson, C., Rögnvaldsson, T., Lönnblad, L. (1994): JETNET 3.0 – A Versatile Artificial Neural Network Package. Comput. Phys. Commun. **81**, 185–220.

Peterson, K. L. (1990): Classification of Cm I Energy Levels Using Counterpropagation Neural Networks. Phys. Rev. A **41**, 2457–2461.

Peterson, K. L. (1991): Classification of Cm II and Pu I Energy Levels Using Counterpropagation Neural Networks, Phys. Rev. A **44**, 126–138.

Peterson, K. L. (1998): Prediction of Isotope Shifts and Lande g Factors of Curium (I) Energy Levels Using Counter-Propagation Neural Networks. Talk presented at the 194th Heraeus Seminar, "Scientific Applications of Neural Nets," Physikzentrum, Bad Honnef, May 13–15.

Pineda, F. J. (1987): Generalization of Back-Propagation to Recurrent Neural Networks. Phys. Rev. Lett. **59**, 2229–2232.

Pineda, F. J. (1989): Recurrent Back-Propagation and the Dynamical Approach to Adaptive Neural Computation. Neural Computation **1**, 161–172.

Plaut, D. C., Nowlan, S. J., Hinton, G. E. (1986): Experiments on Learning by Back-Propagation. Carnegie–Mellon University Computer Science Technical Report CMU-CS-86-126.

Poggio, T., Girosi, F. (1990a): Networks for Approximation and Learning. Proc. IEEE **78**, 1481–1497.

Poggio, T., Girosi, F. (1990b): Regularization Algorithms for Learning That Are Equivalent to Multilayer Networks. Science **247**, 978–982.

Pomerleau, D. A. (1989): ALVINN: An Autonomous Land Vehicle in a Neural Network. *Advances in Neural Information Processing Systems I* (Denver 1988), edited by Touretzky, D. S. (Morgan Kaufmann, San Mateo), 305–313.

Psaltis, D., Venkatesh, S. S. (1988): Information Storage in Fully Connected Networks. *Evolution, Learning and Cognition*, edited by Lee, Y. C. (World Scientific, Singapore), 51–89.

Qian, M., Gong, G., Clark, J. W. (1991): Relative Entropy and Learning Rules. Phys. Rev. A **43**, 1061–1070.

Qian, N., Sejnowski, T. J. (1988): Predicting the Secondary Structure of Globular Proteins Using Neural Network Models. J. Mol. Biol. **202**, 865–884.

Rauschecker, J. P., Singer, W. (1981): The Effects of Early Visual Experience on the Cat's Visual Cortex and Their Possible Explanation by Hebb Synapses. J. Physiol. (London) **310**, 215–239.

Rezcko, M., Martin, A. C. R., Bohr, H., Suhai, S. (1995): Prediction of Hypervariable CDR-H3 Loop Structure in Antibodies. Protein Engineering **8**, 389–395.

Richard, M. D., Lippmann, R. P. (1991): Neural Network Classifiers Estimate Bayesian *a posteriori* Probabilities. Neural Computation **3**, 461-483.

Richards, F. M. (1991): The Protein Folding Problem. Scientific American **264**(1), 54–63.

Ripley, B. D. (1994): Neural Networks and Related Methods for Classification (with Discussion). J. Roy. Statis. Soc. B **56**, 409-456.

Ritter, H., Schulten, K. (1988): Convergence Properties of Kohonen's Topology Conserving Maps: Fluctuations, Stability, and Dimension Selection. Biol. Cybern. **60**, 59–71.

Rosenblatt, F. (1958): The Perceptron: A Probabilistic Model for Information Storage and Organization in the Brain. Psychological Review **65**, 386–408.

Rosenblatt, F. (1962): *Principles of Neurodynamics: Perceptrons and the Theory of Brain Mechanisms* (Spartan Books, Washington, D.C).

Rost, B., Sander, C. (1992): Jury Returns on Structure Prediction. Nature **360**, 540.

Rost, B., Sander, C. (1993a): Secondary Structure Prediction of All-Helical Proteins in Two States. Protein Engineering **6**, 831–836.

Rost, B., Sander, C. (1993b): Prediction of Protein Secondary Structure at Better than 70% Accuracy. J. Molec. Biol. **232**, 584–599.

Rost, B., Sander, C. (1993c): Improved Prediction of Protein Secondary Structure by Use of Sequence Profiles and Neural Networks. Proc. Nat. Acad. Sci. USA **90**, 7558–7562.

Rost, B., Sander, C., Schneider, R. (1993): Progress in Protein Structure Prediction? TIBS **18**, 120–123.

Rost, B., Sander, C. (1994): Combining Evolutionary Information and Neural Networks to Predict Protein Secondary Structure. Proteins **19**, 55–72.

Rost, B., Sander, C., Schneider, J. (1994): Redefining the Goals of Protein Secondary Structure Prediction. J. Molec. Biol. **235**, 13–26.

Ruck, D. W., Rogers, S. K., Kabrisky, M., Oxley, M. E., Suter, B. W. (1990): The Multilayer Perceptron as an Approximation to a Bayes Optimal Discriminant Function. IEEE Trans. Neural Networks, **1**, 296-298 (1990).

Rumelhart, D. E., Hinton, G. E., McClelland, J. L. (1986a): A General Framework for Parallel Distributed Processing. *Parallel Distributed Processing: Explorations in the Microstructure of Cognition*, Vol. 1, edited by Rumelhart, D. E., McClelland, J. L., and the PDP Research Group (MIT Press, Cambridge, MA), 45–76.

Rumelhart, D. E., Hinton, G. E., Williams, R. J. (1986b): Learning Internal Representations by Error Propagation. *Parallel Distributed Processing: Explorations in the Microstructure of Cognition*, Vol. 1, edited by Rumelhart, D. E., McClelland, J. L., and the PDP Research Group (MIT Press, Cambridge, MA), 318–362.

Rumelhart, D. E., Hinton, G. E., Williams, R. J. (1986c): Learning Representations by Back-Propagating Errors. Nature **323**, 533–536.

Sandler, D., Barrett, T. K., Palmer, D. A., Fugate, R. Q., Wild, W. J. (1991): Use of a Neural Network to Control an Adaptive Optics System for an Astronomical Telescope. Nature **351**, 300–302.

Sandler, D. G., Cuellar, L., Lefebvre, M., Barrett, T., Arnold, R., Johnson, P., Rego, A., Smith, G., Taylor, G., Spivey, B. (1994a): Shearing Interferometry for Laser-Guide-Star Atmospheric Correction at large D/r_0, J. Opt. Soc. Am., **A11**, 858-873.

Sandler, D. G., Stahl, S., Angel, J. R. P., Lloyd-Hart, M., McCarthy, D. (1994b): Adaptive Optics for Diffraction-Limited Infrared Imaging with 8-m Telescopes. J. Opt. Soc. Am., **A11**, 925-945.

Sanger, D. (1989): A Technique for Assigning Responsibilities to Hidden Units in Connectionist Networks. Connection Science **1**, 115.

Sanger, T. D. (1989): Optimal Unsupervised Learning in a Single-Layer Linear Feedforward Neural Network. Neural Networks **2**, 459–473.

Sasai, M., Wolynes, P. G. (1990): Molecular Theory of Associative Memory Hamiltonian Models of Protein Folding. Phys. Rev. Lett. **65**, 2740-2743.

Schrieffer, J. R. (1964): *Theory of Superconductivity* (W. A. Benjamin, New York).

Schuur, J. H., Selzer, P., Gasteiger, J. (1996): The Coding of the Three-Dimensional Structure of Molecules by Molecular Transforms and its Application to Structure-Spectra Correlations and Studies of Molecular Activity. J. Chem. Inf. Comput. Sci. **36**, 334–344.

Sejnowski, T. J., Rosenberg, C. R. (1987): Parallel Networks that Learn to Pronounce English Text, Complex Systems **1**, 145–168.

Seung, H. S., Sompolinsky, H., Tishby, N. (1992): Statistical Mechanics of Learning from Examples, Phys. Rev. A **45**, 6056–6091.

Sherrington, D., editor (1989): *Special Issue in Memory of Elizabeth Gardner, 1957-1988*, J. Phys. A **22**(12), 1953–2273.

Smith, M. (1993): Neural Networks for Statistical Modeling (Van Nostrand Reinhold, New York).

Sompolinsky, H. (1988): Statistical Mechanics of Neural Networks. Physics Today **41**(12), 70–80.

Sompolinsky, H., Tishby, N., Seung, H. S. (1990): Learning from Examples in Large Neural Networks. Phys. Rev. Lett. **65**, 1683–1686.

Specht, D. F. (1990): Probabilistic Neural Networks. Neural Networks **3**, 109–118.

Stahl, S. M., Sandler, D. G. (1995): Optimization and Performance of Adaptive Optics for Imaging Extrasolar Planets, Ap. J. **454**, L153-L156.

Staudt, A., Bender, E., Muto, K., Klapdor-Kleingrothaus, H. V. (1990): Second Generation Microscopic Predictions of Beta-Decay Half-Lives of Neutron-Rich Nuclei. At. Data Nucl. Data Tables **44**, 79–132.

Stein, D. (1985): A Model of Protein Conformational Substates. Proc. Nat. Acad. Sci. USA **82**, 3670–3672.

Stimpfl-Abele, G., Garrido, L. (1990): Fast Track Finding with Neural Networks. Comput. Phys. Commun. **64**, 46-56.

Stimpfl-Abele, G., Yepes, P. (1993): Higgs Search and Neural Net Analysis. Comput. Phys. Commun. **78**, 1–16.

Stolorz, P., Lapedes, A., Xia, Y. (1991): Predicting Protein Secondary Structure Using Neural Net and Statistical Methods. J. Molec. Biol. **225**, 363–377.

Sumpter, B. G., Getino, C., Noid, D. W. (1992): A Neural Network Approach to Energy Flow in Molecular Systems. J. Chem. Phys. **97**, 293–306.

Sutton, R. S., Barto, A. G. (1981): Toward a Modern Theory of Adaptive Networks: Expectation and Prediction. Psychological Review **88**, 135–170.

Thomsen, J. U., Meyer, B. (1989): Pattern Recognition of the ^{1}H-NMR Spectra of Sugar Alditols Using a Neural Network. J. Magn. Res. **84**, 212-217.

van Hemmen, J. L. (1986): Spin-Glass Models of a Neural Network. Phys. Rev. A **34**, 3435–3445.

Wade, R. C., Bohr, H., Wolynes, P. G. (1992): Prediction of Water Binding Sites on Proteins by Neural Networks. J. Am. Chem. Soc. **114**, 8284–8286.

Wan, E. A. (1990): Neural Network Classification: A Bayesian Interpretation. IEEE Trans. Neural Networks **1**, 303-305.

Watkin, T. L. H., Rau, A., Biehl, M. (1993): The Statistical Mechanics of a Learning Rule. Rev. Mod. Phys. **65**, 499-556.

Weisbuch, G., Fogelman-Soulié, F. (1985): Scaling Laws for the Attractors of Hopfield Networks. J. Physique. Lett. **46**, L623–L630.

Werbos, P. J. (1974): *Beyond Regression: New Tools for Prediction and Analysis in the Behavioral Sciences*, Ph.D. Thesis, Harvard University, Cambridge, MA.

Widrow, B., Hoff, M. E. (1960): Adaptive Switching Circuits. *Institute of Radio Engineers, Western Electronic Show and Convention, Convention Record*, Part 4 (IRE, New York), 96–104.

Widrow, B. (1962): Generalization and Information Storage in Networks of Adaline "Neurons." *Self-Organizing Systems 1962*, edited by Yovits, M. C., Jacobi, G. T., Goldstein, G. D. (Spartan, Washington, D.C.), 435–461.

Wilcox, G. L., Poliac, M., Liebman, M. N. (1990): Neural Network Analysis of Protein Tertiary Structure. Tetrahedron Comp. Meth. **3**, 191–211.

Williams, R. J., Zipser, D. (1989a): A Learning Algorithm for Continually Running Fully Recurrent Neural Networks. Neural Computation **1**, 270–280.

Williams, R. J., Zipser, D. (1989b): Experimental Analysis of the Real-Time Recurrent Learning Algorithm. Connection Science **1**, 87–111.

Willshaw, D. J., Buneman, O. P., Longuet-Higgins H. C. (1969): Non-Holographic Associative Memory. Nature **222**, 960–962.

Witt, J. C., Clark, J. W. (1990): Experiments in Artificial Psychology: Conditioning of Asynchronous Neural Network Models, Math. Biosci. **99**, 77–104.

Zemel, R. S., Hinton, G. E. (1995): Developing Population Codes by Minimizing Description Length. Neural Computation **7**(3), 549-564.

Zemel, R. S, Dayan, P. (1997): Combining Probabilistic Population Codes. *International Joint Conference on Artificial Intelligence 1997* (Morgan Kaufmann, Denver, CO).

Zemel, R. S., Dayan, P., Pouget, A. (1997): Probabilistic Independence Networks for Hidden Markov Probability Models. Neural Computation **2**. 227–269.

Zemel, R. S., Dayan, P., Pouget, A. (1998): Probabilistic Interpretation of Population Codes. Neural Computation **10**(2), in press.

Zupan, J., Gasteiger, J. (1991): Neural Networks: A New Method for Solving Chemical Problems or Just a Passing Phase? Analytica Chimica Acta **248**, 1–30.

Zupan, J., Gasteiger, J. (1993): *Neural Networks for Chemists: an Introduction* (VCH, Weinheim).

Adaptive Optics: Neural Network Wavefront Sensing, Reconstruction, and Prediction

Patrick C. McGuire[1], David G. Sandler[1,2], Michael Lloyd-Hart[1], and Troy A. Rhoadarmer[1]

[1] University of Arizona, Steward Observatory,
  Center for Astronomical Adaptive Optics,
  Tucson, AZ 85721, U.S.A.
[2] Thermotrex Corporation
  10455 Pacific Center Court
  San Diego, CA 92121, U.S.A.

Abstract. We introduce adaptive optics as a technique to improve images taken by ground-based telescopes through a turbulent blurring atmosphere. Adaptive optics rapidly senses the wavefront distortion referenced to either a natural or laser guidestar, and then applies an equal but opposite profile to an adaptive mirror. In this paper, we summarize the application of neural networks in adaptive optics. First, we report previous work on employing multi-layer perceptron neural networks and back-propagation to learn how to sense and reconstruct the wavefront. Second, we show how neural networks can be used to predict the wavefront, and compare the neural networks' predictive power in the presence of noise to that of linear networks also trained with back-propagation. In our simulations, we find that the linear network predictors train faster, they have lower residual phase variance, and they are much more tolerant to noise than the non-linear neural network predictors, though both offer improvement over no prediction. We conclude with comments on how neural networks may evolve over the next few years as adaptive optics becomes a more routine tool on the new large astronomical telescopes.

1 Principles of Adaptive Optics

Astronomy has the purpose of exploring the heavens, and in so doing, requires observations be made with ever-increasing clarity, usually meaning that the images be as sharp as possible (blurriness is to be avoided), and that the images have the highest possible contrast (so that dim features can be studied). Military reconnaissance has similar requirements of the images taken from earth-orbiting satellites. The first requirement, called the 'high-resolution imaging' requirement, has repercussions on the second requirement, called the 'faint imaging' requirement – usually when the image is made sharper, then it is possible to see fainter features of the image. Adaptive optics has the main purpose of providing much higher-resolution images for ground-based optical/infrared telescopes, that would otherwise be severely limited by the turbulent blurring of the atmosphere.

There are many motivating factors and many challenges associated with observing astronomical objects or man-made satellites from the surface of the Earth at near-visible (i.e., visible, near- to mid-infrared, and ultra-violet) wavelengths. The motivating factors include:

(1) Discovery and imaging of planets and brown dwarfs orbiting nearby stars;

(2) Study of star-formation in nearby star-forming regions;

(3) Searching for dark matter in our Galactic halo in the form of stellar remnants (i.e. neutron stars) or substellar objects (brown dwarfs) by their gravitational micro-lensing of background field stars;

(4) Verification, discovery, and study of gravitationally-lensed Active Galactic Nuclei (AGNs) in order to constrain the Hubble Constant;

(5) Mapping of the inner realms of star-forming galaxies ("starbursts") at moderate red-shift ($z \equiv \frac{\Delta \lambda}{\lambda} = 1$) and possible associated AGNs to constrain galaxy evolution models;

(6) High-resolution mapping of planetary bodies like Jupiter's moon Io to determine the spatio-temporal properties of its volcanoes, as well as the asteroid Vesta to study its mineralogy;

(7) High-accuracy determination of the time-dependence of the two- and three- dimensional shape of the Sun in order to determine solar structure and to constrain theories of gravitation;

(8) Spectroscopic studies of the rotation curves of nearby galaxies in order to infer dark matter content;

(9) Spectroscopic studies of distant galaxies and protogalaxies which measure redshifts to infer distances and also to determine primordial element/isotope/molecular abundances in order to better understand big-bang nucleosynthesis;

(10) Direct imaging of satellites orbiting the Earth in order to infer their capabilities.

In order to achieve the above goals for high-resolution imaging with ground-based telescopes, there are many technical challenges. These include the following:

(1) The long-exposure blurring of images by the turbulent atmosphere at *all* near-visible wavelengths, producing seeing-limited images with resolution (1 arcsecond), which are not limited by the diffraction due to the aperture size;

(2) The near- to mid-infrared thermal background of the atmosphere and the telescope;

(3) Light pollution at visible wavelengths due to encroaching cities;

(4) Wind-buffetting of the telescope;

(5) Selective atmospheric absorption at various infrared and near-UV wavelengths;

(6) The motion or apparent motion of the science object due to Earth rotation or satellite revolution-about-the-Earth;

(7) The diurnal rising of the Sun and Moon;

(8) Weather shutdowns (humidity, thick clouds, snowstorms) or weather annoyances (high-altitude cirrus);

(9) The construction of telescopes of ever-increasing size in order to observe dimmer objects and also to attain higher spatial resolution;

(10) The construction of CCD and Infrared Detectors of ever-increasing pixel number (e.g. 4096×4096), focal-plane area, quantum efficiency, electronic speed, spectral coverage, and ever-decreasing read-noise and thermal 'dark-current' noise.

A great deal of progress, perhaps a 'quantum leap', can be made on all of the motivating science factors if practical solutions can be found to Challenge 1, which is highly-linked to the effective usage of large ground-based telescopes (Challenge 9).

A much more expensive solution to all the Challenges facing ground-based astronomical and military observing is to place telescopes in orbit around the Earth. The absence of atmospheric blurring and other ground-based problems makes images with resolution better than 0.1 arcsecond available to astronomers (for telescope diameters greater than 2.4 m diameter at a wavelength of 1μm). For example, the Hubble Space Telescope (HST) serves *many* astronomers in their near-visible research; therefore observing time on the HST is relatively hard to obtain. Astronomers would be well-suited to have 5-10 different HST's, but at a price tag of over US$2 billion/HST, this will not happen. The current plan of US space-based near-visible astronomy (the Next Generation Space Telescope (NGST)) is to put into orbit by rocket by 2007 a single large telescope (8-meter class) at less cost than the HST. A ground-based 8-meter class telescope (e.g. Gemini, LBT, Keck) costs an order of magnitude less than the HST – hence, at the cost of 10-20 large ground-based telescopes, a single large space-based telescope can be put into orbit to produce images with resolution of 0.025 arcsecond (at 1 micron). Is such a space-based effort worth the high cost? Can we make ground-based competitive with the HST and NGST at a much lower cost?

One relatively inexpensive solution to (time-dependent) atmospheric blurring (Challenge 1) is first to record a series of very short exposures (1-20ms) of the science object or of a bright 'guidestar' near the science object in order to 'freeze' the turbulence. Each of these short-exposure images will predominantly have a high-resolution bright spot, called a 'speckle', somewhere in the image, and the resolution of this bright speckle is limited only by the telescope size, and not by the atmospheric turbulence. With the long series of short exposure images, during offline processing, the observer's computer program would first shift each image so that the dominant speckle of that image is coincident with all the other images' dominant speckle, after which the computer program would proceed to co-add all the images. This 'Shift-

and-Add' technique and its application to speckle images was developed by astronomers in the 1980's and achieved near diffraction-limited imagery (0.1 arcsecond resolution for a 2.4 m diameter telescope at a wavelength of 1 μm), but required very high signal-to-noise ratios. Hence, speckle-imaging is limited to relatively bright stars and their small neighborhoods. Unfortunately, only a limited amount of science can be produced by imaging near the bright stars that are suitable for speckle imaging.

Following up on military research that was declassified in the early 1990's, most major observatories around the world are developing a technique called 'adaptive optics' (AO). An AO-equipped telescope continually takes short-exposure measurements of the turbulence-distorted wavefronts of light coming from a guidestar above the large telescope, and then rapidly applies the conjugate of the turbulence's phase-map to a deformable mirror (DM) in either the AO instrument or telescope in order to cancel the blurring effect of the turbulence on the image of the science object. Long-exposures of the science object can easily be obtained just by integrating the image continuously during the application of the ever-changing shape of the DM (the AO system operates in 'closed-loop' servo). An AO system costs much less than the large telescope itself and costs significantly more than speckle-imaging offline processing. With the diffraction-limited performance obtained with large ground-based telescopes equipped with AO at only a small fraction of the cost of similar performance from a large space-based telescope, one begins to understand the allure and ongoing renaissance of ground-based telescope construction. In the remainder of this Section, we present a beginner's introduction to adaptive optics; we defer the astronomers to the final two Sections where we discuss old and new results pertaining to the use of neural networks in adaptive optics.

1.1 Large Ground-Based Telescopes

Since Newton and Galileo, astronomers and telescope builders have been entranced by the large-telescope mystique for two compelling reasons: more photons and higher spatial resolution. The rate of photon collection is proportional to the area of the primary mirror and hence the square of the telescope primary mirror diameter. Therefore, the 10-meter reflective telescopes of today afford a factor of 100 higher sensitivity to dim stars and dim galaxies than the 1-meter refractive telescopes of the nineteenth century.

Due to the diffraction theory of wave-like propagation of light developed by Fraunhofer and Airy, perfect 'stigmatic' focussing of light by a mirror or lens of finite size was shown to be impossible (see Meyer-Arendt 1984), as was naively predicted by geometric optics. The half-width of the perfectly diffraction-limited spot was shown to be very close to the ratio $1.22\lambda/D$, where λ is the wavelength of light and D is the diameter of the telescope aperture. For example, for a wavelength of $\lambda = 0.5$ μm (blue-green light), and a lens of aperture $D = 1$ m, we obtain a diffraction-limited resolution of

0.61 μRadian $\equiv$ 0.12 arcseconds. From the inverse-proportional dependence of resolution upon aperture size, astronomers and opticians like Herschel and Fraunhofer immediately realized that by increasing the aperture size, one enhances the ultimate resolution of the optic. Unfortunately, the art of that period was refracting telescopes instead of reflecting telescopes, so chromatic aberration of refractive elements prevented diffraction-limited performance of these telescopes. It was not until the early twentieth century that the trend shifted, and the world's cutting edge 2 to 5 meter telescopes were constructed from reflective components.

Fifty to seventy years ago a few 5-meter-class reflecting telescopes were constructed (e.g. Hale). But until the 1980's, construction of such large telescopes had stagnated, with much of the telescope building effort going towards 2 to 3 meter telescopes. One reason for this was that light-weight mirror-making technology had not yet blossomed, and 5-meter telescopes meant very heavy, very high thermal time-constant primary mirrors. The high thermal time-constants meant that the telescopes did not cool to ambient temperatures during the night, causing some dome turbulence, but more destructively, adding significant thermal emission and making infrared astronomy much more difficult. Another important reason for the delay of the large primary mirror renaissance until the 1990's, was that large, shallow mirrors demanded long telescopes, which are difficult to construct and stabilize from a mechanical engineering standpoint. In the 1980's and 1990's, methods were perfected using spin-cast techniques (Hill & Angel 1992), stressed-lap computer-controlled polishing (Martin *et al.* 1997), and also actively-controlled segmented primary mirrors (Cohen, Mast & Nelson 1994), to construct large, deep mirrors with small focal ratios ($f/D \sim 1.1$), which allowed compact telescope structures.

But the primary reason for the stagnation was that without adaptive optics technology to correct for atmospheric blurring, the payoff in spatial resolution from building larger telescopes was small. In the 1970's the Multiple Mirror Telescope (MMT) (a co-mounted array of six 1.8-meter diameter telescopes, spanning 6.9 meters) was constructed on Mt. Hopkins in Arizona (see Figure 1). Not equipped with adaptive optics until the 1990's, the main advantage of this telescope was the simultaneous pointing of 6 moderate-sized telescopes with a common focal-plane at the science object of interest, thus increasing the number of photons collected by a factor of 6 over 'easily' constructed telescopes of the day. The main disadvantage of the MMT is that when compared to other telescopes of similar size, the 41% filling factor means a factor of 2.5 less light collected than the state-of-the-art 6.5 meter telescope which is replacing the old MMT in the winter of 1998 (see Figure 2).

In Table 1, we list the large telescopes in order of size being built around the world that will be or already are incorporating adaptive optics systems. The 6.5 meter MMT upgrade $f/D = 1.25$ primary mirror was spin-cast in a

Fig. 1. The Multiple Mirror Telescope (MMT) (before March 1998) atop Mt. Hopkins in Arizona (after Serge Brunier of *Ciel et Espace*).

huge rotating oven at the University of Arizona Steward Observatory Mirror Laboratory in the early 1990's, and was also polished at the Mirror Lab in the mid 1990's, and will be taken to the telescope site on Mt. Hopkins in November 1998. A twin of the 6.5 meter MMT primary mirror is now being polished at the Mirror Lab for installation at the Magellan telescope in Chile. The largest telescope mirror in the world (one of the Large Binocular Telescope (LBT) 8.4 meter $f/D = 1.14$ primaries destined for Mt. Graham in Arizona) was successfully spun-cast at the Mirror Lab in 1997. The Keck I and II 10 meter $f/D = 1.75$ telescopes, built and maintained by the University of California, have been installed and are operating on Mauna Kea in Hawaii, and

Fig. 2. The MMT-upgrade (6.5 meters diameter), as it will appear in January 1999 (after Serge Brunier of *Ciel et Espace*).

each consists of 36 separately controlled close-packed segments. The Gemini I and II 8 meter $f/D = 1.5$ telescopes will operate independently on Mauna Kea and in Chile, and are operated by a British/US/Canadian consortium. The Very Large Telescope Array, operated by the European Southern Observatory, consists of four 8.2 meter telescopes which can operate together in interferometric mode, of which one telescope became operational in 1998 in Cerro Paranal in northern Chile. Starfire Optical Range, run by the US Air Force Research Lab, in New Mexico has a 1.5 meter and a new 3.5 meter telescope, the primary mirrors being made at the University of Arizona. The CFHT (Canada-France-Hawaii telescope) is a 3.6 meter telescope operating

TELESCOPE	DIAMETER	LOCATION
SOR I	1.5 m	Starfire Optical Range, New Mexico
MW 100"	2.5 m	Mt. Wilson, California
Shane	3.0 m	Mt. Hamilton, California
SOR II	3.5 m	Starfire Optical Range, New Mexico
Calar Alto	3.5 m	Calar Alto, Spain
Apache Point	3.5 m	Sacramento Peak, New Mexico
La Silla	3.6 m	La Silla, Chile
CFHT	3.6 m	Mauna Kea, Hawaii
WHT	4.5 m	La Palma, Canary Islands
MMT upgrade	6.5 m	Mt. Hopkins, Arizona
Magellan	6.5 m	Las Campanas, Chile
Gemini I	8.0 m	Mauna Kea, Hawaii
Gemini II	8.0 m	Cerro Pachon, Chile
VLT I-IV	8.2 m	Cerro Paranal, Chile
Subaru	8.3 m	Mauna Kea, Hawaii
LBT I & II	8.4 m	Mt. Graham, Arizona
Keck I & II	10.0 m	Mauna Kea, Hawaii

Table 1. List of telescopes that are using or will use adaptive optics

on Mauna Kea, and the SUBARU telescope is a Japanese 8.3 meter telescope soon to be operating on Mauna Kea. And last but not least among this non-exhaustive list of large telescopes, the 3 meter Shane telescope is run by the University of California's Lick Observatory on Mt. Hamilton in California. All of these large telescopes have been or will be outfitted with adaptive optics systems soon after commissioning.

1.2 The Blurring Effects of the Atmosphere

Newton (1730) recognized early on that the atmosphere was a major limiting factor to high resolution imaging through the atmosphere, and he recommended telescopic observations in the 'rarefied' air found on high mountaintops. Kolmogorov (1961) and Tatarskii (1961) developed a statistical model for the phase variations of light passing through turbulent air caused by refractive index variations. From his turbulence theory, Kolmogorov derived the following structure function:

$$D\left(r\right) \equiv E\left([\phi(\mathbf{r}+\mathbf{r}_0) - \phi\left(\mathbf{r}_0\right)]^2\right) = 6.88\left(\frac{r}{r_0}\right)^{\frac{5}{3}}, \qquad (1)$$

where $D(r)$ represents the spatial structure function for phase variations as a function of the spatial separation r and parametrized by the Fried parameter r_0 (see Fried 1965). The Fried parameter is calculated as an average over altitude of turbulence strength, and depends on zenith angle and wavelength

(Beckers 1993), and is the maximum aperture through which diffraction-limited resolution can be obtained. This structure function equation is valid for distances between an empirically-determined inner scale and outer scale ($l < r < L$). The structure function is the mean-square phase error between two points in the aperture, and leads to a form for the residual RMS phase over an aperture which behaves as $(D/r_0)^{5/6}$. The Kolmogorov structure function states that for spatial separations $r < r_0$, there is relatively little phase variation, or in other words, there is spatial coherence of the wavefront of light passing through a patch of atmosphere of size r_0, whereas for separations of $r > r_0$, the wavefront of light is uncorrelated and hence the structure function is large, but limited by the outer scale. The Kolmogorov theory predicts (Noll 1976) that there will be large-scale components which grow with aperture, including large components of tilt and focus. For very rapid images of a bright star taken with a telescope of diameter D, greater than the Fried coherence length r_0, there are typically D^2/r_0^2 bright speckles in the image, caused by the small coherence length. The Fried coherence length varies with time for a given site, but above Mt. Hopkins, the median r_0 at visible wavelengths (0.5 μm) is ~ 0.15 meters. Due to the $\lambda^{6/5}$ dependence of r_0 on wavelength λ, this translates to a median r_0 in the near-infrared (2.2 μm) of ~ 0.9 meters. Hence, the number of speckles (in the image plane) or coherence cells (in the pupil plane) for a short-exposure image taken with a large telescope (e.g. $D = 6.5$ meters) in the near-infrared (52 speckles at 2.2 μm) is 36 times less than in the visible (1900 speckles at 0.5 μm), thus making adaptive optics possible in the infrared and next-to-impossible in the visible.

The atmosphere is often composed of either 2 or 3 discrete layers of turbulence (Hufnagel 1974) (see Figure 3). Each of these layers can in crude approximation be considered 'frozen' on timescales much longer than the time it takes the turbulence to be blown across the telescope aperture. This 'Taylor picture' (Taylor 1935) of winds blowing a fixed pattern of turbulence past the observer is complemented by the prescription of the coherence time-scale τ or 'Greenwood frequency' f_0 (Greenwood 1977, Fried 1994). For temporal sensing frequencies below f_0, there is constant and large correlation of the wavefront of light impinging on a telescope (rigorously, the total phase error equals 1 radian at the Greenwood frequency, and the closed-loop error is 0.3 radians and one needs to sample about 10 times more frequently, so the sampling rate could be 30 times higher than the Greenwood frequency). For frequencies above f_0, the correlation plummets with a $f^{-8/3}$ power law dependence. The Greenwood frequency is given as:

$$f_0 \approx \frac{0.4v}{\sqrt{\lambda z}} = \frac{0.134}{\tau}, \tag{2}$$

where v is the bulk velocity of the wind in the turbulent layer of air and z is the altitude of the layer. For $v = 40$ m/s, $\lambda = 2.2$ μm, and $z = 10000$

meters, the Greenwood frequency is 100 Hz. Hence, for imaging in the near-infrared, adaptive optics system bandwidths need to be from several hundred Hz to 1kHz in order to effectively correct for the varying turbulence. This high bandwidth practically defines adaptive optics; lower bandwidth systems are called 'active optics' systems. The high bandwidth demands that only sufficiently bright adaptive optics guidestars be used; otherwise there will be insufficient photons to accurately measure or sense the shape of the distorted wavefront of light.

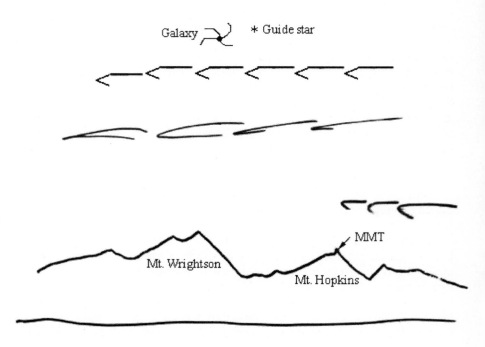

Fig. 3. A qualitative (not to scale) sketch of the main problem facing ground-based near-visible astronomy: atmospheric turbulence; shown here as three layers above Mt. Hopkins.

1.3   Resulting AO-Corrected Images: the Meaning of Strehl Ratios and Strehl Reduction by Non-Zero Wavefront Variance

The main purposes of any adaptive optics system are to improve resolution and concomitantly to increase the peak brightness of images of point-like sources taken beneath the blurring atmosphere. The fundamental limit to both of these objectives is given by the diffraction due to the finite diameter of the primary mirror in the telescope (see section 1.1). Adaptive optics merit-functions are therefore derived with respect to this diffraction limit. The diffraction-limited (DL'ed) on-axis intensity is given by the brightness of the emitting source, the wavelength of the light, the diameter of the aperture, and the focal length of the telescope. For a fixed source and telescope, this DL on-axis intensity is *fixed* at I_0, and all aberrated performance is measured relative to the DL (see Figure 4), giving a relative on-axis intensity of:

$$S \equiv \frac{I}{I_0}, \tag{3}$$

which is called the Strehl ratio. In the pupil plane, the atmosphere and optics will aberrate the phase-map of light coming from a point source, and the variance of this distortion (summed over the entire pupil), σ^2 (in rad^2), quantifies the disturbance. If the variance is given in terms of distance (i.e. μm^2), the conversion to radians is accomplished by multiplying by $(2\pi/\lambda)^2$. Born & Wolf 1975 show that for moderate aberrations ($\sigma^2 < 1\,\mathrm{rad}^2$), the Strehl ratio is:

$$S \approx \exp\left(-\sigma^2\right) \approx 1 - \sigma^2. \tag{4}$$

The DL resolution is given by Rayleigh's criterion ($0.251\lambda/D$ arcseconds, where λ is given in μm and D is given in meters), and seeing-limited resolution is often given in multiples of the DL resolution. However, for weak aberrations, the resolution may be very close to the DL, but the Strehl ratio may be as low as 0.6 or 0.7. Therefore, the Strehl ratio is sometimes a better performance criterion than the resolution. In the best of worlds, a full spatial transfer function (OTF/MTF) study should characterize an AO system, but we do not include such sophistication in the neural net prediction analysis reported here (Section 3). There are several contributions to the phase-variance σ^2 of wavefronts of light incident upon and going through an adaptive optics system (see Tyson 1991, Sandler *et al.* 1994b, Fried 1994). These include fitting error, wavefront sensor error (due to finite number of photons and CCD read noise), servo lag error (discussed in detail in section 3), tilt anisoplanatism error, focus anisoplanatism (for laser guidestars), and reconstruction error.

1.4   Wavefront Sensing

The most common wavefront sensing technique used in adaptive optics systems was put forward by Shack & Platt (1971) and Hartmann (1900). Alternatives include the curvature sensor of Roddier (1988). A Shack-Hartmann sensor consists of an array of lenslets that is put at a pupil in the adaptive optics beam train and a camera put at the focus of the lenslet array. The

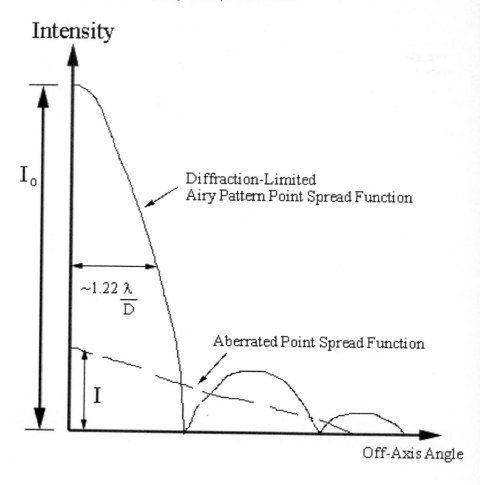

Fig. 4. A qualitative sketch showing the unaberrated and aberrated point spread functions for imaging. The ratio of the aberrated peak intensity to the unaberrated peak intensity is called the Strehl ratio, which is a key indicator of image quality.

most common lenslet and (CCD) camera geometry is called the Fried geometry, and consists of a square array of square lenslets, with each lenslet focussing the unperturbed wavefront to a spot that is centered on the intersection of four CCD pixels, called a quad-cell (see Figure 5). The centers of the quad-cells in the lenslet-array pupil are made to be coincident with the corresponding positions of the actuators in the deformable mirror pupil. If the atmospherically-perturbed wavefront has tilt over a particular subaperture's lenslet in the x-direction, then there will be an imbalance of the signal registered by the CCD quad-cell in the x-direction (see Figure 6). Likewise,

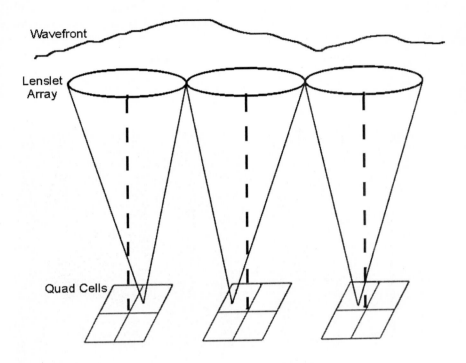

Fig. 5. A qualitative sketch showing the principle of Shack-Hartmann wave-front-sensing with a toy 1×3 lenslet array and 3 quad-cells. the wavefront slope is measured by the imbalance in the signals in the 4 CCD pixels that comprise a quad-cell.

a tip in the y-direction will register a y-direction quad-cell imbalance. The array of tips and tilts over the pupil are often called the slopes, and can be represented as both a matrix or a vector. Read-out noise and dark-current in the CCD detector (represented by n = RMS number of equivalent electrons of noise per pixel) and photon-counting noise (finite number of photons, N, caused by finite subaperture size, the required high temporal bandwidth, and the use of dim guidestars) will only allow imprecise determinations of the slopes. The RMS wavefront sensor centroid error is $N^{-1/2}$ waves, for pure Poisson noise – thus for 100 photons, there is 0.1 waves of centroid error contribution to the error in wavefront determination. For subapertures (on the sky) of 0.5×0.5 m^2, CCD integration times of 5 milliseconds, CCD read noise of $n = 3$ electrons, guidestars brighter than a visual magnitude of ~ 10 are considered bright and offer good adaptive correction (with an average of better than 3000 photons per subaperture per exposure and high signal to noise ratio), and guidestars dimmer than a visual magnitude of ∼ 15 are considered dim (with less than 30 photons per subaperture per exposure and a signal to noise ratio of less than one).

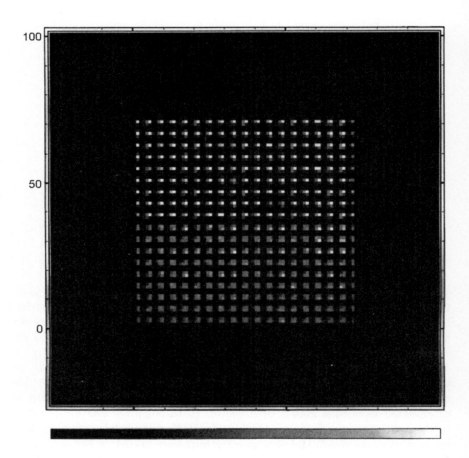

Fig. 6. An image of the CCD data output from the 13 × 13 subaperture Shack-Hartmann wavefront sensor camera being prepared for use with the MMT-upgrade AO system. The upper half of the array was not being clocked properly, so its signal was not centered on quad-cells, as in the lower half of the array (this problem has been fixed).

1.5 Laser Guidestars

For purposes of accurate wavefront sensing and subsequent adaptive correction of the wavefront, guidestars of visual magnitude of at least 10 are required. Also, the guidestar is required to be within 20-40 arcseconds of the desired science object in order for the measured wavefront distortions in front of the guidestar to correspond closely to the wavefront distortions in front of the science object. Otherwise, due to the non-proximal guidestar, there is 'tilt anisoplanatism', and adaptive correction suffers. Rarely is there a guidestar of magnitude 10 or brighter within 20-40 arcseconds of the science object. In Figure 7, Quirrenbach and collaborators have found statistically that only 0.15% of the sky is within 30 arcseconds of a magnitude 10 guidestar (near the Galactic pole), and even when observing in the plane of the Milky Way, only 0.6% of the sky is 'covered'. The scarcity of bright natural guidestars severely limits the quantity and perhaps the quality of astronomical studies which can be conducted with the improvements in resolution and brightness-sensitivity offered by adaptive optics.

Therefore, within the secret auspices of U.S. military research in the early 1980's, Happer (1982) proposed a key extension to the then nascent military artificial guidestar program, namely resonant back-scattering of a finely-tuned laser beam (projected from the ground) off a 10 km thick layer of naturally occurring atomic sodium at an altitude of 90 km and using the excitation of sodium from the hyperfine-split ground state to the first excited state as the resonance (Kibblewhite 1997). They found that there is sufficient sodium column density and sufficient sodium absorption cross-section of yellow D2 light (589 nm), that a projected laser of moderate power (1-10 Watts) could produce a guidestar of magnitude greater than 10. Therefore, with such a laser co-projected from the observing telescope, a bright guidestar could be projected anywhere on the sky, thus allowing adaptive-optics-fostered science to be pursued on any astronomical or man-made object of interest. These sodium guidestars have been developed and used by several research groups (Jacobsen 1997, Carter et al. 1994, Martinez 1998, Roberts et al. 1998, Shi et al. 1996, Friedman et al. 1995, Friedman et al. 1997, Avicola et al. 1994); science has been produced with sodium guidestar correction (Max et al. 1997, Lloyd-Hart et al. 1998a); and most of the world's AO groups will be outfitting their respective large telescopes with sodium laser guidestar systems in the coming years. The technology of tuning a laser to precisely the right wavelength (589.0 nm) is blossoming, with the past use of dye lasers (Martinez 1998) and the present development of solid state Raman-shifted lasers (Roberts et al. 1998).

Previous to the invention and development of the sodium laser guidestar, the US Air Force at Starfire Optical Range in New Mexico successfully developed and is still primarily using another type of laser guidestar: the Rayleigh back-scattering guidestar (Fugate et al. 1991, Primmerman et al. 1991, Sandler et al. 1994a). Such guidestars rely on the λ^{-4} Rayleigh scattering off of

Sky coverage or probability of finding a natural guide star with
a given magnitude within 30 and 10 arcsec from a random point
in the galactic plane and at the pole.

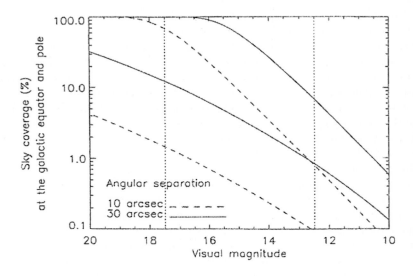

Fig. 7. A graph (after Quirrenbach 1997) showing the fraction of the sky that has
a star brighter than the given magnitude within 10 or 30 arcseconds. The upper
solid and dashed lines are for directions towards the Galactic plane, and the lower
solid and dashed lines are for directions towards the Galactic poles.

aerosols and dust in the atmosphere at altitudes of 10-30 km, and is pulsed
and subsequently range-gated to select a sufficiently high-altitude to form the
guidestar. The beauty of Rayleigh guidestars is that they can be made very
bright (V magnitude < 5) with copper vapor lasers due to the non-necessity
of fine-tuning of the laser wavelength; such bright guidestars are needed to
ensure sufficient resolution in the corrected visible images of the satellites in
question. The main drawback of Rayleigh guidestars is that they are low altitude, and hence for two reasons do not sample the same turbulence as seen by
the light emanating from the science object, but the US Air Force uses these
low-altitude beacons because of the unavailability of magnitude 5 sodium
laser guidestars. The first reason is that for laser guidestar altitudes of less
than 20 km there can be unsensed turbulence *above* the Rayleigh guidestar.
The second reason is that even if there was no atmosphere above the Rayleigh

guidestar, 'focus anisoplanatism' or the 'cone effect' allows proper sampling of the turbulence within a cone emanating from the laser guidestar to the telescope aperture, but does not allow *any* sampling of the turbulence outside this cone, but within the wider cone (or practically a cylinder) of light subtended by the science object at the apex and the telescope as the base. This cone effect also applies to the higher altitude sodium guidestars, but to a much lesser extent.

One problem does remain, even with the development of laser guidestars. This problem arises because the laser guidestar is projected from the ground, usually by a projection telescope attached to the large observing telescope. This means that the upgoing laser samples the same turbulence as the down-going resonant backscatter laser light. Therefore, when the telescope aperture is divided into subapertures for wavefront sensing, each subaperture will only be able to measure the tilt *relative* to the upgoing projected laser beam, which is itself experiencing a *global* deflection or tilt relative to a 'fixed' direction towards the science object. This global tilt would go unsensed unless auxiliary countermeasures are taken. The countermeasure that is most frequently applied is using a third camera (the first being the science camera, the second being the wavefront sensor camera) to measure the global tilt, using light from a natural guidestar collected by the *whole* telescope aperture instead of the fine spatial subdivision of the light as done by the wavefront sensor. This still relies on using a natural guidestar, but allows the use of much dimmer natural guidestars (perhaps 300 times dimmer, or magnitude 16), which are *much* more plentiful and allow full sky coverage. Such coverage would not be possible without the laser guidestars used for differential tilt measurements.

1.6   Control of the Adaptive Mirror

The standard method of adaptive correction relies on inserting a reflective element in the adaptive optical system, prior to the science camera, and then from the signals derived from the wavefront sensor camera, deforming the mirror so as to null the wavefront sensor signals as accurately as possible. Of course, for accurate correction of the turbulence, three characteristics are required of the adaptive element:

(1) sufficient stroke;
(2) sufficient speed;
(3) sufficient spatial resolution.

By the 1970's, low-order active control of optical systems had been demonstrated (Mikoshiba & Ahlborn 1973, Bridges *et al.* 1974, Bin-Nun & Dothan-Deutsch 1973, Feinleib, Lipson & Cone 1974, Hardy, Feinleib & Wyant 1974, Muller & Buffington 1974), either correcting for tilt, focus or a small number of segments/zones; and high-order segmented mirrors and continuously-deformable mirrors (DM's) were in their nascency (Ealey 1989, Ealey & Wheeler 1990, Freeman *et al.* 1977).

In the 1990's, after military declassification of adaptive optics technology, astronomers used this technology to develop AO systems for more heavenly purposes. One such low-order system was developed for the Multiple Mirror Telescope (FASTTRAC2) (Lloyd-Hart *et al.* 1997), and was innovative in the sense that the adaptive element replaced a mirror that was an essential part of the telescope without adaptive optics. In FASTTRAC2, the MMT's 6-facetted (static) beam combiner was replaced with a 6-facetted beam-combiner built by Thermotrex with rapid control of the tip/tilt of each of its facets. By making an essential telescope mirror the adaptive element, a performance-degrading beam train of perhaps 6 additional mirrors before the science camera was avoided. More standard AO systems which have these additional mirrors suffer considerably because of the additional reflection losses, and more importantly for infrared-imaging, the additional thermal emission from each of the uncooled mirrors. This concept of making an essential telescope mirror the adaptive one is currently being developed for the new single 6.5m mirror MMT (Foltz *et al.* 1998, West *et al.* 1996): this rather bold program (Lloyd-Hart *et al.* 1998b) replaces the solid $f/D = 15$ secondary with a 2 mm thick, 64 cm diameter thin-shell (high-order) adaptive secondary, with 336 voice-coil actuators behind the thin-shell (see Figure 8). The adaptive secondary is being built by the University of Arizona and an Italian consortium and should see first light in a complete AO system (being built by the University of Arizona and Thermotrex, Inc.) at the MMT at the end of 1999. The main problem that we have encountered in the development of the MMT adaptive secondary is that the deformable mirror is very 'floppy' and requires the damping effects of a thin air gap (40 microns) between the back of the secondary and a solid reference plate in order to eliminate resonances (see Figure 9, as calculated by T. Brinkley at Thermotrex). Another main problem that our Italian collaborators discovered in their prototype tests was that standard local Proportional-Integral-Derivative (PID) control of each actuator was much too slow to actuate high-stress high-order modes, sometimes taking 20-40 milliseconds to reach the commanded state. Therefore, a 'feed-forward' (not to be confused with feed-forward perceptrons) semiglobal control algorithm was developed in which the required forces for the next commanded mirror-shape are calculated and applied to all the actuators within 1 millisecond, obviating the slow PID control.

1.7   Wavefront Reconstruction

With each measurement of the atmospherically-aberrated wavefront from a natural or laser guidestar, a high-speed computer needs to calculate the commands to send to the actuators that deform the mirror. The wavefront reconstruction is most standardly accomplished by a matrix multiplication:

$$\phi(t + \tau) = \mathbf{R}\mathbf{s}(t), \tag{5}$$

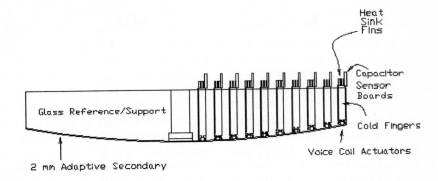

Fig. 8. A sketch showing the adaptive secondary for the MMT upgrade AO system, complete with a 2 mm deformable mirror made of Zerodur (low-expansion glass), voice coil actuators, and a glass reference plate.

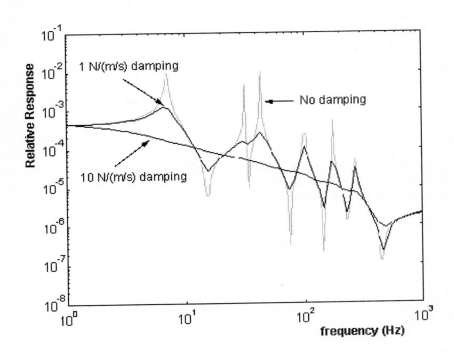

Fig. 9. A frequency response graph for the adaptive secondary shows the abundance of resonances at different sub-kiloHertz frequencies without damping, and how the resonances are eliminated with damping (after T. Brinkley, Thermotrex).

where $\mathbf{s}(t)$ is a vector of all the wavefront sensor slopes (tip/tilts) (dimension $= S$) for an integration time of τ centered at the latest time t, $\phi(t+\tau)$ is the vector of P actuator commands ('pokes') to be applied to the mirror for the next integration time centered at $t+\tau$, and $\mathbf{R}$ is the $P \times S$ reconstructor matrix that transforms from slope to phase. This reconstruction calculation for *all* the actuators has to be completed within a fraction of the wavefront sensor integration time. Since wavefront sensors for adaptive optics take exposures at a 100-1000 Hz rate, this demands wavefront reconstruction within 0.2-2 milliseconds. For high-order adaptive optics with hundreds of actuators over the aperture and an equal or greater number of wavefront sensor subapertures, such complex calculations continuously demanded within such a short time interval would tax even the fastest general purpose workstation computer. Therefore, special purpose computers with Digital Signal Processing (DSP) boards and matrix multiplication boards are employed to perform this dedicated reconstruction calculation.

In the standard AO scheme, the reconstructor matrix R is calculated prior to the astronomical or military observation and depends only on the subaperture/actuator geometry and a model of the deformable mirror. For the standard Fried geometry with DM actuators at the corners of all the wavefront sensor subapertures, a 'poke' matrix $\mathbf{M}$ is measured in which each matrix element M_{ij} represents the 'influence' of a unit poke by a single actuator j on each of the slope measurements i:

$$M_{ij} \equiv \frac{\delta s_i}{\delta \phi_j}. \tag{6}$$

Usually, these matrix elements are highly local in nature, with large numbers only for subapertures bordering an actuator. For $P \leq S$, the reconstructor matrix $\mathbf{R}$ is most often calculated by the Gaussian inverse of $\mathbf{M}$:

$$\mathbf{R} = \left(\mathbf{M}^T \mathbf{M}\right)^{-1} \mathbf{M}^T. \tag{7}$$

More sophisticated wavefront reconstruction algorithms have been developed which take into account the wind directions, atmospheric turbulence statistics, and measurement noise (Wallner 1983, Wild *et al.* 1995, Wild, Kibblewhite & Vuilleumier 1995, Angel 1994a), and one of these algorithms employs neural networks (Angel 1994a), which will be reviewed here. Another neural network algorithm, first developed by Lloyd-Hart (1991) & Lloyd-Hart *et al.* (1992), uses phase-diversity (two images one in-focus and one out-of-focus) as the input data and by training performs both the wavefront-sensing and wavefront-reconstruction in one-step, and will be a focus of this review.

2   Neural Network Wavefront Sensing and Reconstruction

2.1   Motivation

Astronomers currently measure atmospherically-perturbed wavefronts by two main techniques: Shack-Hartmann tip/tilt sensing and Roddier curvature sensing. Both of these techniques operate on the wavefront at a pupil of the telescope, thus demanding reimaging relay optics. A technique, described here, that operates on the *image*-plane data (i.e. the point-spread-function (PSF)) of the telescope would act directly on the quantity of interest to the observer. Additionally, Shack-Hartmann and Roddier wavefront sensing both require wavefront reconstruction from slopes or curvatures to phase. This requires integration over a connected pupil, which is not possible for array telescopes (i.e. the MMT), and therefore one begs for a more direct solution. Neural networks are offered as a solution, and due to their non-linear nature, these neural networks can effectively invert the non-linear transformations (PSF= $|FFT\left(e^{i\phi}\right)|^2$) from pupil-plane to image plane that are needed in phase-retrieval using phase-diversity. In the work described here, supervised-learning (back-propagation) was chosen because of its superiority to unsupervised/self-organizing networks in learning complex functional relationships.

2.2   Widrow-Hoff Wavefront Reconstruction

Angel (1994a) trained a network to transform simulated data from a 6 sub-aperture Shack-Hartmann quad-cell wavefront sensor (2 derivatives per quad-cell, which gives 12 inputs to the net) of circular/annular/segmented geometry (see Figure 10) for a 4 meter telescope in the infrared (2.2 microns, $r_0 = 1.18$meters) to reconstruct the amplitudes of the 6 pistons needed to minimize wavefront variance. Angel utilized the simplest of networks, without a hidden layer, using linear transfer functions without thresholds on the output layer, with Widrow-Hoff (WH) delta-rule training. When training with significant noise, the WH net's weights were found to be much smaller than without noise, thus allowing the conclusion that the WH net was able to learn that in high-noise situations, it is better to either do nothing or to simply average the data from all the inputs so as to reduce noise and determine global slope. In the absence of noise, Noll (1976) predicts that with 12 measurements and corresponding corrections of the wavefront that for this system, the idealized residual wavefront error would be $0.0339(D/r_0)^{5/3} = 0.26\,\text{rad}^2$, and the WH neural net was found to allow an error of only $0.041(D/r_0)^{5/3} = 0.31\,\text{rad}^2$, which is modest considering that the WH net includes both measurement error and fitting error.

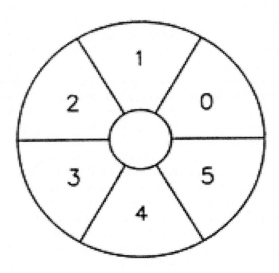

Fig. 10. The subaperture geometry for Angel's Widrow-Hoff linear net (no hidden layer) reconstructor.

2.3   Simulations for the Multiple Mirror Telescope

Angel *et al.* (1990) and Lloyd-Hart (1991) implemented a neural network phase-diversity wavefront sensor and reconstructor for low-order adaptive optics for a simulated Multiple Mirror Telescope. This input layer consisted of 338 input nodes, half being from an in-focus but aberrated 13×13 image of a star at a wavelength of 2.2 μm, and the other half being from a similar, simultaneous but deliberately out-of-focus image of the same star. The output layer of the multilayer perceptron neural network was composed of 18 linear nodes, each representing either tip, tilt, or piston for each of the 6 MMT mirrors. And the hidden layer consisted of 150 sigmoidal nodes (see Figure 11). The training data consisted of 2.5×10^5 image pairs calculated from Kolmogorov turbulence with $r_0 = 1$ m (see Figure 12c-d) on the input and their corresponding best-fit tip,tilt, and pistons in the pupil plane for the outputs, and standard momentum-less back-propagation was employed. A comparison of the unmasked pupil-plane data in Figure 12a and the neural network output in Figure 12e shows the correspondence of the net's output to the 'real' world, and a comparison of the uncompensated image seen in Figure 12b with the neural network corrected image in Figure 12f shows the improvement in image resolution and Strehl ratio offered by a phase-diversity neural network. The long-exposure Strehl ratio, obtained by adding 500 simulated speckle images corrected by the trained neural network, was $S = 0.66$, which is comparable to the Strehl predicted by tilt-corrected Kolmogorov turbulence ($S_{\text{theory}} = 0.70$). This phase-diverse neural network wavefront sensor and reconstructor (which operates in the infrared (K band)) was found to handle photon-counting noise for stars as dim as 10th magnitude (assuming an infrared detector with 10 photo-electron read noise). Similar simulation results have been reported by Vdovin (1995).

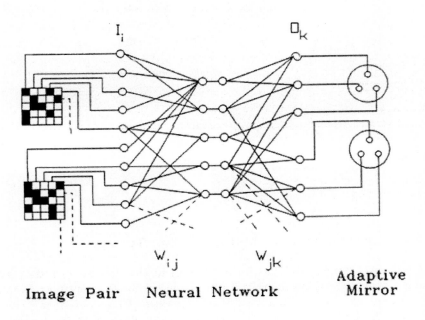

Image Pair   Neural Network   Adaptive Mirror

Fig. 11. The neural network used by Lloyd-Hart (1991) to sense and reconstruct the wavefront from phase-diverse MMT data.

2.4   Transputers at the Multiple Mirror Telescope

Lloyd-Hart (1991) and Lloyd-Hart *et al.* (1992) performed experiments that used a neural network with real starlight at both the Steward 2.3 meter telescope on Kitt Peak and at the old MMT on Mt. Hopkins. The experiments with the 2.3 meter continuous aperture telescope demanded the use of a pupil mask and a shift from K-band (2.2μm) to H-band ($1.6\,\mu$m) to simulate the bigger MMT and the MMT's better seeing conditions. Phase-diverse data was

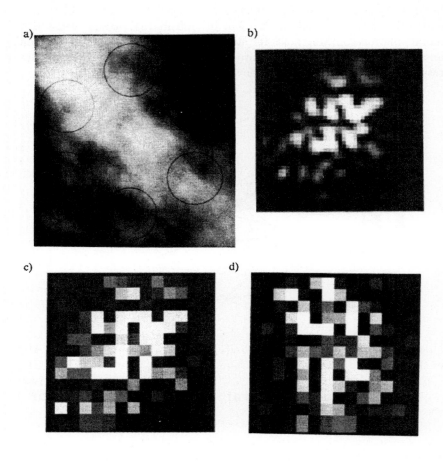

Fig. 12. (after Lloyd-Hart 1991) Sub-figure a) shows the simulated phase wavefront that serves to distort the image formed by the MMT, as seen in b). Sub-figures c) and d) show the in- and out-of-focus downsampled images, respectively, that serve as the input to the phase-diversity neural net wavefront sensor. Sub-figure e) shows the graphically-depicted output of the neural net (tips,tilts, and pistons: compare to the original wavefront in sub-figure a)), and sub-figure f) shows the resulting corrected image, which should be compared with sub-figure b). Sub-figures e) and f) see next page.

e)

f)

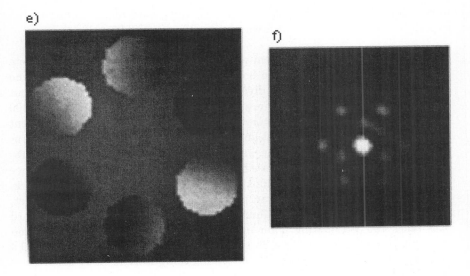

then taken at the 2.3 meter telescope, and then later in off-line mode, the pairs of images were shown to the neural net that was trained on simulated data for the MMT, and the net then outputted the wavefront (tips, tilts, and pistons) for the six apertures. From this neural net wavefront, speckle images were computed, and found to be in decent agreement with the actual speckle phase-diverse images, with some discrepancies caused by the high sensitivity of the image calculation to small changes in the net's output. By averaging 388 wavefronts derived by the net from the input image pairs, the mean wavefront was found not to be zero, but to have a significant (about half a wave) defocus aberration. A similar offline experiment was performed at the MMT, with non-phase-diverse infrared focal plane data taken with two of the six MMT mirrors at the same time as temporal visible centroid measurements were made of the images of each of the two mirrors separately. A neural net was then trained on simulated infrared images to output the relative tip/tilt and piston for the speckle image. With the trained neural net, the actual infrared data was passed through it to derive the tip output; when the net's infrared tip output was compared to the measured visible centroid tip, a significant correlation was determined but with 'a fair bit of scatter'. Prior to the completion of the MMT adaptive system, D. Wittman demonstrated, for the first time in the laboratory, closed-loop wavefront correction by a neural network with a HeNe laser and pin hole and a two-mirror aperture mask. Ten thousand phase-diverse random images were presented for training to the net by randomly changing the piston, tip, and tilt of the 2 element segmented mirror, and one thousand random images were used for testing. The control of the neural net effectively cancelled the random user-imposed mirror deformations, improving a very poor integrated image to one with a superb Strehl ratio of 0.98, or a wavefront variance of $0.02\,\text{rad}^2$.

A six-element adaptive system was built for the MMT in 1991, complete with transputers to serve as the hardware for a neural net that sensed the wavefront from a pair of phase-diverse images and controlled the wavefront with the adaptive mirror (see Figure 13). The neural net was trained on 30,000 image pairs taken in the laboratory in Tucson with the adaptive instrument itself supplying known wavefront aberrations. Of course, the adaptive instrument was unable to aberrate the wavefront on a scale smaller than the sub-aperture size, so 10,000 computer simulated images with fine-scale turbulence *and* fixed trefoil MMT aberrations were used to further train the network. With 16 Inmos T800-25MHz transputer modules serving as the neural net, training was accomplished in 33 minutes. Since the neural net's feed-forward pass takes 160,000 floating point operations, and since the integration and read out time of the two 26×20 sub-arrays of the infrared detector is 9.4 milliseconds, a machine capable of sustaining computations at 17 Mflops was required; the transputer array was benchmarked at a sufficient 25 Mflops. Lloyd-Hart took 15×10 subframes that included most of the energy from the pair of 26×20 images from the infrared camera. Therefore the network had 300 inputs. After the wavefront is calculated by the transputer net, global tilt was subtracted from the wavefront to ensure that the image always stays centered, and the mean phase was subtracted so that the actuators stay centered in its range. Prior to the run, nets were trained on laboratory data, using either 36 or 54 hidden neurons, and 6 or 10 output neurons, depending on whether 2 or 3 telescope apertures were used. An extra output was found useful, so that instead of predicting phase which has a discontinuity at π, nets outputting the continuously-valued sine and cosine of the phase were trained. A three mirror net was trained with laboratory data to a phase error of $0.05\,\text{rad}^2$ after 60,000 laboratory image pairs (50 minutes), but a net with 22 outputs for control of 6 mirrors did not achieve a low enough error, which was probably due to a local minimum. Neural network real-time adaptive control of *two* of the mirrors of the MMT was achieved in 1991, with training in the Tucson lab. For most of the 1991 observing run, the seeing was very poor (1 arcsecond at 2.2μm which is equivalent to an r_0 of 45cm, half the normal value), and since the net was trained for data of the normal character, sky correction with the neural network was impossible. When seeing improved later in the run, a two-mirror net achieved the first on-sky success in adaptive correction of atmospheric phase distortion on the image of the bright star ψ Pegasi (K magnitude of 0.0), see Figure 14, improving the seeing-limited resolution from 0.62 arcseconds to a diffraction-limited 0.1 arcseconds. By taking the summed product of the hidden and output weights $(V_{ij} = \Sigma_j W_{ij} W_{jk})$, Sandler (1991) claimed to see the Zernike modes reflected in this spatial product for his work on a continuous aperture neural network (Sandler *et al.* 1991). Lloyd-Hart (1991) performed the same analysis and

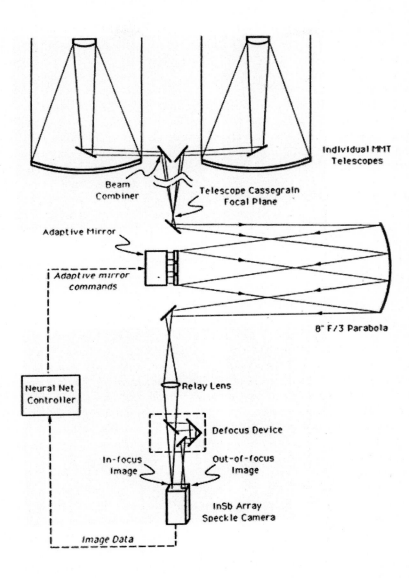

Fig. 13. Complete AO system (after Lloyd-Hart *et al.* 1992) including the phase-diversity neural network wavefront sensor, as implemented at the MMT.

a)

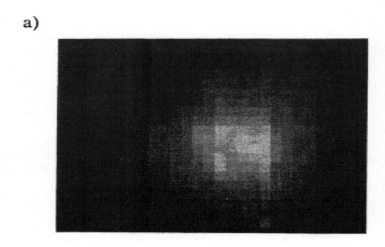

b)

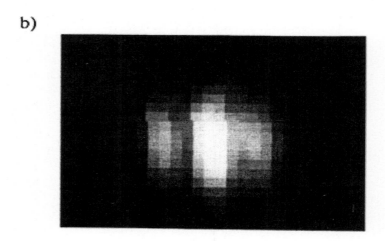

Fig. 14. (After Lloyd-Hart 1991) With and without neural-network AO control of two MMT primary mirrors. The top figure shows the star ψ-Pegasi at 2.2μm wavelength without AO correction during a 10 second exposure (1000 10 millisecond images co-added) with only 2 of the 6 MMT primaries uncovered, giving a resolution of FWHM= 0.62 arcseconds, and a Strehl ratio of 0.15. The bottom figure shows the same star with neural net phase-diversity 67 Hz closed-loop AO correction, giving a fringe resolution of 0.1 arcseconds, and a Strehl ratio of 0.27.

found encouraging results, shown in Figure 15. From the bottom to top of Figure 15, the panels represent V_{ij} for the tip and tilt for mirror A, tip and tilt for mirror F, and the sine and cosine of the phase; the images on the left side are for the in-focus image, and the images on the right side are for the out-of-focus image. The tip/tilt images of mirror A appear to be negative images of the tip/tilt images of mirror F, which is expected because positive tip on one mirror causes relative negative tip on the other mirror. These patterns, like the Zernike patterns in the weights seen for Sandler's continuous aperture net, are only expected in the case of a net with linear transfer functions on the hidden units. Since sigmoidal transfer functions were used, the appearance of this dominant structure in the weights is somewhat of a mystery. Perhaps the nets are not utilizing their non-linearities extensively.

2.5   Simulations for Single-Mirror High-Order Adaptive Optics Systems

Sandler *et al.* (1991), under the auspices of the Air Force Phillips Laboratory, at Starfire Optical Range, implemented a similar phase-diversity neural network wavefront sensor and reconstructor, but for a single-mirror (continuous aperture) and with data obtained from phase-diverse measurements of a *real* atmosphere (and a real star). Historically this work was performed prior to and was the inspiration for the work at the MMT outlined in the last two sections (despite similar publication dates). The 16×16 (I-band: $\lambda = 0.85\mu$m) images of Vega had 0.412-arcsecond pixels, an integration-time of 2 milliseconds, and were taken with the 1.5m Starfire Optical Range telescope in New Mexico; and the out-of-focus image was 1.1μm out-of-focus. A Shack-Hartmann wavefront sensor accumulated slope wavefront data simultaneously for later comparison to the wavefront derived off-line from the neural network which was trained with simulated data. The perceptron neural network consisted of 512 input neurons, a sigmoidal hidden layer, and 8 linear output neurons (see Figure 16). Each of the 8 outputs was trained via standard back-propagation to respond to the two input images (see the bottom pair of images in the inset of Figure 17) with the amplitude of a different Zernike mode's contribution to the atmospheric distortion. (Zernike polynomials (Born & Wolf 1975) are like Hermite polynomials in that they are orthogonal, but they are applied to a unit disk, giving products of angular functions and radial polynomials (e.g., focus, astigmatism, coma, spherical aberration)). The neural-net reconstructed Zernike amplitudes were then used to derive a phase-map of the total atmospheric aberration (by summing all the Zernike modes, weighted by the NN amplitudes), and in Figure 17 for *one* atmospheric realization, the resulting contour-like NN phase-map ('interferogram') can be compared with the phase-map derived by the Shack-Hartmann wavefront sensor and a standard reconstructor. Sandler *et al.* (1991) analyze 107 independent snapshots of the atmosphere with the off-line phase-constructing neural net, and find that the neural net would

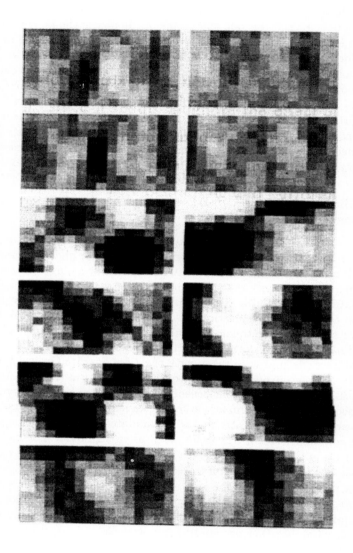

Fig. 15. The product of the hidden and output weights for the phase-diversity neural net for the MMT (after Lloyd-Hart *et al.* 1992). The left column corresponds to the in-focus images, and the right column corrsponds to the out-of-focus images. From bottom to top, the six panels represent tip and tilt for mirror A, tip and tilt for mirror B, and the sine and cosine of the relative phase between the two mirrors. These images represent the filters which would be applied by the outputs to the input image pair if the net were linear.

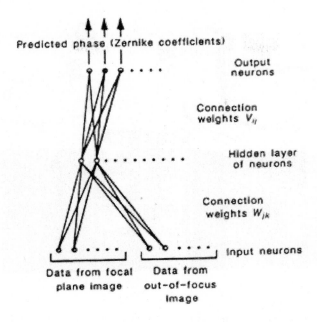

Fig. 16. A diagram showing the neural network used to recover phase for the continuous aperture phase-diversity telescope data (after Sandler *et al.* 1991).

improve the wavefront error to 0.78 rad^2 (the uncorrected wavefront error was 1.77 rad^2), thus improving image resolution by a factor of 3. The use of neural networks for phase diversity wavefront sensing has been described in detail in Barrett & Sandler (1996). This same technique was applied by Barrett & Sandler (1996) to Hubble Space Telescope (HST) data before the Hubble optics were fixed. Using the real HST stellar images at different focal positions, the NN technique was compared to slow off-line Fourier based phase-retrieval methods, and the two methods were found to predict basically the same amount of spherical aberration.

3   Neural Network Wavefront Prediction

Wavefront sensor measurements of the atmospheric distortion need to be made with CCD exposures that have non-zero exposure time, in order to collect sufficient photons to characterize the wavefront. For typical laser guidestars of visual magnitude 10, exposure times of 1 millisecond will give about 600 photons per $0.5 \times 0.5\,\mathrm{m}^2$ subaperture, which gives high (but not

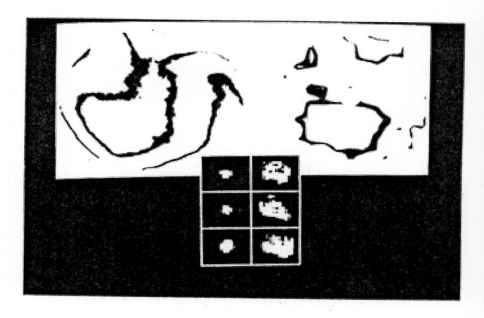

Fig. 17. An example of neural network performance (after Sandler *et al.* 1991), showing three pairs of in- and out-of-focus far-field patterns, and two interferograms. The lowest pair of images is the actual camera data obtained at Starfire Optical Range and which are then input to the neural network. The middle set of images is produced by numerically simulating the in- and out-of-focus stellar images using the phase predicted by the neural network. The top set of images is produced by numerically simulating camera data corresponding to the phase reconstructed from the Shack-Hartmann sensor. The interferogram on the left represents the phase predicted by the neural network, and the interferogram on the right represents the phase measured by the Shack-Hartmann sensor.

infinite) signal-to-noise ratios. Laser guidestar exposure times much shorter than 1 millisecond would diminish performance. If a laser guidestar is not available, or if the researcher wishes to avoid laser guidestar focus anisoplanatism, then a natural guidestar may be used for the adaptive reference. In this case, the natural guidestar may not be optimally bright (dimmer than magnitude 10), so in order to maximize wavefront sensor signal-to-noise ratio, longer exposures are required. For a magnitude 12 natural guidestar, there would be about 475 photons per $0.5 \times 0.5 \, m^2$ subaperture for 5 millisecond exposures. With these non-zero exposure times and the additional time needed to reconstruct the wavefront and apply the actuator commands (0.5-2.5 milliseconds, using eight C40 digital signal processors), winds will have blown the 'old' measured turbulence some fraction of a subaperture across the pupil

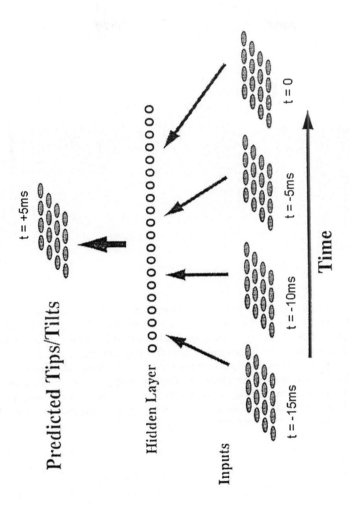

Fig. 18. A figure (adapted from Jorgenson & Aitken 1994) showing the neural and linear network architecture used to predict wavefront slopes for a simulated 2 meter telescope into the future using the past arrays of slopes. For this case, there are 128 inputs, 60 hidden units, and 32 outputs (each wavefront sensor pixel corresponds to an x-tilt and a y-tip).

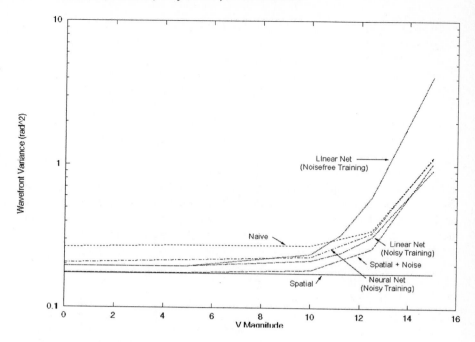

Fig. 19. A comparison of the prediction wavefront variance results of both the neural net and the linear net, after training for 178,000 5-millisecond Shack-Hartmann snapshots of a simulated $\tau = 14$ millisecond atmosphere. The bottom curve shows the simulated spatial (reconstructor and fitting) errors, the next curve adds photon-counting noise to the spatial error. The top curve shows the "naive" prediction error, computed by predicting that the slope will not change from one timestep to the next. And the remaining three curves are the predictor curves, discussed in the text. The performance is shown as a function of the guidestar brightness or magnitude, the dimmer guidestars being on the right.

and new turbulence will take its place, either coming in from outside the aperture or from another subaperture. Concomitantly, the turbulent phase-screen may not remain static (as is assumed by Taylor's frozen flow hypothesis) due to the relative motion of turbulent eddies or due to the relative motion of the multiple turbulent layers of the atmosphere. For a single-layer atmosphere, blowing across at possible jet stream speed of 20m/s, the turbulent phase-screen will have shifted by 0.1 meters in a wavefront sensor CCD exposure time of 5 milliseconds, which is potentially a change of 20% in the turbulent phase-screen above a given $0.5 \times 0.5 \, \text{m}^2$ subaperture. For a dynamic atmosphere, the phase-variance between one wavefront and co-located wavefront t milliseconds later increases by $\sigma_t = (t/\tau)^{5/3}$, where $\tau = 0.31 r_0/v_w$, and v_w is the turbulence-weighted wind-velocity. In the infrared (e.g. $\lambda \sim 2.2 \mu\text{m}$), where the science is done, the Fried coherence length is $r_0 \sim 1\,\text{m}$ and the wavefront decorrelation time is $\tau \sim 16\,$milliseconds. Therefore, in the infrared, there would be an 'unavoidable' temporal variance of $\sim 0.14\,\text{rad}^2$ for 5 millisecond exposures, unless auxiliary measures are employed. This decorrelation is a factor of 4 higher than for 1 millisecond exposures and is similar in magnitude to the fitting and reconstruction errors (Sandler *et al.* 1994b), so temporal decorrelation begins to become significant for these 5 millisecond exposures that might be demanded for AO with magnitude 12 natural guidestars.

Of course, the obvious solution to this temporal decorrelation and servo lag problem is the application of predictive technology to anticipate the wavefront changes that occur during the CCD exposures and actuator/mirror control. In order to apply prediction, the atmosphere needs to be predictable, not random in nature – indeed, Jorgenson, Aitken & Hege (1991) showed that centroids of a star's position followed a chaotic temporal trajectory with correlation dimension near 6. This implies that the atmosphere has a deterministic component that can be predicted, and thus some of the temporal decorrelation of the wavefront can be negated.

Several groups set out to use the past history of the measured wavefront slopes to predict the wavefront for the next WFS exposure. Jorgenson & Aitken (1994) used two-dimensional wavefront data from the COME-ON AO system in a neural network that incorporated both temporal and *spatial* information to predict the future array of slopes. In good seeing their wavefront variance with neural network prediction was a factor of two better than without prediction, with equivalent performance for their linear predictors. For poorer seeing and longer-term predictions, the neural net predictor's wavefront variance was half that of the linear predictor and one-sixth that of no prediction. Lloyd-Hart & McGuire (1996) also thoroughly investigated the efficacy of least-squares linear predictors for both simulated and real data as a function of lookback depth, signal-to-noise ratio and turbulence timescale, and they also did an initial foray into neural network prediction, capitalizing on McGuire's experience in currency futures market prediction with neural networks. The neural network was found to perform many times better than the linear predictors for low signal-to-noise experiments after the predictors had been trained without noise. Training without noise was the lore McGuire learned in his finance prediction studies, as well as the histogram equalization preprocessing used successfully in the neural network studies of Lloyd-Hart & McGuire (1996). The histogram equalization did *not* work well for the linear least-squares prediction. Other groups that have worked on spatio-temporal modeling and prediction of the atmosphere for AO include Schwartz, Baum, & Ribak (1994), Bonaccini, Gallieni, & Giampieretti (1996), and Dessenne, Madec, & Rousset (1997).

Aitken & McGaughey (1996) have published significant work suggesting that the atmosphere is a linear filtering process of Brownian motion with a

Hurst exponent of 5/6, and that the predictability stems from the spatial averaging done by the wavefront sensor. They suggest that linear predictors "should be able to extract *all* of the available information of the linear process", despite evidence that non-linear predictors such as neural nets perform better than linear predictors. Part of the rationale for the new work that we present here is to study better the differences between linear and non-linear predictors, spurred by Aitken & McGaughey's result of linearity.

We follow Jorgenson & Aitken (1994) in the basic network architecture, but use only a single hidden layer. We use the past four $2 \times 4 \times 4$ arrays of open-loop slopes from a simulated 2 meter telescope and AO system to predict the next array of x & y slopes 5 milliseconds into the future (see Figure 18). For our neural network, we have linear transfer functions on the outputs and sigmoidal transfer functions on the hidden units; for our linear network, the hidden units are made linear. MinMax preprocessing is used for the input and output data, due to its linear nature and the linear output transfer functions (in Lloyd-Hart & McGuire 1996, sigmoids were used on the output layer, which allowed the non-linear preprocessing ("Uniform preprocessing") histogram equalization to work well). Standard back-propagation was the chosen training method for both the linear and non-linear nets, using 178,000 5 millisecond Shack-Hartmann wavefront sensor snapshots of a simulated atmosphere with a correlation time of $\tau = 14$ milliseconds, and guidestars of varying magnitudes. This training finally arrived at wavefront variance values which we depict in Figure 19 (for $\sigma^2 < 1$, the Strehl ratio is $\mathcal{S} \approx \exp(-\sigma^2)$). The bottom curve is the combination of spatial errors like reconstructor error and fitting error; when noise is added, we arrive at the next curve; and without any prediction we arrive at the curve that we call the "naive prediction" error, which is the error resulting from the servo lag that assumes that the best prediction is 'the slopes will not change' that comes from assuming a random-walk type evolution. The intermediate curves are those with prediction. Noiseless training with noisy recall performed much worse for the linear predictor than noisy training with noisy recall, with errors that are almost twice the naive prediction error for a magnitude 15 guidestar. The linear net predictor with noisy training does better than the non-linear neural net predictor for all levels of noise (the higher the guidestar magnitude, the more photon counting noise). In high noise situations, the linear net predictor even does better than the spatial and noise errors combined (which might have been considered a floor in the possible performance). Both the linear net and the neural net predictors outperform the naive prediction at low noise levels, but at high noise levels only the linear net's performance is admirable, with the neural net's wavefront error approaching the naive prediction error. The sub-noise-level linear net performance means that the linear net is successfully temporally averaging out the noise. The outperformance of the neural net by the linear net may be caused by several things, first of which is the natural accordance of the linear net with McGaughey

& Aitken's linear atmosphere, but possibly also the choice of linear output transfer functions and the decision not to use histogram equalization as in previous studies. We have also attempted recursive linear least squares (RLS) prediction (Strobach 1990), but for this simulated data set, performance was poor compared to our linear net and our neural net. However, RLS prediction performance for a real data obtained from a high-order AO system on the 1.5 meter telescope at Starfire Optical Range Rhoadarmer (1996) was *much* better than our initial attempts at neural net prediction for this much higher dimensional real data set.

4   Future Work

Further network prediction studies, both linear and non-linear are warranted, and a careful comparison with other techniques (e.g., Wild 1996, Dessenne, Madec, & Rousset 1997, RLS and non-RLS) should be undertaken. The dimensionality curse needs solution, in order to extend the predictive networks' domain of excellence from low-order AO systems to high-order AO systems. This solution requires speeding up the training by 1-2 orders of magnitude, or it requires limiting the size of the networks by using only local, instead of global, neighborhoods of past influence. Once the dimensionality problem is solved, then the predictive networks may be of use in real-time observations at the telescope. Such real-time observations will require the development of these networks in our Adaptive Solutions VME computer or perhaps in a more sophisticated dedicated computer.

The phase-diversity wavefront-sensing neural networks are an inexpensive and rapid phase retrieval method that can be applied to both static optical deformation determination, as well as the rapid atmospheric deformation sensing. This approach requires only two sets of images to determine the wavefront, and so it can be applied at almost any telescope with little difficulty. With a deformable mirror in the system, the phase-diversity neural network will form a complete, inexpensive adaptive optics system perhaps capable of the ultra-high-order adaptive correction (that may allow extra-solar planet imaging) currently in adaptive opticians' plans and dreams (Angel 1994b, Stahl & Sandler 1995).

Acknowledgements

The new work outlined in Section 3 was supported by the US Air Force Office of Scientific Research under grants #F49620-94-1-0437 and #F49620-96-1-0366. The authors thank G. Angeli, J.R.P. Angel, C. Shelton, and A. Krogh for helpful discussions, R.Q. Fugate for collaboration at SOR, and Angeli and Angel for reviewing this paper.

References

Aitken, G.J.M. and McGaughey, D. (1996): "Predictability of Atmospherically-distorted Stellar Wavefronts", *Proc. European Southern Observatory Conf. on Adaptive Optics* **54**, Ed. M. Cullum, Garching, Germany, p. 89.

Angel, J.R.P., Wizinowich, P., Lloyd-Hart, M., and Sandler, D.G. (1990): "Adaptive Optics for Array Telescopes using Neural network Techniques", *Nature* **348**, 221.

Angel, J.R.P. (1994a): "Wavefront Reconstruction by Machine Learning Using the Delta Rule", *Proc. SPIE Conf. on Adaptive Optics in Astronomy* **2201**, Kona, Hawaii, 629.

Angel, J.R.P. (1994b): "Ground Based Imaging of Extrasolar Planets Using Adaptive Optics", *Nature* **368**, 203.

Avicola, K., Brase, J.M., Morris, J.R., Bissinger, H.D., Friedman, H.W., Gavel, D.T., Kiefer, R., Max, C.E., Olivier, S.S., Rapp, D.A., Salmon, J.T., Smanley, D.A., and Waltjen, K.E. (1994): "Sodium Laser Guide Star System at Lawrence Livermore National Laboratory: System Description and Experimental Results" *Proc. SPIE Conf. on Adaptive Optics in Astronomy*, Eds. M.A. Ealey, F. Merkle **2201**, 326.

Barrett, T.K. and Sandler, D.G. (1993): "Artificial Neural Network for the Determination of the Hubble Space Telescope Aberration from Stellar Images", *Appl. Optics* **32**, 1720.

Barrett, T.K. and Sandler, D.G. (1996): "Neural Networks for Control of Telescope Adaptive Optics", in *Handbook of Neural Computation*, Eds. E. Fiesler & R. Beale, Oxford University Press, p. G3.1:1.

Beckers, J.M. (1993): "Adaptive Optics for Astronomy: Principles, Performance, and Applications", *Annu. Rev. Astron. Astrophys.* **31**, 13.

E. Bin-Nun & F. Dothan-Deutsch (1973), *Rev. Sci Instrum.* **44**, 512.

Bonaccini, D., Gallieni, D., and Giampieretti, R. (1996): "Prediction of Star Wander Path with Linear and Non-Linear Methods", *Proc. European Southern Observatory Conf. on Adaptive Optics* **54**, Ed. M. Cullum, Garching, Germany, p. 103.

Born, M. and Wolf, E. (1975): *Principles of Optics*, 5th edition, Pergamon Press, Oxford.

Bridges, W.D., Brunner, P.T. Lazzara, S.P., Nussmeier, T.A., O'Meara, T.R., Sanguinet, J.A., and Brown, W.P. (1974): *J. Appl. Opt.* **13**, 2.

Carter, B.J., Kibblewhite, E.J., Wild, W.J., Angel, J.R.P., Lloyd-Hart, M., Jacobsen, B.P., Wittman, D.M., and Beletic, J.W. (1994): "Sodium Beacon used for Adaptive Optics on the Multiple Mirror Telescope", *Proc. NATO/ASI School on Adaptive Optics for Astronomy*, Eds. D.M. Alloin, J.-M. Mariotti, Cargese, Corsica.

Cohen, R.W. ,Mast, T.S., and Nelson, J.E. (1994): "Performance of the W.M. Keck Telescope Active Mirror Control System", *Proc. SPIE Conference on Advanced Technology Optical Telescopes*, Ed. L.M.Stepp **2199**, 105.

Dessenne, C., Madec, P.Y., and Rousset, G. (1997): "Modal Prediction for Closed-Loop Adaptive Optics", *Optics Letters* **22**, 1535.

Ealey, M.A. (1989): Proc. Soc. Photo-opt. Instrum. Eng. **1167**.

Ealey, M.A. and Wheeler, C.E. (1990): *Proc. 3rd Int. Cong. Opt. Sci. and Eng.* **1271**, Paper 23.

Feinleib, J., Lipson, S.G., and Cone, P.F. (1974): *Appl. Phys. Lett.* **25**, 311.

Foltz, C. *et al.* (1998): http://sculptor.as.arizona.edu/foltz/www/mmt.html

Freeman, R.H., Garcia, H.R., DiTucci, J., and Wisner, G.R. (1977): *High-Speed Deformable Mirror System*, U.S. Air Force Weapons Laboratory Report AFWL-TR-TR-76-146.

Fried, D.L. (1965):"The Effect of Wavefront Distortion on the Performance of an Ideal Optical Heterodyne Receiver and an Ideal Camera", *Conf. on Atmospheric Limitations to Optical Propagation*, U.S. Nat. Bur. Stds. CRPL.

Fried, D.L. (1994): "Atmospheric Turbulence Optical Effects: Understanding the Adaptive-Optics Implications", *Proc. NATO/ASI School on Adaptive Optics for Astronomy*, Eds. D.M. Alloin & J.M. Mariotti, Cargese, Corsica, p.25.

Friedman, H.W., Erbert, G.V., Kuklo, T.C., Salmon, J.T., Smauley, D.A., Thompson, G.R., Malik, J.G., Wong, N.J., Kanz, K., and Neeb, K. (1995): "Sodium beacon laser system for the Lick Observatory" *Proc. SPIE Conf. on Adaptive Optical Systems and Applications*, Eds. R.K. Tyson & R.Q. Fugate, **2534**, 150.

Friedman, H., Gavel, D., Max, C., Erbert, G., Thompson, G., Kuklo, T., Wong, N., Feldman, M., Beeman, B., and Jones, H. (1997): *Laser Guidestar System: Diagnostics and Launch Telescope Critical Design Review Report*.

Fugate, R.Q., Wopat, L.M., Fried, D.L., Ameer, G.A., Browne, S.L. Roberts, P.H., Tyler, G.A., Boeke, B.R., and Ruane, R.E. (1991): "Measurement of Atmospheric Wavefront Distortion using Scattered Light from a Laser Guide-star", *Nature* **353**, 144.

Ge, J. (1998): "Sodium Laser Guide Star Technique, Spectroscopy and Imaging with Adaptive Optics", Ph.D. dissertation, University of Arizona.

Greenwood, D.P. (1977): "Bandwidth Specifications for Adaptive Optics Systems", *J. Opt. Soc. Am.* **67**, 390.

Hardy, J.W., Feinleib, J., and Wyant, J.C. (1974): *Real-Time Correction of Optical Imaging Systems*, OSA Meeting on Optical Propagation through Turbulence, Boulder, CO.

Happer, W. (1982): Princeton U. & member of JASON's advisory group, private report; W. Happer, G.J. MacDonald, C.E. Max, & F.J. Dyson (1994), *J. Opt. Soc. Am.* **A11**, 263.

Hartmann, J. (1900): *Zt. Instrumentenkd.* **24**, 1.

Hill, J.M. and Angel, J.R.P. (1992): "The Casting of the 6.5m Borosilicate Mirror for the MMT Conversion", http://medusa.as.arizona.edu/lbtwww/tech/mirror.htm, *Proceedings of the ESO Conference on Progress in Telescope and Instrumentation Technologies*, Garching, Germany, ed. M-H. Ulrich, 57.

Hufnagel, R.E. (1974): *Proc. Topical Mtg. on Opt. Propagation through Turbulence*, Boulder, CO.

Jacobsen, B.P. (1997): "Sodium Laser Guide Star Projection for Adaptive Optics", Ph.D. dissertation, University of Arizona.

Jorgenson, M.B., Aitken, G.J.M., and Hege, E.K. (1991): "Evidence of a Chaotic Attractor in Star-wander Data", *Optics Letters* **16**, 2.

Jorgenson, M.B. and Aitken, G.J.M. (1994): "Wavefront Prediction for Adaptive Optics", *European Southern Observatory Conf. on Active and Adaptive Optics*, Ed. F. Merkle, Garching, Germany, p. 143.

Kibblewhite, E. (1997): "Physics of the Sodium Layer", *Proc. of the NATO/ASI School on Laser Guide Star Adaptive Optics for Astronomy*, Ed. J.C. Dainty, Cargese, Corsica.

Kolmogorov, A. (1961): in *Turbulence, Classic Papers on Statistical Theory*, Eds. S.K.Friedlander & L. Topper, Interscience, New York.

Lloyd-Hart, M. (1991): "Novel Techniques of Wavefront Sensing for Adaptive Optics with Array Telescopes Using and Artificial Neural Network", Ph.D. Dissertation, University of Arizona.

Lloyd-Hart, M., Wizinowich, P., McLeod, B., Wittman, D., Colucci, D., Dekany, R., McCarthy, D., Angel, J.R.P., and Sandler, D. (1992): "First Results of an On-line Adaptive Optics System with Atmospheric Wavefront Sensing by an Artificial Neural Network", *Astrophysical Journal* **390**, L41-44.

Lloyd-Hart, M. and McGuire, P.C. (1996): "Spatio-Temporal Prediction for Adaptive Optics Wavefront Reconstructors", *Proc. European Southern Observatory Conf. on Adaptive Optics* **54**, Ed. M. Cullum, Garching, Germany, p. 95.

Lloyd-Hart, M., Angel, R., Groesbeck, T., McGuire, P., Sandler, D., McCarthy, D., Martinez, T., Jacobsen, B., Roberts, T., Hinz, P., Ge, J., McLeod, B., Brusa, G., Hege, K., and Hooper, E. (1997): "Final Review of Adaptive Optics Results from the Pre-conversion MMT", *Proc. SPIE conference on Adaptive Optics and Applications* **3126**, San Diego.

Lloyd-Hart, M., Angel, J.R.P., Groesbeck, T.D., Martinez, T., Jacobsen, B.P., McLeod, B.A., McCarthy, D.W., Hooper, E.J., Hege, E.K., and Sandler, D.G. (1998a): "First Astronomical Images Sharpened with Adaptive Optics Using a Sodium Laser Guide Star" *Astrophys. J.* **493**, 950;
http://athene.as.arizona.edu:8000/caao/m13.html

Lloyd-Hart, M., Angel, R., Sandler, D., Barrett, T., McGuire, P., Rhoadarmer, T., Bruns, D., Miller, S., McCarthy, D., and Cheselka, M. (1998b), "Infrared adaptive optics system for the 6.5 m MMT: system status and prototype results," *Proc. SPIE Conference on Adaptive Optical System Technologies*, ed. D.Bonaccini & R.K.Tyson **3353**, Kona, Hawaii.

Martin, H.M., Burge, J.H., Ketelson, D.A., and West, S.C. (1997): "Fabrication of the 6.5-m primary mirror for the Multiple Mirror Telescope Conversion", *Proc. SPIE Conference on Optical Telescopes of Today and Tomorrow*, Ed. A.L. Ardeberg **2871**, 399.

Martinez, T. (1998): "Continuous Wave Dye Laser for Use in Astronomical Adaptive Optics", Ph.D. dissertation, University of Arizona.

Max, C.E., Olivier, S.S., Friedman, H.W., An, J., Avicola, K., Beeman, B.V., Bissinger, H.D., Brase, J.M., Erbert, G.V., Gavel, D.T., Kanz, K., Macintosh, B., Neeb, K.P., and Waltjen, K.E. (1997): "Image Improvement from a Sodium-layer Laser Guide Star Adaptive Optics System", *Science* **277**, 1649;
http://ep.llnl.gov/urp/science/lgs_www/lgs_lick.html

Meyer-Arendt, J.R. (1984): *Introduction to Classical and Modern Optics.* 2nd Edition (Englewood Cliffs, New Jersey: Prentice-Hall), 214–223.

Mikoshiba, S. and Ahlborn, B. (1973): *Rev. Sci. Instrum.* **44**, 508.

Muller, R.A. und Buffington, A. (1974), *J. Opt. Soc. Am.* **64**,9.

Newton, I. (1730): *Opticks*, 4th edition, Dover, New York (1979).

Noll, R.J. (1976): "Zernike Polynomials and Atmospheric Turbulence", *J. Opt. Soc. Am.* **66**, 207.

Primmerman, C.A., Murphy, D.V., Page, D.A., Zollars, B.G., and Barclay, H.T. (1991): "Compensation of Atmospheric Optical Distortion Using a Synthetic Beacon", *Nature* **353**, 141.

Quirrenbach, A. (1997): "Laser Guidestars", *Proc. of the NATO/ASI School on Laser Guide Star Adaptive Optics for Astronomy*, Ed. J.C. Dainty, Cargese, Corsica.

Roberts Jr., W.T., Murray, J.T., Austin, W.L., Powell, R.C., and Angel, J.R.P. (1998): "Solid-State Raman Laser for MMT Sodium Guide Star", *Proc. SPIE Conf. on Adaptive Optical System Technologies* **3353**, Kona, Hawaii.

Rhoadarmer, T.A. (1996): unpublished.

Roddier, F. (1988): *Appl. Opt.* **27**, 1223.

Sandler, D.G., Barrett, T.K., Palmer, D.A., Fugate, R.Q., and Wild, W.J. (1991): "Use of a Neural Network to Control an Adaptive Optics System for an Astronomical Telescope", *Nature* **351**, 300.

Sandler, D.G. (1991): unpublished.

Sandler, D.G., Cuellar, L., Lefebvre, M., Barrett, T., Arnold, R., Johnson, P., Rego, A., Smith, G., Taylor, G., and Spivey, B. (1994): "Shearing Interferometry for Laser-guide-star Atmospheric Correction at Large D/r_0", *J. Opt. Soc. Am.* **A11**, 858.

Sandler, D.G., Stahl, S., Angel, J.R.P., Lloyd-Hart, M., and McCarthy, D. (1994): "Adaptive Optics for Diffraction-Limited Infrared Imaging with 8-m Telescopes", *J. Opt. Soc. Am.* **A11**, 925.

Schwartz, C., Baum, G., and Ribak, E.N. (1994): "Turbulence-degraded Wavefronts as Fractal Surfaces", *J. Opt. Soc. Am.* **A11**, 444.

Shack, R.B. and Platt, B.C. (1971): abstr. *J. Opt. Soc. Am.* **61**, 656.

Shi, F., Smutko, M.F., Chun, M., Larkin, J., Scor, V., Wild, W., and Kibblewhite, E. (1996): poster paper at January 1996 American Astronomical Society meeting, http://astro.uchicago.edu/chaos/poster3/poster3.html

Stahl, S.M. and Sandler, D.G. (1995): "Optimization and Performance of Adaptive Optics for Imaging Extrasolar Planets", *Astrophys. Journal* **L454**, 153.

Strobach, P. (1990): *Linear Prediction Theory: a Mathematical Basis for Adaptive Systems*, Springer-Verlag, Berlin.

Tatarskii, V.I. (1961): *Wave Propagation in a Turbulent Medium*, McGraw-Hill, New York.

Taylor, G.I. (1935): "Statistical Theory of Turbulence", *Proc. R. Soc. Lond.* **A151**, 421.

Tyson, R.K.(1991): *Principles of Adaptive Optics*, Academic Press, San Deigo.

Vdovin, G. (1995): "Model of an Adaptive Optical System Controlled by a Neural Network", *Optical Engineering* **34**, 3249.

Wallner, E.P. (1983): "Optimal Wavefront Correction Using Slope Measurements", *J. Opt. Soc. Am.* **73**, 1771.

West, S.C., Callahan, S., Chaffee, F.H., Davison, W., DeRigne, S., Fabricant, D., Foltz, C.B., Hill, J.M., Nagel, R.H., Poyner, A., and Williams, J.T. (1996): "Toward first light for the 6.5-m MMT telescope", *Proc. of SPIE conference on Optical Telescopes of Today and Tomorrow* **2871**,
http://medusa.as.arizona.edu/lbtwww/tech/light.htm

Wild, W.J., Kibblewhite, E.J., Vuilleumier, R., Scor, V., Shi, F., and Farmiga, N. (1995): "Investigation of Wavefront Estimators Using the Wavefront Control Experiment at Yerkes Observatory", *Proc. SPIE Conf. on Adaptive Optical Systems and Applications*, Eds. R.K. Tyson & R.Q. Fugate **2534**, 194.

Wild, W.J., Kibblewhite, E.J., and Vuilleumier, R. (1995): "Sparse Matrix Wave-front Estimators for Adaptive-Optics Systems for Large Ground-based Tele-scopes", *Optics Letters* **20**, 955.

Wild, W.J. (1996): "Predictive optimal estimators for adaptive-optics systems", *Optics Letters* **21**, 1433.

Nuclear Physics with Neural Networks

Klaus A. Gernoth

[1] Department of Physics, UMIST, PO Box 88, Manchester M60 1QD, UK [**]
[2] Department of Physical Sciences, Theoretical Physics, University of Oulu, FIN–90570 Oulu, Finland

Abstract. This article surveys modeling and prediction of various nuclidic properties with feedforward artificial neural networks. Special emphasis is placed on neural network modeling of nuclear ground state masses, the cleanprop algorithm for training neural networks on data with error bars, global neural network models of nuclear stability and branching ratios of decay of unstable nuclides, and higher-order probabilistic perceptrons for classifying nuclides as stable or unstable. The various network architectures and training algorithms devised for and successfully tested in these applications are discussed in detail and the best of numerical neural network results are presented.

1 Introduction

Since the beginning of this decade custom-tailored artificial neural networks containing an input, an output, and one or several hidden layers of analog units are being applied successfully to a variety of scientific problems in nuclear physics. With the proton and the neutron number serving as input variables, global neural network models have been constructed of the stability/instability dichotomy (Gazula, Clark, and Bohr 1992; Gernoth et al. 1993; Clark et al. 1999), of ground-state spins (Gernoth et al. 1993), of atomic mass excesses (Gernoth et al. 1993), and of branching ratios in decay of unstable nuclides (Gernoth and Clark 1995a). Trained with error backpropagation strategies based on steepest gradient descent minimization of appropriate objective functions, the mature networks display high performance, competitive with the accuracy achieved in conventional approaches, in both modeling the training data as well as in quantitatively predicting nuclidic properties for input patterns outside the training set.

The numerical results obtained with neural network techniques in nuclear physics being of interest in their own right, an equally important purpose of this research is to develop and test novel network architectures and training algorithms as are most appropriate for a given type of problem. For instance, in work connected with neural net modeling of the atomic mass table, presented in Sect. 2, a modified backpropagation algorithm, dubbed cleanprop (Gernoth and Clark 1995b) was designed on the basis of the maximum likelihood criterion specifically for training neural networks on data

---

[**] Permanent address

contaminated with Gaussian-like noise, a situation frequently encountered in statistical analysis of experimental results. Section 3 is devoted to an exposition of the cleanprop algorithm.

In neural network modeling of nuclear decay (Gernoth and Clark 1995a), suitably renormalized squashing functions for the non-linear responses of the output neurons are adopted, whereby it is ensured that a network will produce a probability distribution at its output layer for any pattern impressed on its input units. A backprop training scheme is implemented in which the relative entropy of the known target probability distribution with respect to the actual output probability distribution of the network is used as pattern-specific error measure. This work is described in Sect. 4.

In recent work (Clark et al. 1999) higher-order probabilistic perceptron (HOPP) networks, containing an input and an output layer only and no hidden units, have been applied to the nuclear stability/instability discrimination problem. In contrast to the multi-layer nets with exclusively pairwise interneuronal connections, used in all other work mentioned above, the HOPP architectures allow for feedforward interactions between input and output units up to any order. An account of the research on the HOPP classifiers is given in Sect. 5.

In Sects. 2, 4, and 5 a number of important, very recent neural network results are reported that are not yet published elsewhere. Section 6 summarizes some important conclusions and future prospects of neural network modeling of nuclidic properties, taking into account the most recent, and highly motivating, developments of ongoing research in this field. Looking ahead, it can be anticipated that materials science will emerge in the foreseeable future as a field affording ample opportunities for applying neural network methods developed in the context of nuclear and high-energy physics.

2   Neural Network Modeling of the Atomic Mass Table

In modeling the atomic mass table (Clark et al. 1992; Clark and Gernoth 1992; Gazula, Clark, and Bohr 1992; Gernoth et al. 1993) feedforward neural networks with various coding schemes for the independent input variables have been extensively explored. Common to all these nets is that they contain one or more hidden layers with analog units and an output layer with just a single analog neuron. The activity of the output unit represents, suitably coded, the atomic mass excess that the network produces in response to a nuclide impressed at its input interface.

For an input pattern, labeled r, the activity $a_k^{(r)}$ of a neuron k in a hidden or output layer is given by the logistic response function

$$a_k^{(r)} = \frac{1}{1 + e^{-u_k^{(r)}}} , \tag{1}$$

where the stimulus $u_k^{(r)}$ to unit k is a linear combination of the activities $a_l^{(r)}$ of the units in the respective preceding layer plus a threshold weight w_{k0},

$$u_k^{(r)} = \sum_l w_{kl} a_l^{(r)} + w_{k0} \ . \tag{2}$$

Here, the symbol w_{kl} denotes the strength of the connection from a unit l in the input or a hidden layer to a unit k in the next layer. Note that the nets used here do not possess connections that skip layers or feedback interneuronal links. The thresholds w_{k0} may be viewed as the strengths of connections to the neural network from an external unit with a perpetual maximum activity of unity. For a given input pattern the activities of the output neurons may then be obtained by successively applying (2) and (1), starting with the first hidden layer and proceeding forward from layer to layer, until the output layer is reached. The output activities represent the values computed by the neural net for the dependent quantities.

In its most basic version the backpropagation training algorithm (Rummelhart et al. 1986; Hertz, Krogh, and Palmer 1991; Müller, Reinhardt, and Strickland 1995) aims at minimizing the square cost function

$$E = \sum_r E_r \ , \tag{3}$$

where the pattern-specific error E_r is given by

$$E_r = \frac{1}{2} \sum_k \left(t_k^{(r)} - a_k^{(r)} \right)^2 \ . \tag{4}$$

The sum in (3) extends over all patterns r in the training set and the one in (4) over all output units. In (4) the quantities $t_k^{(r)}$ represent the known target values for the dependent variables, with which the actual neural network outputs $a_k^{(r)}$ must be compared.

In the present context the output interface possesses just a single neuron, for which the pattern-specific target values $t_1^{(r)}$ are obtained from the nuclidic mass excesses

$$\Delta m(r) = m(r) - A m_u \ , \tag{5}$$

wherein $m_u = 931.48\,\mathrm{MeV}$ is the atomic mass unit, $m(r)$ the mass, and A the atomic mass number of the nucleus r currently fed into the neural network at its input interface. The target values $t_1^{(r)}$ are produced by suitably mapping the mass excesses (5) onto the interval $(0, 1)$. The mass excesses computed by the neural network are then obtained by inverting the coding scheme for the output activities $a_k^{(r)}$.

The procedure for minimizing the objective function (3) in turn is rooted in a steepest gradient descent technique, in which upon presentation of a pattern r the current weights and thresholds w_{kl} are altered by incremental amounts $\Delta w_{kl}^{(r)}$ formed by

$$\Delta w_{kl}^{(r)} = -\eta \frac{\partial E_r}{\partial w_{kl}} + \xi \Delta w_{kl}^{(r-1)} = \eta \delta_k^{(r)} a_l^{(r)} + \xi \Delta w_{kl}^{(r-1)} \ . \tag{6}$$

When updating a threshold w_{k0} via prescription (6), the corresponding activity $a_0^{(r)}$ of the external neuron must be set to unity. According to (6) weights and thresholds move in multi-dimensional weight space in the direction in which the pattern-specific error E_r decreases. Equation (6) deviates somewhat from this strict gradient descent rule by the addition of the momentum term $\xi \Delta w_{kl}^{(r-1)}$, where $\Delta w_{kl}^{(r-1)}$ denotes the changes for the last pattern, $r-1$, that was presented to the network prior to the current one, r. Numerically, the adjustments $\Delta w_{kl}^{(r)}$ in (6) are governed by the learning rate η and the momentum parameter ξ. The momentum term helps to damp fluctuations in weight space about the (local) minimum in the total error E, reached towards the end of the training procedure (Rummelhart et al. 1986; Hertz, Krogh, and Palmer 1991; Müller, Reinhardt, and Strickland 1995). For the output units the error signals $\delta_k^{(r)}$ in formula (6) are given by

$$\delta_k^{(r)} = a_k^{(r)} \left(1 - a_k^{(r)}\right) \left(t_k^{(r)} - a_k^{(r)}\right) \ . \tag{7}$$

All other error signals may be obtained by backpropagating the output error signals (7) according to

$$\delta_k^{(r)} = a_k^{(r)} \left(1 - a_k^{(r)}\right) \sum_l \delta_l^{(r)} w_{lk} \tag{8}$$

from layer to layer, until the first hidden layer is reached. The sum in (8) comprises all units l to which unit k extends a connection.

In the work of the early 90s (Clark et al. 1992; Gernoth et al. 1993) the experimental ground state masses of 2291 nuclei, which have been available then from the National Nuclear Data Center (NNDC) at the Brookhaven National Laboratory (BNL), have been used. The mass excesses and a number of other ground state and excited state nuclidic properties may be accessed on-line through the internet services provided by the NNDC.[1] By random selection a training set of 1719 nuclei and a test set of 572 nuclei have been formed out of the total data base of 2291 input-output patterns. The test set is used to evaluate the predictive capabilities, that a neural network has developed under training, on patterns it has not seen before.

The neural networks are trained by presenting the patterns in the training set in random order and updating weights and biases after every pattern according to (6)–(8). A complete pass through the whole training set is referred to as an epoch. The random order in which the nuclei are presented to a network in learning changes from epoch to epoch. This element of randomness helps to avoid getting trapped in local minima during the early phases of

---

[1] For up-to-date information on the BNL and NNDC internet services see the www-sites *http://www.bnl.gov* and *http://www.nndc.bnl.gov*.

training. The weights and biases are initialized by random sampling of a uniform distribution on the interval $[-0.5, 0.5]$. At the beginning every neuron in a layer is connected to every other neuron in the respective next layer. In a training run a prescribed maximum number of epochs – in the atomic mass problem typically a few thousands – is performed and the objective function (3) evaluated after each epoch. The training program is instructed to store those weights and thresholds for which the smallest value of E was found. These values are then taken as the starting configuration for a consecutive run. In this manner series of up to a few tens of successive training runs are carried out. Connections are deleted regularly from the network after every few training runs according to by how much omitting a connection increases the total error (3). The learning rate η is suitably decreased between runs, as long as no connections are removed, and increased again after deletion of connections. In retraining a skeletonized network the learning rate is then decreased again between training cycles. Such a decrease of the learning rate permits closer investigation of the weights in the vicinity of a minimum of the cost function (3). In the mass problem the learning rate η varied between 0.1 and 0.5, whereas the momentum parameter ξ was held at a fixed value of 0.9. The training procedure involves also more frequent presentation of hard-to-learn patterns. These and other details of training neural nets on nuclidic properties are discussed more thoroughly in the context of neural network modeling of nuclear stability and decay in Gernoth and Clark (1995a).

The best results (Gernoth et al. 1993) have been obtained with a network architecture in which eight input units are used for encoding the proton number Z in binary form, another eight neurons for representing the neutron number N as a binary number, and another two units for analog coding of the atomic mass number $A = Z + N$ and the neutron excess number $I = N - Z$. These nets possess three hidden layers with ten neurons in each of them. Such a network architecture will be denoted symbolically by $^h(18 + 10 + 10 + 10 + 1)_a[P]$, where the superscript h refers to the hybrid coding, binary and analog, of the input variables and the subscript a to the analog coding of the output variable. The number P of non-zero weights and thresholds remaining after skeletonization is given in square brackets.

Table 1 compares the best neural network models with two of the best conventional theoretical calculations of nuclidic mass excesses. The model by Masson and Jänecke (1988) is a statistical fit, based on inhomogeneous partial difference equations, with 471 adjustable parameters to a data base of 1504 experimental ground state masses. Due to its strong physical motivation and quantum-mechanical input the macroscopic-microscopic model by Möller and Nix (1994) needs far fewer, namely 14, parameters than the statistical approaches provided by neural networks or the mass fit by Masson and Jänecke (1988). All models exhibit comparable performance on the respective (training) sets to which the adaptable parameters have been fitted. The accuracy achieved with neural networks in predicting the masses in the test set is only slightly inferior to the performance on the training set. Although, owing to the nature of the test set, making predictions for these patterns is mainly a task of interpolation, these figures nonetheless demonstrate the remarkable power of neural networks as an alternative to more elaborate approaches for theoretically predicting atomic ground state masses. The macroscopic-microscopic model by Möller and Nix (1994) was constructed on grounds of the 1323 experimental masses used already for an earlier calculation (Möller and Nix 1981). The prediction error shown in Table 1 is the performance measure of the Möller and Nix (1994) model for another 351 nuclei, whose masses have been added to the data base since 1981. As these nuclei lie mainly at the very edges of the 1981 data base, prediction in this case is predominantly a task of extrapolation. The excellent performance of the Möller and Nix (1994) model on the 351 new nuclei attests to the superb quality of their calculations.

Table 1. Comparison of neural network models of atomic masses with results of conventional theoretical approaches. The neural net results are taken from Gernoth et al. (1993). The root mean square (rms) and mean deviations of the calculated mass excesses from the known target values are given in units of MeV. The learning error refers to the training set of 1719 nuclei and the prediction error to the test set of 572 nuclei, formed from a total of 2291 masses by random selection, as explained in the text. The theoretical model of Möller and Nix (1994) is fitted to the (training) set of 1323 masses used by Möller and Nix (1981) already in an earlier calculation. The prediction error in Möller and Nix (1994) was obtained on a (test) set of 351 nuclei, for which ground state masses have been measured experimentally since 1981. The data base used in Masson and Jänecke (1988) contains 1504 nuclei. The model by Masson and Jänecke (1988) uses 471 adjustable parameters and the one by Möller and Nix (1994) a mere 14. See text for more details.

	Learning error		Prediction error	
	rms	mean	rms	mean
$^h(18+10+10+10+1)_a[353]$	0.629	−0.001	0.736	−0.055
$^h(18+10+10+10+1)_a[363]$	0.564	0.015	0.725	0.006
Möller and Nix 1994	0.673		0.735	
Masson and Jänecke 1988	0.346	0.014		

To make possible a more direct comparison with the theory of Möller and Nix (1994), neural networks have been trained also on their 1981 data base with 1323 patterns and tested on the 351 new nuclei. The outcome of this project is displayed in Table 2. The first neural net results for this problem (Gernoth et al. 1993) have been somewhat discouraging, which may be attributed to the fact that here making predictions for the test nuclei absent from the training set is a task of mainly extrapolation rather than interpolation. Kalman (1994), adopting an input coding, training strategy, and network architecture differing from Gernoth et al. (1993), was able to improve noticeably on these previous attempts. A brief account of this unpublished work of Kalman is given by Gernoth and Clark (1995b).

Employing a modified input coding scheme for the independent variables Z and N and their parities and a more sophisticated backpropagation learning rule and training strategy than used before, very recent work (Athanassopoulos et al. 1998) succeeded in constructing a neural network model that predicts the masses of the 351 new nuclei with a remarkably small root mean square error of 1.209 MeV. The learning and prediction errors for this net on the Möller and Nix (1994) data base, along with its architecture, input coding, and number of weights, are also shown in Table 2. It is worthwhile to point out that this net is also one of the most efficient of all neural net models of atomic masses in use of parametric resources. To an extent, this explains the amazing success of this net in extrapolating to the 351 new nuclei. The details of the more elaborate training procedure mentioned above and of the neural network models of nuclear ground state masses obtained therewith will be published in a forthcoming paper (Athanassopoulos et al. 1998).

Table 2. Neural network results for learning the Möller and Nix (1981) data base containing 1323 nuclei and predicting the mass excesses for the 351 nuclei (Möller and Nix 1994) added to this data base since 1981 in comparison to the macroscopic-microscopic model of Möller and Nix (1994). Listed are the root mean square deviations of the theoretical masses from the experimental ones. Units are MeV. The performance figures shown in the first two rows are taken from Gernoth et al. (1993) and Gernoth and Clark (1995b). The net in the second row was constructed and trained by Kalman (1994). The superscript s refers to the single-value decomposition used in the input coding scheme. See Gernoth and Clark (1995b) for more details on the Kalman (1994) net. The net in the third row was constructed and trained by Athanassopoulos et al. (1998). This net employs a hybrid input coding scheme, in which two units are used for analog coding of Z and N and another two neurons for binary representation of the parities of Z and N.

	Learning	Prediction
$^h(18 + 10 + 10 + 10 + 1)_a[421]$	0.828	5.981
$^s(4 + 40 + 1)_a[241]$	1.068	3.036
$^h(4 + 10 + 10 + 10 + 1)_a[272]$	0.617	1.209
Möller and Nix 1994	0.673	0.735

3   Cleanprop Algorithm for Training Neural Nets on Noisy Data

The measured ground state masses on which the neural networks are trained are unavoidably afflicted with experimental uncertainties. There is a danger that a neural net, like any other statistical or semi-statistical theoretical model in which parameters have to be fitted to data contaminated with noise, learns under training some of these experimental errors. It is therefore desirable to have a method available that allows for the experimental errors in the target data to be eliminated, as much as is possible, from the theoretical model being built on the basis of these noisy data. Such a method is provided by the maximum likelihood criterion (Mathews and Walker 1970; von Mises 1964) for estimating adjustable parameters and has been applied successfully by Möller and Nix (1988) to assess the influence of the experimental errors on their macroscopic-microscopic model of atomic ground state masses. Inspired by this and other work (Möller and Nix 1994), further research by Gernoth and Clark (1995b) succeeded in implementing the maximum likelihood criterion into the neural network approach, leading to a modified backpropagation algorithm, dubbed cleanprop (Clark and Gernoth 1995; Gernoth and Clark 1995b; Gernoth and Clark 1995c), for training neural nets on noisy target data.

One commences with the deviations of the experimental and thus noisy target data $t_1^{(r)}$, on which the neural nets are trained, from the true values $Y_1^{(r)}$, which, however, are not known. The pertinent relation reads

$$t_1^{(r)} = Y_1^{(r)} + \gamma_r \ , \tag{9}$$

where the pattern-specific experimental error $\gamma_r \in N\left(0, \sigma_r^2\right)$ for the quantity represented by the single output unit, labeled by the subscript $k = 1$, is assumed to be normally distributed with zero mean and standard deviation σ_r. Likewise, one further assumes that the actual neural network outputs $a_1^{(r)}$ differ from the true values $Y_1^{(r)}$ according to

$$a_1^{(r)} = Y_1^{(r)} + \epsilon_r \tag{10}$$

with the theoretical neural network model error $\epsilon_r \in N\left(\mu, \sigma^2\right)$ for the single output unit being a random variable with mean μ and standard deviation σ. In the approach followed here the quantities μ and σ are the appropriate error measures by which the quality of the neural network model may be assessed.

Combining (9) and (10), the target values $t_1^{(r)}$ and the network responses $a_1^{(r)}$ are related via

$$t_1^{(r)} = a_1^{(r)} + \epsilon_r + \gamma_r \ , \tag{11}$$

where the discrepancy $\epsilon_r + \gamma_r$ is normally distributed with mean μ and standard deviation $\sqrt{\sigma^2 + \sigma_r^2}$. The conditioned probability density for obtaining the target value $t_1^{(r)}$ is then given by the convolution of the two normal distributions $N\left(0, \sigma_r^2\right)$ and $N\left(\mu, \sigma^2\right)$, yielding

$$
f\left(t_1^{(r)} | a_1^{(r)}; \mu; \sigma_r^2 + \sigma^2\right) = \frac{1}{\sqrt{2\pi}} \frac{1}{\sqrt{\sigma_r^2 + \sigma^2}}
$$

$$
\times \exp\left[-\frac{\left(t_1^{(r)} - a_1^{(r)} - \mu\right)^2}{2\left(\sigma_r^2 + \sigma^2\right)} \right] . \quad (12)
$$

The likelihood function L is formed as the product of the probability densities (12) over all patterns r. The parameters of the neural network theory with respect to which, invoking the maximum likelihood criterion, the likelihood function L needs to be maximized are the weights and biases w_{kl} and the standard deviation σ and the mean μ of the theoretical neural network model. If P is the number of weights and thresholds, this results in P equations of the form

$$
\sum_r \frac{t_1^{(r)} - a_1^{(r)} - \mu}{\sigma_r^2 + \sigma^2} \frac{\partial a_1^{(r)}}{\partial w_{kl}} = 0 , \quad (13)
$$

an equation optimizing the standard deviation σ,

$$
\sum_r \frac{\left(t_1^{(r)} - a_1^{(r)} - \mu\right)^2 - \left(\sigma_r^2 + \sigma^2\right)}{\left(\sigma_r^2 + \sigma^2\right)^2} = 0 , \quad (14)
$$

and another equation optimizing the mean μ,

$$
\mu = \frac{\sum_r \left[\left(t_1^{(r)} - a_1^{(r)}\right) \big/ \left(\sigma_r^2 + \sigma^2\right)\right]}{\sum_r \left(\sigma_r^2 + \sigma^2\right)^{-1}} . \quad (15)
$$

The summations in (13)–(15) extend over all patterns r in the training set. As before, the symbol w_{kl} in (13) denotes the weight of the connection of a unit l in the input or a hidden layer to a unit k in the next layer or with $l = 0$ the bias of unit k. Note that formula (19) that is given in Gernoth and Clark (1995b) for μ is wrong. However, the numerical results presented there (and in other papers) and reported also in the present publication are not affected, since in all applications of cleanprop the assumption $\mu = 0$ was made from the start and its validity checked with the mature networks as explained further below in this section.

At this stage, the connection to training algorithms for neural networks may be made by observing that (13) (and (15)) derive also from minimizing a modified cost function of the form (3) w.r.t. w_{kl} (and μ). However, the pattern-specific error E_r is now given by

$$
E_r = \frac{1}{2} \frac{\left(t_1^{(r)} - a_1^{(r)} - \mu\right)^2}{1 + \left(\sigma_r/\sigma\right)^2} . \quad (16)
$$

Following the same steps that led to the usual backpropagation formulas (6)–(8) in Sect. 2 – with now (16) for E_r in (6) instead of (4) – yields the corresponding recipe for training feedforward neural nets on data with error bars. As before, the logistic function (1) for the activity $a_k^{(r)}$ of a unit k in response to the stimulus $u_k^{(r)}$, given by (2), is adopted. Equation (6) needs to be changed to

$$\Delta w_{kl}^{(r)} = \eta_r \delta_k^{(r)} a_l^{(r)} + \xi \Delta w_{kl}^{(r-1)} \ ,$$

(17)

where now a pattern-specific effective learning rate

$$\eta_r = \frac{\eta}{1 + (\sigma_r/\sigma)^2}$$

(18)

must be used. Equation (7) now reads

$$\delta_1^{(r)} = a_1^{(r)} \left(1 - a_1^{(r)}\right) \left(t_1^{(r)} - a_1^{(r)} - \mu\right) \ ,$$

(19)

determining the output error signals $\delta_1^{(r)}$ for the single output unit.

Equations (14)–(19) now suggest the following cleanprop training strategy in the presence of normally distributed noise, characterized by the error bars σ_r, in the target data $t_1^{(r)}$. Similarly to ordinary backprop, upon presentation of a training pattern r the corresponding output error signal $\delta_1^{(r)}$ is computed by (19) and backpropagated from layer to layer according to (8). Weights and biases are updated now according to (17) with the modified learning rate η_r being given by (18). Prior to the first epoch and then after every completed epoch the model errors σ and μ are computed by iteratively solving (14) and (15), starting with $\mu^{(0)} = J^{-1} \sum_r \left(t_1^{r)} - a_1^{(r)}\right)$ and the rms error

$$\sigma^{(0)} = \sqrt{\frac{1}{J} \sum_r \left(t_1^{(r)} - a_1^{(r)} - \mu^{(0)}\right)^2}$$

(20)

as input values. Here, the integer J denotes the number of training patterns. Equation (14) is employed in a form conducive to iterative solution for σ (Möller and Nix 1988; Gernoth and Clark 1995b),

$$\sigma^2 = \frac{\sum_r \left\{\left[\left(t_1^{(r)} - a_1^{(r)} - \mu\right)^2 - \sigma_r^2\right] \Big/ \left(\sigma_r^2 + \sigma^2\right)^2\right\}}{\sum_r \left(\sigma_r^2 + \sigma^2\right)^{-2}} \ .$$

(21)

The iteration process alternates between (21) and (15) until convergence is reached. It is worthwhile to point out that, as is evident from (18), the training patterns enter the cleanprop algorithm with a relative importance which is the smaller, the more uncertain the corresponding target values $t_1^{(r)}$ or, put differently, with an importance which is the larger, the smaller the experimental error bars σ_r.

It is illuminating to consider two limiting cases of the cleanprop algorithm outlined above. Firstly, in the absence of experimental noise, i.e. $\sigma_r = 0$ for all r, and with $\mu = 0$ the ordinary backprop rules (6) and (7) (for $k = 1$) are readily retrieved from (17)–(19). In this case the learning rate is the same, η, for all patterns and the pattern-specific cost function (16) reduces to (4) (for a single output neuron). From (14) it then follows that the standard deviation σ becomes the ordinary root mean square error.

The other limiting case is the more interesting one. As one can infer from (18) the learning rate η_r decreases with decreasing σ, i.e. the better the neural network model becomes, the smaller the weight and threshold changes (17) get. In the extreme limit of the neural network model becoming a perfect one with $\sigma \to 0$ and $\mu \to 0$ also the objective function (16) and the effective learning rate (18) vanish. The latter in particular has the advantage of learning to cease altogether, whereby the already perfect neural net model is prevented from being corrupted by any spurious weight changes, which, in such a situation, would be present still in ordinary backprop. In ordinary backprop, learning would stop only if a net reproduces exactly the target data $t_1^{(r)}$, including the experimental errors!

The cleanprop algorithm was first tested on a problem for which also the true values $Y_1^{(r)}$ are known (Gernoth and Clark 1995b; Gernoth and Clark 1995c). To this end theoretical values $Y_1^{(r)}$ have been generated by suitably mapping onto the interval $(0, 1)$ the mass excesses $\Delta m(r)$ provided by the liquid-drop model of Myers and Swiatecki (1966). These are obtained by inserting in (5) the atomic masses

$$m(r) = Z m_{\mathrm{H}} + N m_{\mathrm{n}} - a_1 A + a_2 A^{2/3} + a_3 Z^2 / A^{1/3} + a_4 (Z - N)^2 / A \quad (22)$$

of the liquid-drop model. Here, $m_{\mathrm{H}} = 938.77\,\mathrm{MeV}$ and $m_{\mathrm{n}} = 939.55\,\mathrm{MeV}$ are the masses of the hydrogen atom and the neutron. The coefficients in the Bethe-Weizsäcker semi-empirical mass formula (22) read $a_1 = 15.68\,\mathrm{MeV}$, $a_2 = 18.56\,\mathrm{MeV}$, $a_3 = 0.717\,\mathrm{MeV}$, and $a_4 = 28.1\,\mathrm{MeV}$ (Myers and Swiatecki 1966).

The true mass excesses of the theoretical liquid-drop model are then artificially contaminated with white noise. For a total of 2251 nuclei, a subset of the 2291 nuclei used in the applications described in Sect. 2, error bars σ_r are drawn at random from a normal distribution of zero mean and given half-width Γ. To generate noisy target data $t_1^{(r)}$ by means of (9), for each pattern r in turn a value γ_r is chosen randomly from a pool of 400 values that are normally distributed with zero mean and standard deviation σ_r. A 551 patterns are selected at random from the 2251 nuclei to form the test set, leaving a training set of 1700 patterns.

For this application of cleanprop to the liquid-drop model as outlined above it is further assumed that the mean μ appearing in (12)–(16) and (19) and (21) vanishes. This entails that the errors ϵ_r in (10) which a net makes in predicting the true values $Y_1^{(r)}$ must be normally distributed with zero mean and a standard deviation σ which is determined self-consistently along the lines sketched in the context of (20) and (21). Because the true values $Y_1^{(r)}$ are known by construction, the validity of the assumptions surrounding (10), which are essential to the cleanprop algorithm, can be tested with the mature networks. Such checks have shown (Gernoth and Clark 1995b) that the neural network errors ϵ_r are indeed normally distributed with zero mean and with the standard deviation σ resulting from solving (14) with $\mu = 0$. Furthermore, one is in a position to compute also the rms deviation

$$\sigma^{(\text{true})} = \sqrt{\frac{1}{J} \sum_r \left(Y_1^{(r)} - a_1^{(r)} \right)^2} \qquad (23)$$

of the network responses $a_1^{(r)}$ from the true values $Y_1^{(r)}$. In real-world applications, where the true values $Y_1^{(r)}$ would not be known, it is nonetheless useful to check the validity of (10) on only the patterns with the smallest error bars σ_r, as was done in Möller and Nix (1988).

Some of the results of training neural networks with cleanprop on the liquid-drop ground state masses are displayed in Table 3 in comparison to training with ordinary backprop (Gernoth and Clark 1995b). A good many more such results may be found in Gernoth and Clark (1995b). Two input neurons are used for analog coding of Z and N. Else, the nets possess three hidden layers with three units each and a single output unit. The nets have been trained for a maximum number of epochs of 20000 with a bulk learning rate $\eta = 0.5$ and a momentum parameter $\xi = 0.9$ adopted in (18) and (17). No further attempts of improving or skeletonizing the mature nets have been made.

For all of the nets listed in Table 3 the theoretical model error σ, determined self-consistently by solving (14) with $\mu = 0$, is invariably very close to the rms deviation $\sigma^{(\text{true})}$, defined in (23), of the actual neural net outputs $a_1^{(r)}$ from the true values $Y_1^{(r)}$. This is true not just for the nets trained with cleanprop but also for the nets trained with only ordinary backprop. In a real-world application, in which due to lack of knowledge of the true values $Y_1^{(r)}$ it would not be possible to evaluate $\sigma^{(\text{true})}$ exactly, an approximate estimate for $\sigma^{(\text{true})}$ may be obtained by using only the patterns with the smallest error bars σ_r in (23) and setting $Y_1^{(r)} = t_1^{(r)}$ for these samples. The usual rms error $\sigma^{(0)}$, from (20) with $\mu^{(0)} = 0$, deviates the more from σ and $\sigma^{(\text{true})}$, the larger the error bars σ_r. These findings support the view that for problems meeting the assumptions made in the context of (9)–(11) the quantity σ is indeed the more incisive performance measure than the usual rms error $\sigma^{(0)}$. Owing to the paucity of adjustable parameters, network performances may vary noticeably with the initial configuration in weight space and with the stochastic path in pattern space.

Table 3. Performance figures for learning and prediction of neural networks trained with cleanprop (cp) or ordinary backprop (bp) on the liquid-drop model (22) of atomic ground state masses. The nets are all of the type $^a(2 + 3 + 3 + 3 + 1)_a$[37] with two input neurons for analog coding of Z and N and a single output unit for analog representation of the mass excess values. Units are MeV. The training set contains 1700 and the test set 551 nuclei. Error bars σ_r (cf. (9)) are generated by random sampling of a normal distribution of zero mean and given half-width Γ, shown in square brackets in the first column. For every pattern r an error γ_r is randomly selected out of a set of 400 values normally distributed with zero mean and standard deviation σ_r. The Roman numeral in the first column specifies different choices of errors γ_r and the arabic letter different choices of initial weights and thresholds and of the stochastic path in pattern space. The rms deviation $\sigma^{(0)}$ of the neural net responses from the noisy target data is obtained from (20), applied to the training or test set as appropriate with $\mu^{(0)} = 0$. In the fourth and fifth (seventh and eighth) column the network model error σ, obtained from solving (14) with $\mu = 0$, is compared to the rms deviation $\sigma^{(\text{true})}$ (cf. (23)) of the net outputs from the true values for the training (test) set. The data shown here are a collection of the results published in Gernoth and Clark (1995b).

Run	Rule	Learning error			Prediction error		
		$\sigma^{(0)}$	σ	$\sigma^{(\text{true})}$	$\sigma^{(0)}$	σ	$\sigma^{(\text{true})}$
Ia [5.0]	cp	6.316	4.575	4.709	6.430	4.521	4.645
Ia [5.0]	bp	7.084	5.573	5.728	7.148	5.428	5.433
IIb [5.0]	cp	4.506	1.430	1.574	4.369	1.676	1.757
IIIc [10.0]	cp	8.532	1.366	1.960	8.513	1.706	2.086
IIIc [10.0]	bp	11.012	6.188	6.385	10.822	5.752	6.128
IIId [10.0]	cp	8.432	0.949	1.482	8.429	1.122	1.324
IVe [15.0]	cp	13.263	0.717	1.504	12.178	1.021	1.447
IVe [15.0]	bp	14.138	5.762	6.117	13.495	5.753	5.936
Ve [15.0]	cp	12.957	0.470	1.284	12.198	1.056	1.109
VIf [25.0]	cp	22.026	1.126	1.867	20.184	2.018	1.875

As is evident from the bp-results for nets Ia, IIIc, and IVe in Table 3, also the usual backpropagation algorithm is able to filter out some of the noise in the target data $t_1^{(r)}$. However, the same nets trained with cleanprop generally exhibit the better performance in both learning and prediction. Since the atomic masses provided by the liquid-drop model are a fairly smooth function of the independent input variables Z and N, fitting and predicting these masses is a much easier task for neural networks than modeling the real, experimental masses (Gernoth and Clark 1995b). It is nonetheless remarkable that the cleanprop algorithm is performing so well even on the noisiest data, corresponding to a half-width $\Gamma = 25$ Mev, used for run VIf in Table 3. To conclude, a comparison of the cp-trained nets with the bp-trained ones (pairs Ia, IIIc, and IVe in Table 3) confirms that cleanprop is much more efficient in suppressing the detrimental effects of errors in the target data on the neural network models than ordinary backprop.

Networks of the type $^h(18 + 10 + 10 + 10 + 1)_a[421]$ have been trained with the cleanprop algorithm also on the masses and error bars of 1323 nuclei selected at random from the Möller at al. (1992) data base of 1654 patterns (Gernoth and Clark 1995b). The predictive performance was evaluated for the complementary test set of 331 nuclei, also from the Möller et al. (1992) data base. The cleanprop results are only marginally better than the performance achieved with ordinary backprop learning, which may be ascribed to the fact that the experimental errors are indeed quite small. The cleanprop training strategy was applied also to exactly the same 1323 nuclei used by Möller and Nix (1994) to determine the parameters in their macroscopic-microscopic model and the resulting networks tested on the 351 new nuclei (see Tables 1 and 2 for this partitioning of the Möller and Nix (1994) data base). Although in this case, in comparison to ordinary backprop training, the improvement in predictive performance on the 351 new nuclei gained by cleanprop learning of the 1323 masses is somewhat more noticeable than in the previous situation, also these nets still fall short of being competitive with the model of Möller and Nix (1994) in extrapolating to the 351 nuclei.

4   Neural Network Models of Nuclear Stability and Decay

Constructing neural network models involving nuclear stability and decay provides further challenging opportunities for developing and testing network architectures and training strategies in applications to problems for which a large data base is available and which are bound to be of continuing interest to both experimental and theoretical nuclear physicists. Most naturally, the first question to ask is whether a nuclide is stable at all or not. As a matter of fact, this classification task was one of the first problems in nuclear physics tackled with fully connected feedforward neural networks (Gazula, Clark, and Bohr 1992) of the type described in Sect. 2 and recently was resumed with perceptrons possessing also higher-order interneuronal connections (Clark et al. 1999). Proceeding from the stability/instability dichotomy, the next topic of interest would be to make neural nets classify further the unstable nuclides according to by which modes or combinations of modes they decay. However, it is possible to do much more than that and have the nets compute also the branching ratios for the various modes of decay (Gernoth and Clark 1995a).

For any given nuclide, the corresponding branching ratios are just the relative probabilities for the different channels by which that nuclide may decay. This section focuses on reviewing and updating the neural net research done on the stability and branching ratios problem (Clark, Gernoth, and Ristig 1994; Clark, Gernoth, and Ristig 1995; Gernoth and Clark 1995a). Other ongoing work is devoted to training neural nets to predict the half-lives and energies released on decay of unstable nuclides of certain classes (Mavrommatis et al. 1998). So far, all these studies are restricted to nuclei in their ground states.

The training and test sets in Gernoth and Clark (1995a) are based on the experimental data on stability and decay provided by the NNDC in 1990. The total data base employed in this work contains 252 stable nuclides (st) and 1319 nuclides which are unstable with respect to emission of an α particle, β^- decay, electron capture (ec), the latter including β^+ decay, or spontaneous fission (sf) or with respect to combinations of two of these four decay channels. The relevant categories are all listed in Table 4. Out of every subset about a 20 % of nuclides have been reserved by random selection to form the test set, resulting in a total of 1256 training patterns and of 315 test patterns. Table 4 displays for each category the total number of nuclides along with the number of training and test patterns. More details on how the training and test sets have been chosen from the Brookhaven data may be found in Gernoth and Clark (1995a). The patterns used there will be referred to as the 1990 data base.

Table 4. The 1990 data base used by Gernoth and Clark (1995a) in modeling nuclear stability and decay with neural networks. The figures shown for the stables (st) and for each category of mono-unstables $\left(\alpha, \beta^-, \text{electron capture } (ec), \text{ including}\right.$ β^+ decay, spontaneous fission (sf)$\left.\right)$ and of bi-unstables ($\alpha/ec, \beta^-/ec, \alpha/\beta^-, \alpha/sf$) are, in this order, the total number of patterns, the number of training patterns, and the number of test patterns in the given class.

st	252/200/52	ec	521/417/104	β^-/ec	32/25/7
α	85/68/17	sf	4/3/1	α/β^-	12/10/2
β^-	517/414/103	α/ec	138/111/27	α/sf	10/8/2

It is clear from the nature of the problem that one is dealing here with five dependent output quantities. The first neuron in the output layer is assigned the task to signal whether a given nuclide is stable or not. Obviously, for a stable nuclide r the corresponding target pattern for the output layer reads $t_1^{(r)} = 1$ and $t_k^{(r)} = 0$ for all other output units k. The probability for a stable nucleus to be stable is unity and the probabilities for it to decay all vanish! For an unstable nuclide the target activity for the stability output unit vanishes, $t_1^{(r)} = 0$, whereas the target data $t_k^{(r)}$ for the other output neurons $k \neq 1$ represent the relative decay probabilities for the four modes α, β^-, ec (including β^+), and sf. For any given nuclide, whether stable or unstable, the target probabilities $t_k^{(r)}$ sum up to unity. If $K+1$ is the number of output quantities, in the present case $K + 1 = 5$, then the target datum $t_{K+1}^{(r)}$ may be written as

$$t_{K+1}^{(r)} = 1 - \sum_{k=1}^{K} t_k^{(r)} . \tag{24}$$

In similar vein, in the present problem setting a neural network must produce at its output layer a probability distribution over the five exhaustive outcomes st, α, β^-, ec, and sf for any pattern r impressed at its input interface. This requirement entails that, regardless of the input, the output activities $a_k^{(r)}$ must all lie in the interval $[0, 1]$ and add up to unity. For feedforward nets of the type introduced in Sect. 2, with the response function (1) for the hidden units and the stimulus (2) for the hidden as well as the output units, only the response function for the output neurons needs to be altered to ensure that also the net outputs $a_k^{(r)}$ will constitute a probability distribution for any input pattern r. For such neural network predictors of probabilities the output activities $a_k^{(r)}$ are determined by (Hertz, Krogh, and Palmer 1991; Lönnblad et al. 1991; Stolorz, Lapedes, and Xia 1991; Gernoth and Clark 1995a)

$$a_k^{(r)} = \frac{e^{u_k^{(r)}}}{1 + \sum_{l=1}^{K} e^{u_l^{(r)}}} \qquad \text{for} \quad 1 \leq k \leq K \tag{25}$$

and

$$a_{K+1}^{(r)} = 1 - \sum_{k=1}^{K} a_k^{(r)} . \tag{26}$$

It is obvious that (25) and (26) meet the necessary requirements for the outputs $a_k^{(r)}$ ($1 \leq k \leq K + 1$) to constitute a probability distribution. Note that because of (26) only K output units are needed, which has the additional advantage of reducing the number of adjustable parameters right from the beginning. In the present case, the activities of the four "real" output units represent, in the given order, the stability/instability dichotomy and the branching ratios for α decay, β^- decay, and electron capture. The probability for spontaneous fission is given by the activity of the $(K + 1)$-th, here the fifth, "virtual" output neuron. This activity is obtained by (26).

The objective function to be minimized in training a neural network on nuclear stability and decay is again a sum of the form (3) of pattern-specific errors E_r. The appropriate measure for comparing the two probability distributions $t_k^{(r)}$ and $a_k^{(r)}$ is the relative entropy of the target probability distribution $t_k^{(r)}$ with respect to the actual neural network output probability distribution $a_k^{(r)}$ (Kullback 1959; Hertz, Krogh, and Palmer 1991; Qian, Gong, and Clark 1991; Gernoth and Clark 1995a). The relative entropy assumes the form

$$E_r = \sum_{k=1}^{K+1} t_k^{(r)} \log\left[\frac{t_k^{(r)}}{a_k^{(r)}}\right] = \sum_{k=1}^{K} t_k^{(r)} \log\left[\frac{t_k^{(r)}}{a_k^{(r)}}\right]$$

$$+ \left[1 - \sum_{k=1}^{K} t_k^{(r)}\right] \cdot \log\left\{\left[1 - \sum_{k=1}^{K} t_k^{(r)}\right] \cdot \left[1 + \sum_{k=1}^{K} e^{u_k^{(r)}}\right]\right\} . \quad (27)$$

The second identity in (27) derives from the first, which is the definition of the relative entropy, by making use of (24)–(26). It may be shown that the relative entropy (27) is positive semi-definite and vanishes iff $a_k^{(r)} = t_k^{(r)}$ for all outcomes $k = 1, 2, \cdots, K + 1$ (Kullback 1959; Qian, Gong, and Clark 1991). Following again the same steps outlined in Sect. 2 with the relative entropy (27) for E_r and exploiting repeatedly (25) it is readily established (Hertz, Krogh, and Palmer 1991; Lönnblad et al. 1991; Gernoth and Clark 1995a) that (6) and (8) remain exactly the same and that the output error signals $\delta_k^{(r)}$ now take the somewhat simpler form

$$\delta_k^{(r)} = t_k^{(r)} - a_k^{(r)} \qquad \text{for} \quad 1 \leq k \leq K . \quad (28)$$

With (6), (8), and (28) the backpropagation algorithm based on the cost function (27) works in exactly the same manner as outlined in Sect. 2 in the context of the square cost function (4).

To impose also on the neural net outputs the mutual exclusiveness of stability and instability and the fact that all unstables decay by at most two of the decay channels, it is useful to consider also postprocessed responses $b_k^{(r)}$ and $c_k^{(r)}$ constructed from the raw output activities (25) and (26) in two stages. If the net deems a nuclide r to be stable, i.e. if of the raw activities $a_k^{(r)}$ the one for the stability unit is the largest, a winner-take-all criterion is applied and the stability output neuron assigned the total activity of unity, $b_1^{(r)} = 1$, and all other output units a vanishing activity, $b_k^{(r)} = 0$ for $k = 2, 3, 4, 5$. Else, the activity for the st unit is set zero, $b_1^{(r)} = 0$, and the total activity of unity redistributed among the four decay modes according to

$$b_k^{(r)} = \frac{a_k^{(r)}}{\sum_{l=2}^{5} a_l^{(r)}} \qquad \text{for} \quad k = 2, 3, 4, 5 . \quad (29)$$

In the second stage the activities $b_k^{(r)}$ are left unchanged for nuclides classified by the network as stable, $c_k^{(r)} = b_k^{(r)}$ for all k. In the complementary case of a pattern put by the network into the category of unstables let $l \in \{2, 3, 4, 5\}$ and $n \in \{2, 3, 4, 5\}$ be the two neurons among the four decay output units with the largest raw activities. The phase-two postprocessed output activities $c_k^{(r)}$ are now formed as

$$c_k^{(r)} = \frac{a_k^{(r)}}{a_l^{(r)} + a_n^{(r)}} = \frac{b_k^{(r)}}{b_l^{(r)} + b_n^{(r)}} \qquad \text{for} \quad k = l, n \qquad (30)$$

and $c_k^{(r)} = 0$ for all other units $k \neq l, n$. Numerical results demonstrate (Gernoth and Clark 1995a) that network performances improve in both learning and prediction from one postprocessing stage to the next. The major gain in performance is effected by moving from the raw outputs $a_k^{(r)}$ to the phase-one responses $b_k^{(r)}$. From there results improve only marginally upon going to the phase-two activities $c_k^{(r)}$.

Although the relative entropy (27) is the appropriate error measure upon which to build an algorithm for training neural nets to predict probabilities, this quantity is hard to comprehend in human terms and might actually be misleading. A single pattern for which a network produces an output $a_k^{(r)}$ close to zero for a corresponding non-vanishing target value $t_k^{(r)}$ suffices to make the relative entropy averaged over all patterns become disproportionately large, although the net might perform very well on all other patterns. It is therefore necessary to assess the performance of the networks also in terms of more illuminating error measures. Perhaps the most incisive performance measure is provided by the average over patterns of the maximum taken over the five output neurons (including the "virtual" one) of the deviation $\left| t_k^{(r)} - c_k^{(r)} \right|$. Explicitly, the definition of this quantity, expressed as a percentage, reads

$$D = \frac{1}{J} \sum_r \max \left\{ \left| t_k^{(r)} - c_k^{(r)} \right| ; 1 \leq k \leq 5 \right\} \times 100\% \ . \qquad (31)$$

To obtain individual performance figures on each one of the nine categories listed in Table 4, the sum in (31) is to be taken over all training or test patterns belonging to that class with J being the total number of patterns involved. Gross figures may be formed by summing all patterns in the training or test set and taking J to be the total of nuclides contained in the respective set. Another useful performance measure is the number of close matches, where here an output pattern is regarded a close match if all output activities $c_k^{(r)}$ are within 5% of their targets $t_k^{(r)}$, i.e. $\max \left\{ \left| t_k^{(r)} - c_k^{(r)} \right| ; 1 \leq k \leq 5 \right\} \leq$ 0.05. The above introduced error measures may, of course, be evaluated also for the raw outputs and the phase-one postprocessed responses by simply inserting $a_k^{(r)}$ or $b_k^{(r)}$ for $c_k^{(r)}$. Note that, when taking the phase-one or phase-two postprocessed outputs, a close match for a stable nuclide is identical to a perfect match.

Along the lines just described a variety of networks have been trained and tested on nuclear stability and decay in the work reported in Gernoth and Clark (1995a). The momentum parameter was fixed at $\xi = 0.9$ and the learning rate η varied between 0.005 and 0.02 with typically a few thousands of epochs per training run and skeletonization applied in the course of the training procedure. The best results are obtained with nets of the type $^b(16 + 10 + 10 + 4)_a[P]$ with respectively eight input neurons for binary coding of the independent variables Z and N. Table 5 displays for the best network the performance measure D, defined in (31), along with the number of close matches for the whole training and test set and for each one of the nine relevant classes of nuclides individually.

Table 5. Performance of the best neural network model of nuclear stability and decay. The figures shown here are taken from Gernoth and Clark (1995a). The network is of the type $^b(16 + 10 + 10 + 4)_a[277]$ with binary coding of the input quantities Z and N and 277 remaining parameters after repeated cycles of skeletonization. Shown for each class of nuclides are the performance measure D, defined in (31), and the number M of close matches for both the training and the test set. The total number of patterns in a given class is shown in square brackets. The first row lists overall performance figures on the whole training and test set. The error measures have been evaluated with the phase-two postprocessed output activities $c_k^{(r)}$ as introduced in the context of (30). An output pattern $c_k^{(r)}$ is regarded a close match, if $\max\left\{\left|t_k^{(r)} - c_k^{(r)}\right|; 1 \le k \le 5\right\} \le 0.05$.

Class	Learning error		Prediction error	
	D	M	D	M
All	4.91	1112 [1256]	13.15	246 [315]
st	1.00	198 [200]	30.77	36 [52]
α	22.35	43 [68]	19.79	11 [17]
β^-	1.70	405 [414]	8.00	94 [103]
ec	3.18	395 [417]	5.60	93 [104]
sf	18.50	1 [3]	23.70	0 [1]
α/ec	13.51	52 [111]	18.87	8 [27]
β^-/ec	16.98	9 [25]	17.72	4 [7]
α/β^-	5.56	9 [10]	16.27	0 [2]
α/sf	47.82	0 [8]	55.19	0 [2]

The network performs best on recognizing the β^- and ec mono-unstables. This task is facilitated to an extent by the large number of β^- and ec unstable nuclides in the data set and by the fact that in many cases the net only has to decide whether a nuclide is below or above the valley of stables in the $N - Z$ plane. The stable nuclides and the β^-, ec, and β^-/ec unstables spread over the entire range of the atomic mass number $A = Z + N$, from light nuclides to heavy ones. A neutron-rich nuclide is prone to decay by the β^- channel, whereas a proton-rich nuclide is likely to take the ec (including β^+) route. In the first case, a neutron (n) decays into a proton (p) and an electron (e^-), which process is accompanied by the emission of an antineutrino ($\bar{\nu}$), n $\longrightarrow$ p+e^-+$\bar{\nu}$. In the latter case, a proton and an electron of the atomic shell combine to produce a neutron and a neutrino, p + e^- $\longrightarrow$ n + ν. In β^+ decay a proton transforms into a neutron and a positron (e^+) with concomitant emission of a neutrino, p $\longrightarrow$ n + e^+ + ν. The β^-/ec patterns lie in between the β^- and the ec mono-unstables and mix with the stable nuclides. In this stripe of the $N-Z$ plane β^- and ec decay compete with each other. Mainly for the reason that the β^- and the ec (including β^+) mode represent opposite, complementary decay channels the β^-/ec bi-unstables must be considered highly problematic and hard to predict accurately within any theory. As demonstrated by the performance figures on the mixed β^-/ec category, the network interpolates remarkably well between the two extremes of pure β^- and pure ec decay. The net seems to use some of the knowledge it gained on the mono-unstables to compute the branching ratios for the β^-/ec bi-unstables.

The α mono-unstables and α/ec bi-unstables are found in the range of medium-heavy up to heavy nuclides and overlap significantly with the ec mono-unstables. An α unstable decays by ejecting a positively charged helium nucleus, $^4_2\text{He}^{2+}$, the α particle. On the proton-rich side of the valley of stability α and ec decay compete with each other. The network performs quite satisfactorily on singling out the α mono-unstables and does even better in computing the branching ratios for the α/ec bi-unstables. In view of the complexity of how the ec, the α, and the α/ec unstables mingle in the $N - Z$ plane the net may be said to handle very successfully the far from trivial task of making valid predictions for these cases. The net produces respectable results also for the few α/β^- patterns, which lie in the region of heavy nuclides and are embedded in the bulk of β^- unstables on the neutron-rich side of the $N - Z$ plane. The rare cases making up the sf and α/sf cases are invariably very heavy nuclides. Predicting the decay probabilities for these patterns necessarily involves a dominant element of extrapolation. Due to this and the paucity of sf and α/sf unstables the comparatively poor performance of the net on these classes isn't too surprising after all.

As in earlier work (Gazula, Clark, and Bohr 1992), the neural networks exhibit somewhat of a weakness in correctly classifying the stable nuclides also in the present application. Referring to the net in Table 5, the almost perfect performance on the stables in the training set drops sharply when trying the net on the stables in the test set. The stables are located in the $N - Z$ plane along the valley of β stability in a complicated pattern which the network apparently finds hard to predict. Moreover, the overwhelming majority of nuclides are unstable, biasing to a degree the net in favor of categorizing an unknown nuclide as unstable. As pointed out by Gazula, Clark, and Bohr (1992), the Mathews coefficient C (Mathews 1975) is a more suitable error measure for assessing the performance of a neural network model on the stability/instability discrimination problem than the number of perfect matches. With n_{st} being the number of nuclides (in the training or test set as appropriate) classified correctly by the network as stable, $n_{\overline{st}}$ the number of correctly classified unstables, $\hat{n}_{st}$ the number of patterns incorrectly classified as unstable, and $\hat{n}_{\overline{st}}$ the number of nuclides put incorrectly by the net into the category of stables, the Mathews coefficient is defined as

$$
C = \frac{n_{st} n_{\overline{st}} - \hat{n}_{st} \hat{n}_{\overline{st}}}{\sqrt{(n_{st} + \hat{n}_{st})(n_{st} + \hat{n}_{\overline{st}})(n_{\overline{st}} + \hat{n}_{st})(n_{\overline{st}} + \hat{n}_{\overline{st}})}} \ . \tag{32}
$$

The Mathews coefficient C assumes values between -1 and $+1$. It is -1 for all patterns put into the incorrect category and $+1$ in case of all stable and unstable nuclides classified correctly. For a net classifying invariably all patterns as either stable or unstable the Mathews coefficient vanishes. The error measure (32) takes into account properly the relative abundance of stables and unstables in the data set. A neural network classifier is the more reliable, the larger the Mathews coefficient. Evaluating (32) numerically with the net of Table 5 yields a Mathews coefficient of 0.93 for learning and of 0.68 for prediction (Gernoth and Clark 1995a). The latter figure is identical to the best predictive performance reported in Gernoth et al. (1993) for the stability/instability classification problem. However, with $C = 0.87$ for learning, the net in Gernoth et al. (1993) did not do as well in classifying the training patterns. So far, a predictive performance with a Mathews coefficient of 0.68 is the best that could be obtained in neural network modeling of the stability/instability dichotomy.

A more thorough discussion of the quality of this best of neural network models of nuclear stability and decay may be found in Gernoth and Clark (1995a). The 1990 NNDC nuclear data table, from which the training and test patterns have been selected, contained another 344 ground state nuclides, doubtless all unstable, with no, incomplete, or only imprecise information on the relative probabilities for decay. The net in Table 5 was tested also on these nuclides. It was found (Gernoth and Clark 1995a) that the neural network responses confirm most of the incomplete or imprecise data that were available at all on the decay mechanisms and branching ratios of the 344 nuclides.

However, for a number of the 344 nuclides omitted originally from the 1990 training and test sets all or at the least some of the missing information on decay has become available in the meantime. In early 1997, when the research reviewed in this section was resumed, complete and precise data on decay mechanisms and branching ratios could be obtained through the NNDC internet services for another 134 nuclides of the 344 data set, of which 132 belong to one of the nine categories listed in Table 4. For another 42 of the 344 nuclides the data of early 1997 were only approximate. A total of 40 of these nuclides belong to one of the categories listed in Table 4. It is most useful to further test the net of Table 5, trained on the 1990 learning set containing 1256 nuclides, on these 132 nuclides as well as on a larger set of 172 nuclides formed by lumping together the 132 and the 40 nuclides. To obtain performance figures for the latter set, the approximate data for the 40 nuclides are taken to be the precise ones. Note that the nuclides not falling into one of the nine categories considered in the original work (Gernoth and Clark 1995a) are excluded. The results of this work are shown in Table 6.

Table 6. Performance of the neural network of Table 5, trained and tested on the 1990 training and test sets listed in Table 4, on two further sets of, respectively, another 132 and 172 nuclides that had to be excluded from the data base in the original work (Gernoth and Clark 1995a). In 1990, information on stability and decay for these nuclides was missing altogether, incomplete, or imprecise. In early 1997, the data on decay mechanisms and branching ratios were complete and precise for the 132 nuclides considered here and approximate for another 40 nuclides. The set with 172 patterns is the union of the 132 and the 40 nuclides. For the sake of obtaining performance figures for this larger set the approximate data for the 40 nuclides are taken to be the precise ones. Shown for each class of nuclides are the performance measure D, defined in (31), and the number M of close matches. The total number of patterns in a given class is shown in square brackets. The first row lists overall performance figures. The error measures have been evaluated with the phase-two postprocessed output activities $c_k^{(r)}$ as introduced in the context of (30). An output pattern $c_k^{(r)}$ is regarded a close match, if $\max\left\{\left|t_k^{(r)} - c_k^{(r)}\right|; 1 \leq k \leq 5\right\} \leq 0.05$.

Class	132 Nuclides		172 Nuclides	
	D	M	D	M
All	6.58	113 [132]	9.65	128 [172]
α	27.58	4 [7]	23.22	6 [12]
β^-	0.11	65 [65]	0.11	66 [66]
ec	5.98	37 [41]	9.48	38 [45]
sf	100.00	0 [1]	100.00	0 [1]
α/ec	18.88	7 [16]	18.57	16 [41]
β^-/ec	14.42	0 [1]	14.42	0 [1]
α/sf	6.21	0 [1]	11.84	2 [6]

Focusing on the four classes α, β^-, ec, and α/ec, for which sufficiently many patterns are available to draw conclusions, the network is doing best again on the β^- and ec unstables. The neural network performance on the β^- unstables must be considered a complete success. Furthermore, it is very encouraging that the performance on the α unstables is only slightly inferior to the one on the original 1990 data set. The remarkably good performance on the class of α/ec unstables also for the new sets of 132 and 172 nuclides speaks for the quality of the neural network model of nuclear stability and decay.

5 Higher-Order Probabilistic Perceptrons

In contrast to multi-layer neural networks with one or more layers of hidden units, perceptrons (Minsky and Papert 1969; Duda and Hart 1973) contain only an input and an output layer. In the elementary perceptron architecture the input neurons extend feedforward connections to the output units. The adjustable parameters in an elementary perceptron are the weights of the binary input-output connections and the thresholds for the output units. However, the class of problems that such simple perceptrons can be taught to solve is rather limited.

A more general perceptron architecture overcoming the deficiencies of the simple binary perceptron may be constructed by supplementing the usual thresholds w_{k0} and binary weights w_{kl} with higher-order couplings from pairs, triplets, quadruplets, etc. of input units to output neurons (Clark, Gernoth, and Ristig 1994; Clark, Gernoth, and Ristig 1995; Clark et al. 1999). The following discussion will concentrate on the interesting case of a higher-order perceptron with on-off (binary) input units. In such a structure the stimulus $u_k^{(r)}$ that output neuron k receives from the input neurons for a pattern r is formed by adding to the usual formula (2) a sum of linear combinations of the products $a_l^{(r)} a_m^{(r)}$, $a_l^{(r)} a_m^{(r)} a_n^{(r)}$, $\cdots$ of the activities of all pairs, triplets, $\cdots$ of input units. The general formula for stimulus $u_k^{(r)}$ in a higher-order perceptron possessing a total of L on-off input units is given by

$$
\begin{aligned}
u_k^{(r)} = {}& w_{k0} + \sum_{l=1}^{L} w_{kl} a_l^{(r)} + \sum_{l<m=1}^{L} w_{klm}\, a_l^{(r)} a_m^{(r)} \\
& + \sum_{l<m<n=1}^{L} w_{klmn}\, a_l^{(r)} a_m^{(r)} a_n^{(r)} + \cdots \\
& + \sum_{l_1<\cdots<l_\nu=1}^{L} w_{kl_1\cdots l_\nu}\, a_{l_1}^{(r)} \cdots a_{l_\nu}^{(r)} \\
& + \cdots + w_{k12\cdots L}\, a_1^{(r)} a_2^{(r)} \cdots a_L^{(r)} \,.
\end{aligned}
\tag{33}
$$

Here, the strength of the ternary, quaternary, etc. connection from input neurons l, m, n, etc. to output unit k is denoted by w_{klm}, w_{klmn}, etc.

Endowed with the soft-max activation functions (25) and (26) for the output units, higher-order perceptrons may be used for classification tasks or taught to predict probability distributions. Neural network structures of the type discussed in this section are referred to as higher-order probabilistic perceptrons (HOPP). It may be shown that with a suitable choice of interneuronal weights $w_{kl_1 \cdots l_\nu}$ of all orders from $\nu = 0$ to $\nu = L$ a HOPP classifier can produce the *a posteriori* probabilities provided by the Bayes optimal classifier (Clark, Gernoth, and Ristig 1994; Clark, Gernoth, and Ristig 1995; Clark et al. 1999). The remainder of this section will concentrate on the practical aspects of training a HOPP and on the results obtained with HOPP architectures trained on classifying nuclei as stable or unstable (Clark et al. 1999).

For training a HOPP with K ("real") output units (and one "virtual" one) the learning rule (6) assumes the form

$$\Delta w_{kl_1 \cdots l_\nu}^{(r)} = -\eta \frac{\partial E_r}{\partial w_{kl_1 \cdots l_\nu}} + \xi \Delta w_{kl_1 \cdots l_\nu}^{(r-1)} = \eta \delta_k^{(r)} a_{l_1}^{(r)} \cdots a_{l_\nu}^{(r)} + \xi \Delta w_{kl_1 \cdots l_\nu}^{(r-1)} \,,$$
(34)

where the pattern-specific error E_r may be taken to be the square error (4) or the relative entropy (27), the difference being the output error signals $\delta_k^{(r)}$. In (34) the symbol $\Delta w_{kl_1 \cdots l_\nu}^{(r)}$ denotes the modification, upon presentation of pattern r, of the strength $w_{kl_1 \cdots l_\nu}$ of the connection from input neurons $l_1, \cdots, l_\nu$ to output unit k and $\Delta w_{kl_1 \cdots l_\nu}^{(r-1)}$ the last parameter changes immediately before the current ones. The integer ν denotes the order of the connection and may assume values from $\nu = 0$ to $\nu = L$, the total number of the on-off input neurons. The case $\nu = 0$ corresponds to training the bias of output unit k, in which case the product of activities $a_{l_1}^{(r)} \cdots a_{l_\nu}^{(r)}$ on the right side of (34) must be replaced by unity.

For the relative entropy cost function (27) the output error signals are given, as before, by the simple expression (28). To derive the corresponding result for the square cost function (4) one may first recast E_r with the help of the sum rules (24) and (26) in the form

$$E_r = \frac{1}{2} \left\{ \sum_{k=1}^{K} \left[t_k^{(r)} - a_k^{(r)} \right]^2 + \left(\sum_{k=1}^{K} \left[t_k^{(r)} - a_k^{(r)} \right] \right)^2 \right\}$$
(35)

and then proceed with taking the partial derivatives in (34). The second term in (35) is the contribution from the $(K+1)$-th, the "virtual" output neuron. One winds up with the expression

$$\delta_k^{(r)} = a_k^{(r)} \left\{ t_k^{(r)} - a_k^{(r)} + \sum_{l=1}^{L} \left[t_l^{(r)} - a_l^{(r)} \right] \left[1 - \sum_{m=1}^{K} a_m^{(r)} - a_l^{(r)} \right] \right\}$$
(36)

for the output error signals $\delta_k^{(r)}$ for training a HOPP using the square error cost function.

The result (36) may be derived also by using in (34) instead of (35) the original form (4) for the pattern-specific square deviation E_r with, however, the index k running from $k = 1$ to $k = K+1$ and assuming the form (25) also for $a_{(K+1)}^{(r)}$. After evaluation of the partial derivatives of E_r w.r.t. the weights $w_{kl_1\cdots l_\nu}$ the target value $t_{K+1}^{(r)}$ and the activity $a_{K+1}^{(r)}$ of the $(K+1)$-th, the "virtual" output unit, may then be eliminated by means of the constraints (24) and (26). It is worthwhile to note that the same applies to the derivation of (28). Also in this case taking the partial derivatives in (6) or (34) and imposing the constraints (24) and (26) may be interchanged. In case of a HOPP with a single output unit, $K = 1$, equation (36) reduces, apart from an irrelevant factor 2, to the usual formula (7) for the output error signal. The additional factor 2 is due to the fact that the cost function (35), from which (36) derives, includes the contribution from the second, the "virtual" output unit. For $K = 1$ this renders the expression (35) to be exactly twice the usual square cost function.

In a first application the HOPP architecture has been put to test on the nuclear stability/instability classification problem, employing both the relative entropy learning rule (28) as well as the square deviation rule (36). The HOPPs explored in this application contain a single output unit, for which reason the rule (36) is identical to (7), expect for the aforementioned factor 2. As in Sect. 4, respectively eight input neurons are used for binary coding of the independent variables Z and N. For a stable nuclide r a target value $t_1^{(r)} = 1$ is assigned to the single output unit, whereas for an unstable one the target value is given by $t_1^{(r)} = 0$. A winner-take-all criterion is imposed at the output layer, i.e. an output activity $a_1^{(r)} > 0.5$ is interpreted as a stability assignment, while an output response $a_1^{(r)} < 0.5$ means that the HOPP classifies the input nuclide as unstable. The training set contains the 200 stable nuclides of Table 4 and, omitting the eleven cases of sf and α/sf unstables, the 1045 unstables of this table. The test set comprises the 52 stables of Table 4 and, leaving out the three examples of sf and α/sf unstables, the 260 unstables of Table 4.

Retaining all connections of all orders from $\nu = 0$ to $\nu = L$ – in the present case $\nu = 16$ – would entail a total of $\sum_{\nu=0}^{16} \binom{16}{\nu} = 2^{16} = 65536$ adjustable weights for nuclear stability/instability HOPP networks. Apart from the computational demands of training that many weights, it would be also futile to even try, as such a net would possess by far more parameters than the 1245 stability/instability assignments to train on. Such a huge net would by sheer memorization very quickly achieve a perfect fit to the training data, but would be utterly useless for making valid predictions for the test patterns it hasn't seen during training. As a matter of fact, this behavior was found in numerical experiments with HOPPs containing all connections of all orders from $\nu = 0$ to only $\nu = 3$, resulting in nets with 697 adaptable parameters.

For the reasons outlined above the number of connections must be cut down substantially prior to any training. A strategy which has proven highly successful for this purpose originates in the observation that for any given pattern r a weight $w_{kl_1 \cdots l_\nu}$ in the expansion (33) comes into effect only if all of the input neurons involved – $l_1, \cdots, l_\nu$ – are simultaneously on for that pattern. Else, the product $a_{l_1}^{(r)} \cdots a_{l_\nu}^{(r)}$ of activities vanishes. One may therefore, prior to training, easily assess the relative importance of any given connection by simply making a count for how many training patterns that connection comes into effect at all. Only connections that come into effect for at least J_{crit} training examples are retained and all others deleted from the HOPP rightaway. The skeletonization scheme just described has the advantage that, irrespective of order, all connections deemed relevant according to the chosen J_{crit} are kept. Once the non-vanishing weights have been preselected, training a HOPP according to the rules (34) and (28) or (36) follows the same strategy sketched in the previous sections. However, parameters are updated only if $\left| t_1^{(r)} - a_1^{(r)} \right| \geq 0.5$. Training consists of several consecutive runs, within each of which 36000 epochs are performed. The configuration of weights corresponding to the best HOPP found in a run serves as input for the next run. The procedure for training HOPPs on the nuclear stability/instability discrimination task is outlined in greater detail in Clark et al. (1999).

From a large number of numerical experiments networks corresponding to $J_{\mathrm{crit}} = 200$, 170, and 165 with, respectively, a total of 109, 127, and 143 adjustable parameters emerged as the most powerful stability/instability classifiers of the HOPP type. Tables 7 and 8 show the number of non-vanishing weights in each order for $J_{\mathrm{crit}} = 200$ and $J_{\mathrm{crit}} = 165$. Not surprisingly, only connections of the few lowest orders are retained. The learning rate η in (34) was typically 0.025 and the momentum parameter ξ varied between zero and 0.9. In many cases all (retained) weights were initially set zero rather than starting out from a random configuration. The HOPPs starting with (important) weights sampled randomly from a uniform distribution on the interval $[-0.5, 0.5]$ did not perform much differently from the HOPPs constructed from an initial configuration with all weights zero. HOPPs trained with the relative entropy prescription (28) and the square error rule (36) turned out to be of very similar quality.

Table 7. Connectivity pattern of HOPP classifiers for discriminating stable and unstable nuclides. Shown in the second column is the number I_{weights} of connections of order ν, listed in the first column, that come into effect for at least $J_{\text{crit}} = 200$ patterns in the training set. All higher orders, for which no connection comes into effect for at least J_{crit} training examples, are omitted. The third column lists the total number $\binom{16}{\nu}$ of possible connections of order ν.

ν	I_{weights}	$\binom{16}{\nu}$
0	1	1
1	14	16
2	89	120
3	5	560
Total	109	697

Table 8. Connectivity pattern of HOPP classifiers for discriminating stable and unstable nuclides. Same as in Table 7 for $J_{\text{crit}} = 165$.

ν	I_{weights}	$\binom{16}{\nu}$
0	1	1
1	14	16
2	90	120
3	38	560
Total	143	697

The best HOPP network constructed by training in the manner described above displays a Mathews coefficient C (cf. (32)) of 0.80 in learning and of 0.63 on the test set, after a total of 224769 training epochs. For this net J_{crit} was set to 165, resulting in the 143 non-vanishing weights listed in Table 8. The net was trained with the relative entropy learning rule (28). The second best HOPP net possesses 109 retained parameters, corresponding to $J_{\text{crit}} = 200$. The number of non-vanishing weights is listed for each order in Table 7. For this HOPP the Mathews coefficient (32) assumes the value 0.68 on the training set and 0.63 in predicting the test patterns. This performance was reached after completion of 142982 epochs. The net was trained with the square error learning rule (36).

6 Conclusions and Prospects

Artificial neural networks with feedforward connections are demonstrably powerful novel tools to compute nuclidic properties which are hard, if not close to even impossible, to predict with conventional theoretical means based on fundamental quantum mechanics. Provided that the network architecture and training algorithm are tuned to the particular problem one is dealing with, these computational structures are capable to yield numerical results that in accuracy compare very favorably with the best of results attainable within conventional approaches.

Neural network models of nuclear masses seem to be at the point of matching the predictive faculty of the macroscopic-microscopic model of Möller and Nix (1994), one of the best of global models of nuclear masses available at the moment. At present, efforts are being undertaken to implement the cleanprop algorithm, described in Sect. 3, in the advanced training strategy that led to the excellent performance of the $^h(4 + 10 + 10 + 10 + 1)_a[272]$ net (Athanassopoulos et al. 1998) (cf. Table 2 in Sect. 2) in extrapolating to nuclides far from the valley of β stability, the benchmark of comparison with the results of Möller and Nix (1994). Applying the cleanprop learning rule instead of ordinary backprop might make a difference, since the experimental errors with which the measured masses of these nuclei are afflicted are, albeit not very large, not entirely negligible either. It remains to be seen in how far this will further improve the capability of trained networks to extrapolate accurately to regions in the $N - Z$ plane far from stability.

The best of neural network models of nuclear stability and decay displays a striking performance in predicting the branching ratios for a large number of nuclides it has not been trained on. This applies both to the 315 nuclides in the original test set (cf. Tables 4 and 5 in Sect. 4), for which all decay data were available at the time this network was constructed and tested, as well as to the 132 nuclides, for which the decay data have become available since then until early 1997 (cf. Table 6 in Sect. 4). The parametric efficiency of this network is astounding as well. With 1256 nuclides in the training set, 315 in the original test set, another 132 additional ones in the extended test set of early 1997, and with five output quantities this network manages to compute accurately a total of 8515 data with a mere 277 parameters. It is an exciting and continuing challenge to have this network compute the branching ratios of decay for nuclides for which these data have not yet been measured and to compare the network predictions to any experimental results becoming newly available.

The higher-order probabilistic perceptrons excel also in parametric efficiency, yielding a Mathews coefficient in prediction of 0.63 with, respectively, 109 and 143 non-vanishing weights (cf. Tables 7 and 8 in Sect. 5). This performance is to be compared with a Mathews coefficient in prediction of 0.68 obtained with multi-layer networks with, respectively, 227 weights (Gernoth et al. 1993) and 277 weights (Gernoth and Clark 1995a; see also the comments surrounding (32) in Sec. 4). Encouraged by the promising successes of HOPP networks on the nuclear stability/instability categorization problem, these architectures are currently employed for classifying materials according to their crystal structure. It will be a rewarding undertaking for future work to train HOPPs to predict not only whether nuclides are stable or not but to compute also, along the lines presented in Sect. 4 in the context of multi-layer networks, the branching ratios of decay for unstable nuclides.

Acknowledgments

Discussions with J. W. Clark, S. Dittmar, V. Dixit, E. Mavrommatis, and M. L. Ristig are gratefully acknowledged. The kind hospitality of the Department of Physical Sciences, Theoretical Physics of the University of Oulu, Finland, where part of this article was completed, is appreciated.

References

Athanassopoulos, S., Mavrommatis, E., Gernoth, K. A., Clark, J. W. (1998): To be published.

Clark, J. W., Gazula, S., Gernoth, K. A., Hasenbein, J., Prater, J. S., Bohr, H. (1992): Collective Computation of Many-Body Properties by Neural Networks. *Recent Progress in Many-Body Theories*, Vol. 3, edited by Ainsworth, T. L., Campbell, C. E., Clements, B. E., Krotscheck, E. (Plenum Press, New York), 371–386.

Clark, J. W., Gernoth, K. A. (1992): Teaching Neural Networks to do Science. *Structure: From Physics to General Systems*, Vol. 2, edited by Marinaro, M., Scarpetta, G. (World Scientific, Singapore), 64–77.

Clark, J. W., Gernoth, K. A. (1995): Statistical Modeling of Nuclear Masses Using Neural Network Algorithms. *Condensed Matter Theories*, Vol. 10, edited by Casas, M., de Llano, M., Navarro, J., Polls, A. (Nova Science Publishers, Commack, NY), 317–333.

Clark, J. W., Gernoth, K. A., Dittmar, S., Ristig, M. L. (1999): Higher-Order Probabilistic Perceptrons as Bayesian Inference Engines. To be published.

Clark, J. W., Gernoth, K. A., Ristig, M. L. (1994): Connectionist Many-Body Phenomenology. *Condensed Matter Theories*, Vol. 9, edited by Clark, J. W., Shoaib, K. A., Sadiq, A. (Nova Science Publishers, Commack, NY), 519–537.

Clark, J. W., Gernoth, K. A., Ristig, M. L. (1995): Connectionist Statistical Inference. *Recent Progress in Many-Body Theories*, Vol. 4, edited by Schachinger, E., Mitter, H., Sormann, H. (Plenum Press, New York), 283–292.

Duda, R. O., Hart, P. E. (1973): *Pattern Classification and Scene Analysis* (Wiley, New York).

Gazula, S., Clark, J. W., Bohr, H. (1992): Learning and Prediction of Nuclear Stability by Neural Networks. Nucl. Phys. A **540**, 1–26.

Gernoth, K. A., Clark, J. W. (1995a): Neural Networks that Learn to Predict Probabilities: Global Models of Nuclear Stability and Decay. Neural Networks **8**, 291–311.

Gernoth, K. A., Clark, J. W. (1995b): A Modified Backpropagation Algorithm for Training Neural Networks on Data with Error Bars. Comput. Phys. Commun. **88**, 1–22.

Gernoth, K. A., Clark, J. W. (1995c): Neural Network Models of Nuclear and Noisy Data. *New Computing Techniques in Physics Research IV*, edited by Denby, B., Perret-Galix, D. (World Scientific, Singapore), 425–430.

Gernoth, K. A., Clark, J. W., Prater, J. S., Bohr, H. (1993): Neural Network Models of Nuclear Systematics. Phys. Lett. B **300**, 1–7.

Hertz, J., Krogh, A., Palmer, R. G. (1991): *Introduction to the Theory of Neural Computation* (Addison-Wesley, Redwood City, CA).

Kalman, B. L. (1994): Private Communication to Clark, J. W., Washington University, St. Louis, MO, USA.

Kullback, S. (1959): *Information Theory and Statistics* (Wiley, New York).

Lönnblad, L., Peterson, C., Pi, H., Rögnvaldsson (1991): Self-Organizing Networks for Extracting Jet Features. Comput. Phys. Commun. **67**, 193–209.

Masson, P. J., Jänecke, J. (1988): Masses from an Inhomogeneous Partial Difference Equation with Higher-Order Isospin Contributions. At. Data Nucl. Data Tables **39**, 273–280.

Mathews, B. W. (1975): Comparison of the Predicted and Observed Secondary Structure of T4 Phage Lysozyme. Biochim. Biophys. Acta **405**, 442–451.

Mathews, J., Walker, J. L. (1970): *Mathematical Methods of Physics* (Benjamin, New York).

Mavrommatis, E., Dakos, A., Gernoth, K. A., Clark, J. W. (1998): Calculations of Nuclear Half-Lives with Neural Nets. *Condensed Matter Theories*, Vol. 13, edited by da Providência, J., Malik, F. B. (Nova Science Publishers, Commack, NY), in press.

Minsky, H., Papert, S. (1969): *Perceptrons* (MIT Press, Cambridge, MA).

Möller, P., Nix, J. R. (1981): Nuclear Mass Formula with a Yukawa-plus-Exponential Macroscopic Model in a Folded-Yukawa Single-Particle Potential. Nucl. Phys. A **361**, 117–146.

Möller, P., Nix, J. R. (1988): Nuclear Masses from a Unified Macroscopic-Microscopic Model. At. Data Nucl. Data Tables **39**, 213–223.

Möller, P., Nix, J. R. (1994): Stability of Heavy and Superheavy Elements. J. Phys. G **20**, 1681–1747.

Möller, P., Nix, J. R., Myers, W. D., Swiatecki, W. J. (1992): The Coulomb Redistribution Energy as Revealed by a Refined Study of Nuclear Masses. Nucl. Phys. A **536**, 61–71.

Müller, B., Reinhardt, J., Strickland, M. T. (1995): *Neural Networks* (Springer-Verlag, Berlin).

Myers, W. D., Swiatecki, W. J. (1966): Nuclear Masses and Deformations. Nucl. Phys. **81**, 1–60.

Qian, M., Gong, G., Clark, J. W. (1991): Relative Entropy and Learning Rules. Phys. Rev. A **43**, 1061–1070.

Rummelhart, D. E., McClelland, J. L., and the PDP Research Group (1986): *Explorations in the Microstructure of Cognition*, Vols. 1 and 2 (MIT Press, Cambridge, MA).

Stolorz, P., Lapedes, A., Xia, Y. (1991): Predicting Protein Secondary Structure Using Neural Net and Statistical Methods. J. Molec. Biol. **225**, 363–377.

von Mises, R. (1964): *Mathematical Theory of Probability and Statistics* (Academic Press, New York).

Using Neural Networks to Learn Energy Corrections in Hadronic Calorimeters

João Seixas

Dept. Física, Instituto Superior Técnico, Av. Rovisco Pais, 1096 Lisboa Codex, Lisbon, Portugal

Abstract. I present two methods to determine the energy correction factors for the ATLAS hadronic calorimeter. In the first we use a recurrent neural network with nearest neighbour feedback in the input layer and, in the second, an event classification step by a competitive network precedes the learning of the correction factor. A comparison with a normal feed-forward net with backpropagation learning scheme is presented for the first method.

1 Introduction

In recent years calorimetry has become a major tool in high-energy experimental particle physics (Sauli 1993). A calorimeter in this context is quite easy to define: it is just an (eventually segmented) instrumented block of material such that, whenever a high energy particle passes through it, part of its energy is transformed in a way that can be observed and measured (*e.g.*, as light). Calorimeters are normally classified in terms of the physical processes they are designed to observe. Some are particularly sensitive to electromagnetic particles (Electromagnetic Calorimeters or EMC's), some are more sensitive to hadrons (Hadronic Calorimeters or HAC's). As for the active material, it can range from liquid noble gases to crystal scintillators. In all cases the choice of material and the calorimeter design are such that the detector responds as much as possible in a manner proportional to the energy of the incident particle. With this definition a calorimeter is thus mainly an energy-measuring device, although, if properly segmented, it can also provide information on position and particle identification through different responses to leptons and hadrons.

There are several reasons for the actual preference for this type of detector, the major one being that in contrast to other detectors, the precision of a calorimeter increases with incident particle energy (the resolution goes like $\sim 1/\sqrt{E}$). In the very high energy realm that will be attained with the future Large Hadron Collider at CERN, this is crucial, but also the size of the detectors (and the corresponding price!) is a very important constraint. Calorimeters are clearly to be favoured since their size, necessary to contain an event, scales with the logarithm of the incident particle energy. This is to be compared with, for instance, a conventional magnetic spectrometer that scales only as the square root of the incident momentum. Furthermore, the understanding of strong interactions among hadrons has shifted the emphasis towards measurements of global properties of events. Jets of particles produced by decaying quarks rather than individual particles are now the principal entities and since calorimeters are sensitive to both charged and neutral particles they are particularly suited to measure the characteristics of such events. Even neutrinos, although they do not interact normally with the detector, can be inferred from the study of momentum balance or missing transverse energy, provided that the device is hermetic.

Calorimeters are also divided into two broad classes, namely homogeneous and sampling. In the first case the detector is made of a material that is sensitive throughout to the energy deposited by the incoming particle, whereas for sampling calorimeters there is a separation between sensitive material and absorption material. In the latter case the sensitive material is sandwiched between layers of absorptive material and one is thus able to *sample* the fraction of energy that is deposited by the incident particle in the detector. Of course, if the absorption fraction is known for both the active and passive parts the total energy of the incident particle can in principle be reconstructed. The main advantage of this type of calorimeter with respect to homogeneous ones is that one can achieve a much more compact design, yet maintaining as much as possible the hermeticity of the device.

The physical mechanism by which the energy of an incident particle is transformed into a measurable quantity is essentially the same for all types of calorimeters and involves the creation of a cascade or shower of secondary particles in the active part of the calorimeter material. The shower energy (or part of it) is then detected in the form of ionization charges, scintillation or Cherenkov light.

For electromagnetic calorimeters, electrons and positrons lose their energy by collision with atomic electrons and by bremsstrahlung in the field of a nucleus, whereas photons undergo Compton scattering, the photoelectric effect and pair production in the field of a nucleus. Bremsstrahlung for e^- and e^+ and pair production are dominant above 10 MeV and energy independent above 1 MeV. Therefore these processes occur at the beginning of the shower formation and it is through a succession of these processes that an electromagnetically interacting particle loses its energy down to the domain dominated by ionization loss.

In hadronic calorimeters the energy of the incident particle is measured in a way similar to the electromagnetic case, *i.e.*, through the production of a shower. However the mechanism for shower production is quite different since it involves multiparticle production in deep inelastic hadron-nucleus collisions at high energies and thus the elementary processes involve strong interactions. Roughly half of the incident particle energy goes into multiparticle production, while the other half is carried away by a few high-energy leading particles. The large number of possible interaction processes make the shower development more complex. This development involves first the interaction of the incoming hadron with a nucleon inside a nucleus producing nucleons and mesons (*spallation*). In turn these produced particles may interact again inside the same nucleus, since the mean free path of pions in nuclear matter is smaller than the nuclear radius. If the energy is large enough, we then have a fast *intranuclear cascade* which produces more secondary particles. Of course some of these can also escape the nucleus and interact with other nuclei, producing an *internuclear cascade*. All these processes occur on a time scale $\sim 10^{-23}$ s, typical for strong interaction processes.

After this initial burst, spallation proceeds by de-excitation of highly excited nuclei with an associated emission of particles (*evaporation mechanism*) on a time scale $\sim 10^{-18}$s. The emitted particles are nucleons or even nuclear fragments. Furthermore, if heavy nuclei are present (like *e.g.*,^{238}U) the highly excited nucleus might undergo a fission process leaving excited nuclei, which in turn evaporate.

The cascade is thus mostly composed of nucleons and pions. However roughly 30% of these particles are π^0's which then decay electromagnetically producing an *electromagnetic* shower. Showers in hadronic calorimeters are thus seen to be composed of two parts, namely a π^0 component followed by a slower and longer range component due to hadronic activity. The energy spent on π^0 production is determined by the kind of the first inelastic interaction and from event to event is subject to large non-gaussian fluctuations. Part of the initial energy is also dissipated in undetectable form (μ's and ν's escape the calorimeter) and part is lost in nuclear binding energy, breakup or excitation of the recoiling nuclei. The fraction of invisible energy lost this way can reach 40 % of the energy dissipated in non-electromagnetic form.

2   Hadronic Calorimeters: the ATLAS Case

The ATLAS experiment (ATLAS stands for "**A** **T**oroidal **LHC** **A**pparatu**S**) is part of the forthcoming CERN LHC *p-p* physics program aiming to find the Higgs particle present in the Standard Model of the strong and electroweak interactions and also to search for new physics that might exist beyond this model. It is a huge device (see Fig. 1) that is made of a variety of specific detectors ranging from the inner detector to muon detectors. In this work we will be concerned with the Tile Calorimeter: this is a large hadronic sampling calorimeter, which makes use of steel as the absorbing medium and of scintillating plates as the active material. The calorimeter itself consists of a cylindrical structure with an inner radius of 2280 mm and an outer radius of 4230 mm; it is subdivided into a 5640 mm long central barrel and two 2910 mm extended barrels (see Fig. 2). The scintillating tiles lie in the (r, ϕ) plane and span the width of the module in the ϕ direction.

The main function of the Tile Calorimeter is to allow for a better reconstruction of the jets produced in the *p–p* interactions and, together with the end-cap and forward calorimeters, to have a good measure of the missing transverse momentum. It is important to mention that the Tile Calorimeter is the return path for the flux of the solenoid (see Fig. 1).

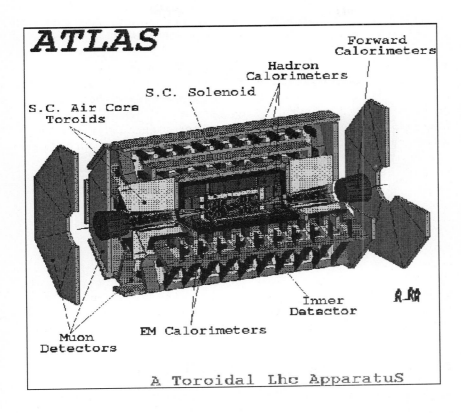

Fig. 1. The ATLAS detector.

3   A Neural Network Approach

As stated above, hadronic showers in calorimeters result from a large variety of multiple particle-production reactions. Fluctuations in the hadronic interactions that occur in each particular event increase the uncertainties and reduce the performance of hadronic calorimeters with respect to electromagnetic calorimeters (Amaldi 1981). Some effort has been devoted to the design of calorimeters which make a hardware compensation of the hadronic shower deficiencies (Brückmann et al. 1986). The main idea is to compensate for the invisible energy by having some active material in the calorimeter that

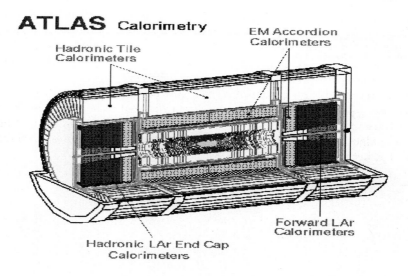

Fig. 2. The ATLAS calorimetry.

would contribute an additional signal, for example by nuclear fission. An e/π factor close to 1 may thus be obtained; however, because the percentage of low-energy neutrons is also subject to large fluctuations, this compensation operates on the average but not at a single-event level. The energy resolution is not improved.

In conclusion: hardware compensation or correction factors in hadronic calorimeters, no matter how accurate they might be on the average, always involve large errors when applied to individual events. Hence, the ideal situation would be to devise a method where a different correction factor is applied to each event, depending on its nature.

In this chapter I present two methods which attempt to adapt the correction factor to the event that is being corrected. They are based on the idea that the nature of each event is reflected in the spatial distribution of the energy density. Then, a correction factor that is sensitive to the energy distribution would be sensitive to the nature of the event. The spatial energy distribution may not give a complete characterization of the hadronic shower but, short of the actual detailed identification of the subprocesses, it is the maximum amount of information that is obtained under normal calorimetry conditions.

The aim of the present work is not only to present new neural network algorithms adapted to the characteristics of the problem, but also to assess the capability of these networks to cope with this new and difficult environment.

In the first method, I use a recurrent neural network with as many input nodes as the number of cells in the calorimeter, space distribution sensitivity being enhanced by the existence of feedback between the nodes associated to nearest-neighbour cells. The feedback is restricted to the input layer, the rest of the network has a feed-forward structure. This allows for a hybrid learning-algorithm which is computationally more efficient than the standard recurrent network-algorithm.

In the second method, a classification of the events is performed by feeding the principal eigenvalues of the normalized energy distribution to a competitive network. For the events of each class the correction factors are then learned by a regular feed-forward network. In the end, the identification system contains the competitive network and as many feed-forward networks as the number of classes.

4   Energy Correction by a Recurrent Network

In the last few years recurrent nets have attracted interest in many fields owing to their generality and efficiency. Although more difficult to implement than conventional feed-forward nets, they allow learning in more general situations, nevertheless at some expense of computational resources. It is thus advisable, whenever possible, to limit this type of architecture to some parts of the network, creating neural nets of mixed architecture. This is the philosophy we will follow in this paper.

Our case study is the correction of the deposited energy in the ATLAS calorimeter. As discussed in the introduction, different amounts of energy with different distributions will be deposited in the calorimeter, depending on the form of the cascade produced. An automatic correction is necessary and any algorithm that aims at performing this correction must include in some way, for each cell of the calorimeter, the information corresponding to the neighbouring cells. This can be already achieved, to some extent, by a simple feed-forward network. However, in this section we introduce a mixed network with an architecture and a learning-algorithm that includes feedback only among the units of the input layer. Furthermore, we restrict feedback for each cell to its next-to-near neighbours. This is not a very serious restriction, as we shall see. As for the other layers of the network, a conventional feed-forward architecture is used together with a normal back-propagation algorithm.

4.1   Architecture and Learning Mechanism for Mixed Networks

The problem consists in gathering information on the energy deposited in the neighbouring cells of the calorimeter and in deciding, on the basis of that information, how to correct the measured value of the total deposited energy. Recurrent networks seem to be a promising candidate for the task, because neighbourhood information may be explicitly introduced in the network architecture.

Recurrent networks are neural networks in which we have feedback; in other words, the output of some or of all the units can act as input for other units in the network, regardless of the layer ordering. It is clear from this that the learning mechanism for this type of systems has to be modified with respect to the one used in usual feed-forward networks. Backpropagation, for instance, can indeed be modified to accommodate feedback (Pineda 1987), (Almeida 1987), (Almeida 1988), although the method becomes slightly more complicated.

In our case we do not need to include feedback for all the units in the network. We use instead a mixed three-layer network (see Fig. 3). Feedback is introduced in the first (input) layer, but the second (hidden) and third (output) layers are designed as normal feed-forward nets. One thus expects to capture the information of the neighbouring cells in the input layer.

We have a single output unit with output values corresponding to the corrected total deposited energy. Feedback is only introduced between neighbouring units. On implementation, however, neighbouring cells of the calorimeter *do not* correspond to neighbouring cells in the actual input layer for the network because of the way cells are referenced. In fact, the cells are counted sequentially and therefore, instead of appearing as a three dimensional matrix on input, they are part of a large input vector. On the other hand, it is important to realize that, although the connections extend only to near neighbours, the *effect* of one unit on the other units of the input layer extends much further than its nearest and next-to-nearest neighbours. We will explore this issue below.

The learning process is also a mixed one. Normal backpropagation is used for the hidden and output layers, but it is modified for the input layer. We thus achieve a drastic simplification, that increases the performance of the net.

4.2   The Learning Mechanism

The starting point is, as usual, the error function

$$E = \frac{1}{2} \sum_j \left(t_j - x_j^{(3)} \right)^2 , \tag{1}$$

where t_j is the teaching input for the output unit j and the sum extends over all possible input patterns. The notation is: $x_i^{(0)}$ $(i = 1, \ldots, n_I)$ is the

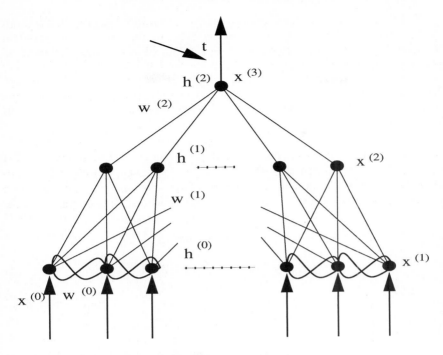

Fig. 3. Architecture of the recurrent net.

input value for unit i of the input layer; $x_i^{(1)}$ is the output of the unit i of the input layer which will serve as input for the units in the hidden layer; $x_i^{(2)}$ is the output of unit i in the hidden layer ($i = 1, \ldots, n_H$); $x^{(3)}$ is the output of the unit in the output layer. For the weight matrices, $w^{(0)}$ corresponds to the weights between units in the input layer, $w^{(1)}$ are the weights of the connections between layers 0 and 1 (input and hidden layers, respectively) and $w^{(2)}$ are the weights for the connections between hidden and output layers. As usual, we also introduce the local fields

$$h^{(2)} = \sum_{i=1}^{N_H} w_i^{(2)} x_i^{(2)} \, , \; h_j^{(1)} = \sum_{i=1}^{N_I} w_{ji}^{(1)} x_i(1) h_j^{(0)} = \sum_{i=1}^{N_I} w_{ji}^{(0)} x_i(1) + x_i(0) \quad . \; (2)$$

The weight-update is made by minimization on the error surface (1) using a steepest-descent method,

$$\Delta w_{ij}^{(i)} = -\eta \frac{\partial E}{\partial w_{ij}^{(i)}} \quad , \tag{3}$$

where η is the learning rate.

The first problem we have to deal with is, to extract the output $x^{(1)}$ from the input units, since they are implicitly defined. We simplify the problem by assuming that we have linear units in the input layer. Then (if there is no self-coupling for the units)

$$x_j^{(1)} = \lambda h_j^{(1)} = \lambda \left(x_j^{(0)} + \sum_{m \neq j} w_{jm}^{(0)} x_m^{(1)} \right) \tag{4}$$

or

$$x_j^{(0)} = \frac{1}{\lambda} x_j^{(1)} + \sum_{m \neq j} w_{jm}^{(0)} x_m^{(1)} = \sum_m \left(\frac{1}{\lambda} \delta_{jm} + w_{jm}^{(0)} \left(1 - \delta_{jm} \right) \right) x_m^{(1)}$$
$$= \left(\mathcal{M} \mathbf{x}^{(1)} \right)_j \quad , \tag{5}$$

where $\mathcal{M}$ is the transfer matrix with components

$$\mathcal{M}_{jm} = \frac{1}{\lambda} \delta_{jm} + w_{jm}^{(0)} \left(1 - \delta_{jm} \right) \; , \tag{6}$$

and $\mathbf{x}^{(0)}$ and $\mathbf{x}^{(1)}$ represent, respectively, the input and output vectors from the input layer. The output from the first layer is thus obtained by solving the system of linear equations

$$\mathbf{x}^{(1)} = \mathcal{M}^{-1} \mathbf{x}^{(0)} \quad . \tag{7}$$

For the matrix inversions we use a simple Gauss-Jordan method with full pivoting. It is certainly not the most efficient method, but is simple and easily implemented. A fully fledged version of the learning mechanism clearly needs to revise this point since much of the learning time is spent in performing the matrix inversions.

The learning mechanism starts by propagating the input values to the output layer. Once this is done, we start back-propagating the weight update:

Step 1: $w^{(2)}$ and $w^{(1)}$: This proceeds by normal back-propagation yielding

$$\Delta w_{ij}^{(2)} = -\eta \delta_i^{(3)} x_j^{(2)} \; , \tag{8}$$

$$\delta_i^{(3)} = \left(t_i - x_i^{(3)} \right) g' \left(h_i^{(2)} \right) \; , \tag{9}$$

$$\Delta w_{ij}^{(1)} = -\eta \delta_i^{(2)} x_j^{(1)} \; , \tag{10}$$

$$\delta^{(2)i} = g' \left(h_i^{(1)} \right) \delta_i^{(3)} w_{ki}^{(2)} \; , \tag{11}$$

where g is the activation function (e.g., a sigmoid function) and

$$g' \left(h_j^{(i)} \right) = \frac{\partial g(h_j^{(i)})}{\partial h_j^{(i)}} \; .$$

Step 2: $w^{(0)}$: We must calculate

$$\frac{\partial E}{\partial w_{ij}^{(0)}} = \delta^3 \sum_{m=1}^{n_H} w_m^{(2)} g'(h_m^{(1)}) \sum_{k=1}^{n_I} w_{mk}^{(1)} \frac{\partial x_k^{(1)}}{w_{ij}^{(0)}} \quad . \tag{12}$$

Assuming (Almeida 1987),(Almeida 1988),(Pineda 1987) that there exists a dynamical law for the system leading to fixed points with $\frac{dx^{(1)}}{dt} = 0$,

$$x_i^{(1)} = \lambda \left(\sum_j w_{ij}^{(0)} x_j^{(1)} + x_i^{(0)} \right) , \tag{13}$$

then

$$x_k^{(1)} = \lambda \left(\sum_{l=1}^{n_I} w_{kl}^{(0)} x_l^{(1)} + x_k^{(0)} \right) , \tag{14}$$

$$\frac{\partial x_k^{(1)}}{\partial w_{ij}^{(0)}} = \lambda \left[\delta_{ik} x_j^{(1)} + \sum_{l=1}^{n_I} n_I w_{kl}^{(0)} \frac{\partial x_l^{(1)}}{\partial w_{ij}^{(0)}} \right] \quad . \tag{15}$$

This implies

$$\lambda \delta_{ik} x_j^{(1)} = \sum_{l=1}^{n_I} \mathcal{L}_{kl} \frac{\partial x_l^{(1)}}{\partial w_{ij}^{(0)}} , \tag{16}$$

$$\mathcal{L}_{kl} = \delta_{kl} - \lambda w_{kl}^{(0)} \quad . \tag{17}$$

Thus

$$\frac{\partial x_l^{(1)}}{\partial w_{ij}^{(0)}} = \lambda \mathcal{L}_{ki}^{-1} x_j^{(1)} \tag{18}$$

and the updating rule for $w^{(0)}$ reads

$$\Delta w_{ij}^{(0)} = \eta \delta_i^{(1)} x_j^{(1)} , \tag{19}$$

$$\delta_i^{(1)} = \lambda \sum_{m=1}^{n_H} \delta_m^{(2)} \sum_{k=1}^{n_I} w_{mk}^{(1)} \mathcal{L}_{ki}^{-1} =$$

$$= \lambda \sum_{k=1}^{n_I} \mathcal{E}_k \mathcal{L}_{ki}^{-1} , \tag{20}$$

$$\mathcal{E}_k = \sum_{m=1}^{n_H} \delta_m^{(2)} w_{mk}^{(1)} \quad . \tag{21}$$

The learning process needs, apart from the matrix inversion needed to obtain the output from the input layer, the inversion of matrix $\mathcal{L}$ at each step. This can be avoided if one introduces an auxiliary network

(Almeida 1987),(Almeida 1988),(Pineda 1987). In this case, the auxiliary matrix variables have the dynamical law

$$\tau \frac{dy_i}{dt} = -y_i + \lambda \sum_{m=1}^{n_I} w_{mi}^{(0)} y_m + \mathcal{E}_i \qquad (22)$$

with

$$\delta_i^{(1)} = \lambda y_i \quad . \qquad (23)$$

Solving Eq. (4.2) allows to find $\delta_i^{(1)}$ without having to invert the matrix $\mathcal{L}$.

It is not clear at the moment which process is the best, since the matrices we have to invert are rather sparse because of the connectivity imposed to the net . This has to be set into relation with the number of steps needed to relax the auxiliary network to a fixed point. A decision upon this point has to be reached for the case of a calorimeter with many units. In our case the two methods are more or less equivalent.

4.3   Network Stability and Computational Implementation

Let us now consider that the units in the input layer are counted sequentially, i.e., for a cell with Cartesian coordinates $(x_1, x_2, x_3) \in \mathbb{N}^3$ we map the coordinates on a linear chain by a coordinate $n = x_1 + x_2 N_1 + x_3 N_1 N_2$. Here, N_i is the extension in the i direction, and we assume that the first cell in Cartesian coordinates is $(0, 0, 0)$. It is clear that nearest neighbours in the original Cartesian configuration *do not* correspond to nearest neighbours in the new one.

We may linearize the dynamical equation

$$\dot{x}_i^{(1)} = -\frac{1}{\tau} x_i^{(1)} + \frac{\lambda}{\tau} \left\{ x_i^{(0)} + \sum_{j \in neighbours} w_{ij}^{(0)} x_j^{(1)} \right\} \qquad (24)$$

around the fixed point 4.2. Introducing the linearized coordinates $x_i^{(1)} = x_i^{(1)^*} + z_i$, Eq. (24) can be written

$$\dot{\mathbf{z}} = \frac{\lambda}{\tau} \mathcal{M} \mathbf{z} \quad , \qquad (25)$$

with

$$\mathcal{M} = \begin{bmatrix} -1 & w_{01}^{(0)} & \cdots \\ w_{10}^{(0)} & -1 & w_{12}^{(0)} & \cdots \\ & w_{21}^{(0)} & -1 & w_{23}^{(0)} \\ \vdots & \vdots & \vdots & \vdots \\ & & \cdots & \cdots & w_{N_1 N_2 N_3, N_1 N_2 N_3 - 1}^{(0)} & -1 \end{bmatrix} \qquad (26)$$

Adopting periodic boundary conditions, $\mathcal{M}$ is replaced by

$$\mathcal{M}' = \begin{bmatrix} -1 & w^{(0)} & & \cdots & & w^{(0)} \\ w^{(0)} & -1 & w^{(0)} & \cdots & & \\ & w^{(0)} & -1 & w^{(0)} & & \\ \vdots & \vdots & \vdots & \vdots & & \\ w^{(0)} & \cdots & \cdots & w^{(0)}_{N_1 N_2 N_3, N_1 N_2 N_3 - 1} & & -1 \end{bmatrix} . \tag{27}$$

To simplify the discussion, let us assume that there is a connection between elements $n-1$, n, and $n+1$ and that these connections have the same weight $w^{(0)}$. The eigenvalues are easily obtained and are $(-1 + 2\lambda w^{(0)} \cos\theta)$. Stability exists only if $|2\lambda| < 1$. Another important result is obtained by inversion of matrix $\mathcal{M}$: the influence of a given unit n on the others connected to it extends *beyond* the units directly connected to it, and its influence goes as $(-\lambda w^{(0)})^n$. It is thus important to have $|\lambda w^{(0)}| < 1$.

A word on the computer implementation. The program for the mixed network is written in C++ and is fully designed in an object-oriented perspective. Most or all of the classes and functions used can be readily included in libraries, but some effort may still be applied to device a more efficient coding, both in terms of memory management and in execution time. This is particularly important if the algorithm is applied to the full calorimeter because of the number of cells involved and the subsequent time for the matrix inversions.[1]

5   Correction Factors by Event Classification

Let $\{E(\alpha)\}$ denote the set of energies deposited in the cells of the calorimeters with 3-dimensional coordinates $\{\vec{x}(\alpha)\}$. For each event we compute the center of mass

$$\vec{X} = \frac{\sum_\alpha E_\alpha \vec{x}(\alpha)}{\sum_\alpha E_\alpha}$$

and the center-of-mass coordinates

$$\vec{y}(\alpha) = \vec{x}(\alpha) - \vec{X} \quad .$$

The center-of-mass moment matrix is

$$M_{ij} = \frac{\sum_\alpha E_\alpha \vec{y}_i(\alpha) \vec{y}_j(\alpha)}{\sum_\alpha E_\alpha} \quad .$$

Diagonalizing this matrix we obtain, for each event, three eigenvalues λ_k, $k = 1, 2, 3$, which are the lengths of the principal axis of the normalized energy distribution. Unless the distribution is Gaussian the principal directions do not give a complete characterization of the shape of the event. For a better

---

[1] The source code can be obtained from joao.seixas at cern.ch

shape characterization in the non-Gaussian case, higher-order moments of the energy distribution should be used. Here, however, we use only the eigenvalues of the second-order moment. A generalization to higher-order moments is straight-forward.

Events of a similar physical nature do not necessarily have exactly the same shape. For example, the sequence of subprocesses might occur in a different order or, if the event occurs near the border of the calorimeter, part of the shower may be lost. Therefore, to recognize classes among events which may differ in their details, a self-organized associative memory is the appropriate device. For this purpose we use a competitive network. We use a single-layer network, the nodes receive as input the set of (three) eigenvalues λ_k , $k = 1, 2, 3$, ordered by increasing size. The output of unit i is

$$y_i = \theta_i f \left(\sum_{k=1}^{3} w_{ik} \lambda_k \right) \quad ,$$

f being a non-decreasing function. The nodes compete among themselves to be the one that is updated. The winner i^* has the largest output

$$\theta_{i^*} f \left(\sum_{k=1}^{3} w_{i^* k} \lambda_k \right) > \theta_j f \left(\sum_{k=1}^{3} w_{jk} \lambda_k \right) \quad , \text{ for } j \neq i^* \quad .$$

The learning (updating) laws for w_{ij} and θ_i are

$$w_{ij}(t + 1) = w_{ij}(t) + \eta_w \delta_{ii^*} (\lambda_j - w_{ij}) \ ,$$

$$\theta_i(t + 1) = \theta_i(t) + \varepsilon \eta_\theta \delta_{ii^*} y_i (1 - y_i) + r_i \ ,$$

where ε is either $+1$ or -1 and $r_i = 0$ if $\varepsilon = +1$.

The patterns λ_j that are presented to the network are drawn at random from the set of eigenvalues corresponding to the set of events that one wants to classify. On the average the same pattern will be presented to the network many times. The learning process eventually stabilizes because the learning intensities η_w and η_θ decrease slowly during the learning period.

This competitive network has two sets of learning parameters, $\{w_{ij}\}$ and $\{\theta_i\}$. The role of adjustable node parameters θ_i in networks and further details on unsupervised competitive learning is discussed in Dente and Mendes (1996). Here we only mention that the node parameters play an important role as modulators of the learning process. The dynamics of the node parameter keeps track of how often each unit is the winner. When $\varepsilon = +1$, if one starts from small θ's and random w_{ij} 's, as soon as a node wins for some pattern λ, it becomes more sensitive and it tends to win again for all patterns close to λ. It means that each winning node becomes assigned to a large neighbourhood in the pattern space. This implies a rough classification of pattern categories but, on the other hand, by avoiding that many nodes are assigned to the same cluster, it maintains many free nodes available to classify rare-event categories. For $\varepsilon = -1$ the effect is the opposite: if a node wins, it becomes less likely that it will win again for a different pattern. This situation is useful to resolve fine structures in a diffuse set of patterns.

For the case of calorimeter data, the hadronic showers are, in general, cigar-shaped. Therefore one of the eigenvalues in $\{\lambda_i\}$ tends to be substantially larger than the others. It is therefore convenient to use $\varepsilon = -1$ to be able to discriminate the fine structure associated with the transverse eigenvalues.

Once the events are separated into classes by the competitive network, the correction factor for the events of each class is obtained by using them to teach a feed-forward network, a different one for each class. After the learning stage with simulated or calibrated data, our correction apparatus is a three-layer device. The first layer determines the eigenvalues associated with the shape of the event. The second is the competitive network that assigns the event to its class, and finally the third is the set of feed-forward networks that outputs the corrected energy of the event.

This approach, close in spirit to modular nets, is still not fully implemented, but we hope to achieve it in the near future. We postpone the presentation of numerical results concerning this method to a forthcoming publication.

6 Numerical Results

As a first test of the recurrent-net algorithm, we checked the convergence. In this case we presented the same sample of 40 events of different energies several times and calculated the average error-function over the events in the sample for each presentation. The number of hidden units in this case is 10 and we use a calorimeter sample containing $3\times4\times5$ calorimeter cells. The results are presented as a function of the parameter λ (see Eq. (4)), the learning rate is fixed, $\eta = 0.9$. We could switch off the lateral connections in the input layer, thus getting a normal feed-forward backpropagation network for comparison. One sees that the backpropagation net performs already quite well (see Fig. 4), although the recurrent net for $\lambda = 0.5$ seems to converge faster and gives slightly more robust results (see Fig. 5).

We performed a more thorough test by using the parameters $\lambda = 0.6$ and $\eta = 0.9$ on a calorimeter sample containing $5\times4\times5$ cells and for a set of 6 event samples at 20, 50, 80, 100, 150 and 300 GeV. The network contains 50 units in the hidden layer and the learning proceeds by a random presentation of events of each energy sample. The results are presented and compared to the feed-forward net obtained by switching off the lateral connections in the input layer of the recurrent net. It is important to mention that the network performance test was done in this case with event samples *not used* during the learning phase.

In Fig. 6 we see the result of the test, that is the average error function divided by the expected value of the energy E_{true} as a function of E_{true}, after

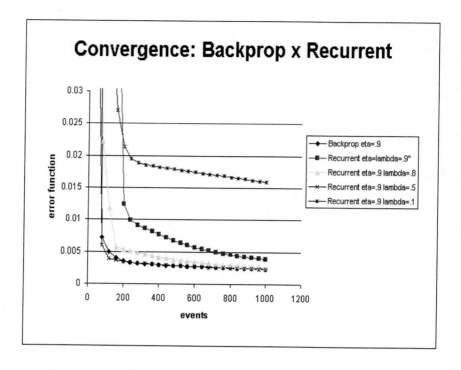

Fig. 4. Learning convergence for a sample of 40 events of various energies using simple and recurrent backpropagation algorithms. Several values of the linear unit's gain are used for comparison. The value of the error-function Eq. (1) averaged over the input sample is represented on the vertical axis.

the presentation of ∼1000 events. Both approaches perform poorly for the samples 20 GeV and 300 GeV and overall the normal backpropagation approach seems to yield better results. However, a more careful analysis shows that the poor results obtained for these energy samples are related to incorrect choices of the amplitude and bias of the sigmoid output functions for each unit. Correcting these values, one sees (Fig. 7) that both algorithms perform equally well and quite nicely (except for the 20 GeV sample), and that introducing the recurrent mechanism does not significantly improve the quality of the net results. Notice that we represent the average relative error for the net energy-value output E_{net}, $\Delta E/E = (E_{net} - E_{true})/E_{true}$, as a function of E_{true} for a simpler comparison. In all, the improvement we obtain by using the recurrent input layer amounts to less than 0.01%. We still do not completely understand why this is so. Further progress can only be achieved by a careful analysis of the event topologies, which we hope to do in the near future.

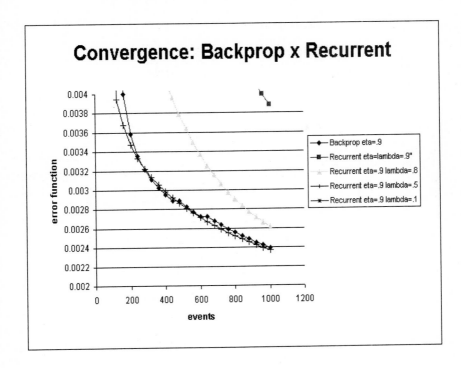

Fig. 5. Zoom of the previous figure showing the details of convergence and increased robustness of the recurrent algorithm.

The poor performance for the 20 GeV sample can be easily understood: in fact we rescale the input and output values to the [0,1] interval because of the output range of the sigmoid functions. This means that the contribution of the energy values of the 20 GeV sample is very small, the induced weight change is also very small and thus that the learning mechanism is very ineffective in this case. The problem can be only overcome, both for the simple and recurrent back-propagation learning schemes, by changing the form of the error function (1) (Hertz *et al.* 1991).

A word on the number of hidden units is necessary, since the question why we are using very few units may be puzzling. In fact we are quite concerned with the generalization capabilities of the network, since it will not perform at fixed energies, but rather in a continuous energy band. If we take a large number of hidden units, the network might fit the input/output relation quite closely and consequently would essentially miss events with energies not used during the learning phase. In fact we tested the network on events with an energy different from the learning samples and the result was very satisfactory. Nevertheless it must also be kept in mind that the strong bottleneck

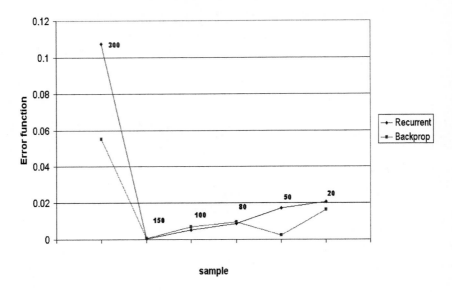

Fig. 6. Performance of a net with normal and recurrent backpropagation algorithms for data samples at 20, 50, 80, 100, 150 and 300 GeV. In this case the parameters of the sigmoid functions are incorrectly chosen.

created by such a small hidden layer might be the reason for the very small influence of recurrency in the network performance.

7   Conclusions

In all, it seems for the moment that the problem of correcting the energy deposited in hadronic calorimeters is quite well solved by conventional feedforward nets with backpropagation. The effect of introducing feedback among the units of the net changes the results very little and only increases the use of computer resources without significantly increasing the performance. The reason for this unexpected result is still not clear and might be related to either the event topologies or to the scarce number of hidden units used. In any case it is quite important to compare this result with the performance of the modular net described above and with more conventional algorithms to assess the validity of using a neural approach for this particular problem.

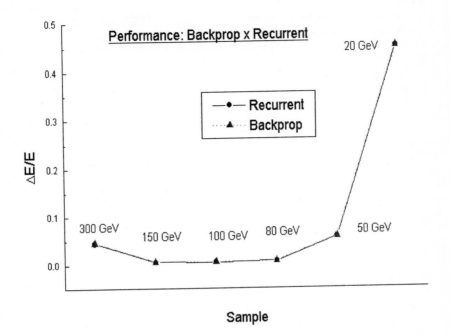

Fig. 7. Performance of a net with normal and recurrent backpropagation algorithms for data samples at 20, 50, 80, 100, 150 and 300 GeV. In this case the parameters of the sigmoid functions are correctly chosen.

Acknowledgements

This work was done in collaboration with R. V. Mendes and A. Amorim. The author would like to thank the Theory Division of CERN for its hospitality during the final writing of this paper.

References

Almeida, L. B. (1987): in "IEEE First International Conference on Neural Networks", Eds. M. Caudill and C. Butler, (IEEE, New York), pp. 609-618.
Almeida, L. B. (1988): in "Neural Computers", Eds. R. Eckmiller and Ch. von der Malsburg (Springer, Berlin), pp. 199-208.
Amaldi, U. (1981): Physica Scripta **23**, 409.
Brückmann, H., Behrens, U., and Anders B. (1986): Nucl. Instrum. Meth. A**263**, 136.

Dente, J. A. and Vilela Mendes, R. (1996): Network: Computation in Neural Systems **7**, 123.

Hertz *et al.* (1991): See, for example, J. Hertz, A. Krogh, and R.G. Palmer: "Introduction to the Theory of Neural Computation" (Addison Wesley, Redwood City, CA), for a complete set of methods and references.

Pineda, F. J. (1987): Phys. Rev. Lett. **59**, 2229; Neural Computation **1** (1989) 161.

Sauli, F. (1993): For a recent review, see: "Instrumentation in High Energy Physics", Ed. F. Sauli, Advanced Series on Directions in High Energy Physics, Vol. 9 (World Scientific, Singapore).

Neural Networks for Protein Structure Prediction

Henrik G. Bohr

Department of Physics, B. 307, DTU, The Technical University of Denmark, DK-2800, Lyngby, Denmark.

Abstract. This is a review of neural network applications in bioinformatics. In particular, the applications to protein structure prediction are discussed here. Examples of such applications are prediction of secondary structures, prediction of surface structure, fold class recognition and prediction of the 3-dimensional structure of protein backbones.

1   Introduction

A brief review of neural network applications in protein and genomic science is given before we discuss specific topics of protein structure prediction. This is in order to point out the many possible applications of neural networks in bioinformatics and especially concerning protein-structure analysis.

The first applications of neural networks in biotechnology was for secondary-structure prediction and appeared in 1988 (Qian and Sejnowski 1988; Bohr et al. 1988a). It was a natural extension into biology of Artificial Intelligence application (such as speech recognition) of neural networks and this methodology was immediately established as a leading-edge technique. Both the successes, the technicalities, and the shortcomings in all further applications are well illustrated in this application.

The next application of neural networks in biotechnology were aimed at splice-site prediction on DNA with database error detection as a spin-off (Brunak et al. 1990) and gave a high score. This spurred a number of successful applications in genomic research, such as to promotor recognition, DNA binding, gene finding etc.

Back to proteins, the next applications were concerned with the more ambitious goal of 3-dimensional, tertiary-structure predictions of proteins but with little success in accuracy (Bohr et al. 1988b). The same situation occurred for protein function prediction. Both applications were based solely on sequence information. More successful were the specific applications concerning ligand binding sites (Wade et al. 1992), fold-class recognition (Reczko et al. 1994), identification of membrane spanning regions, and prediction of the geometry of antibody binding regions (Reczko et al. 1995), to mention some of the more prominent applications of neural networks.

2   Prediction of Secondary Structures by Neural Networks

Historically, prediction of secondary structures in proteins was one of the first application of artificial neural networks in biotechnology. It turned out also to be the most popular one for using neural networks since the appearance of the first papers in 1988 (Qian and Sejnowski 1988; Bohr et al. 1988a) due to the straight-forward applicability and the clearly defined and limited scope of predicting what structure a given amino acid participates in, judged solely from its type.

The secondary structures are some of the most visible structural elements of the proteins which thus could explain the popularity of the endeavour. Furthermore the scope was clearly defined since well-known procedures of how to extract and classify them had been devised (Kabsch and Sander 1983). That procedure was based on hydrogen-bond assignments through a measure of distances calculated from the crystallographic data of atomic coordinates. Hence one could employ a program, the DSSP (Directory of Secondary Structure of Proteins) algorithm, that from a given set of coordinates for a protein, could calculate hydrogen-bond potentials and then assign secondary structures to that protein. Improvements of the network performance can actually be achieved by altering this fundamental procedure.

For most proteins the secondary structures occur only in a few varieties being helices, sheets, and turns, and most commonly only in three specific types, the alpha-helices, the beta-sheets, and the loops. A few percent of the helix occurrences are grouped into helices with another amount of residues per turn than the common α-helices of 3.6 periodicity. Among them are the 3_{10}-and the π-helices. In most prediction schemes for secondary structures all the helix types will be grouped into one class, sometimes denoted by the most frequently occuring α-helix. While the two first classes are well-defined by their hydrogen-bonding pattern the last group of turns is harder to define and will often, for convenience, be grouped together with random coil structure denoted as simply coil and being defined as any structure not belonging to the first two classes. Hence the task of predicting secondary structures in proteins boils down to predicting whether a residue in the protein participates in a structure being either a helix, sheet, or coil and these annotations are defined from the DSSP prescription.

2.1   Neural Networks for the Prediction Task

The type of neural networks most commonly employed for secondary structure prediction were feed-forward networks of the multi-layered perceptron type or more complicated recurrent neural networks equivalent to the networks used with real-time recurrent learning (RTRL) (Williams and Zipser 1989). The former networks have a unique direction of the data stream, such that input is passed through the consecutive layers towards a specific layer of neurons that produces the output while the latter networks have a set of extra feed-back connections. We shall hereafter denote these layers of neurons as, in consecutive order, the input layer, the hidden layers, and the output layer. The reason for choosing this network among many other types is due to its renown ability to generalize speech recognition, image processing and molecular biology data (Qian and Sejnowski 1988; Bohr et al. 1988a; Holley and Karplus 1989; Bohr et al. 1988b) (to mention just a few successes) and its rather simple structure both with respect to processing of data and training via the back-propagation error algorithm (Rumelhart and McClelland) most commonly used. The training procedure is performed during a fixed set of steps until a cost function CF has reached a low value, e.g., by a gradient descent. The cost function CF is normally written as

$$CF = \frac{1}{2} \sum_{\alpha,i} (t_i^\alpha - z_i^\alpha)^2 \qquad (1)$$

that is simply the squared sum of errors, t_i being the correct target value and z_i the actual value of the output neurons.

In each instance of training, a vector of input values (e.g., letters in a text string), the size of the vector (window) representing the correlation among the inputs in time, is to be related to a vector of output values of classes (e.g., secondary-structure classes) corresponding to the value (in the middle of the window) in the input vector. The network is carried out on several types of network architectures, one being for example 60×20 (60 is the window size) input elements, 40 hidden neurons, and 3 output neurons corresponding to the three secondary-structure categories.

In order to evaluate the performance of the network various statistical measures have been proposed. In the case of a dual-valued output one often uses the so-called Mathews coefficient (Mathews 1975). If we denote the two possible output values by 1 and 0 (signifying occurence or absence) and if p is the number of correctly predicted examples of 1s, $\bar{p}$ the number of correctly predicted examples of 0s, q the number of examples of 1s incorrectly predicted and $\bar{q}$ is the number of examples of 0s incorrectly predicted we define the coefficient C as:

$$C = \frac{p\bar{p} - q\bar{q}}{\sqrt{(p+q)(p+\bar{q})(\bar{p}+q)(\bar{p}+\bar{q})}} \; . \qquad (2)$$

For completely correct decisions (ideal performance) the measure is 1 and for completely wrong decisions C is -1. A poor net will give $C = 0$ indicating that it does not capture any correlation in the training set in spite of the fact that it might be able to predict several correct values.

2.2 Implementation and Results for Secondary Structure Prediction

In the first attempts to predict secondary structures by neural networks the bare sequence data of amino acids were represented in the input layer of a multi-layered feed-forward neural network (Bohr et al. 1988a) while the output of secondary structures, classified in three categories, either as helices, sheets, or coils, were represented by three output neurons. 20 neurons were used to represent each type of amino acid in a binary coding such that, for example, alanine was represented by 10000000000000000000 and arginine by 01000000000000000000. This binary coding ensured that there was no initial bias in the representation of the amino acids since the Hamming distance between all of them was equal.

The network would first be trained on roughly 56 crystallographic protein structures from PDB chosen to have high structural resolution. Through each training cycle the synaptic weights and thresholds would be adjusted to give the lowest error in the secondary-structure assignments of the 56 proteins compared to the true assignments taken from the DSSP prescription presented to the network. After a series of training cycles the network would be tested to predict secondary structures from sequences of proteins with a known structure but novel to the network. Chris Sander and Reinhard Schneider have developed a procedure (Sander and Schneider 1991) for constructing databases of known three-dimensional structures of proteins with a sequence similarity that guarantee low structural homology provided that the similarity is below a certain threshold. This is what is needed for construction of a training set that has maximal structural variability and provides the basis for a neural network with possibly the best capacity for predicting new structures by interpolation. The key point of the construction of such a database is the observation that there is a threshold of sequence similarity above which structural similarity can be inferred. This threshold of sequence similarity can be found by making an alignment study of all available sequences corresponding to known three-dimensional structures. The threshold sufficient for structural similarity is (for small length) strongly dependent on the length of the alignment.

Before the introduction of the neural network methodology the traditional statistical methods, of which the Chou and Fasman (1974) method is the most well-known, were able to predict secondary structures correctly up to about 55%. The neural network methods gave a considerable improvement to that, improving the correctness score by at least 10%. This is not surprising since the neural network methods operated within a wider context, i.e., were based on a longer range of interaction in the sequence space. Sometimes windows even as large as 51 residues were employed for the prediction. In the most successful predictions of Rost et al. (1993) the percentage of correctly predicted residues went as high as 73 for a standard protein set, with a Mathews coefficient above 0.5.

3   Surface Prediction

A very useful application of neural networks is the prediction of surface positions on proteins. Such networks are trained to tell from sequence information whether a given residue of a sequence is positioned on the surface or is in the interior of that protein. This information is of course very much related to the problem of where the solvent and, in general, ligands bind on a protein.

In one of these specific applications (Bohr et al. 1992) feed-forward neural networks of the perceptron-type with hidden layers of neurons have been employed to predict the surface structures of proteins. After being trained on known surface structures the network can predict local surface properties for new proteins on the basis of their sequence. The coding is done by assigning a number to each residue in the sequence signifying whether that residue is deeply buried in or positioned on the surface of the protein. Such a network was up to 70% correct in predicting surface structures of proteins, novel to the trained network.

What concerns ligand binding prediction and, especially, water binding prediction is that it has for a long time been an important goal to predict ligand binding sites on proteins since that knowledge gives important information on the surrounding structures of the 3-dimensional folded protein and its functionality. Detailed ligand binding data are quite sparse, except when the ligand is a water molecule, and therefore prediction of water binding sites was the main issue in many studies. There is a great need in biotechnological research to locate water sites fast, accurately, and efficiently, for example for molecular-dynamics calculations, for drug design, and for crystallographic, and NMR studies of protein structure.

Methods for such predictions have been based on tertiary-structure information and hence are only applicable to a limited number of proteins. Therefore, it is of great use if a method for predicting binding sites on proteins can be applied solely on the basis of the protein-sequence data since a much larger class of proteins with unknown structure can be treated.

The neural networks trained on the available data bases of ligand binding site information could reasonably well predict binding sites for novel proteins. The neural networks were using protein sequences as input and producing binding-site locations as output (Wade et al. 1992). In most of these surface predictions the score was not much above 70 % correctness (percentage of residues correctly predicted to be outside or not).

4   Neural Network Prediction of Protein Fold Classes

The next application of neural networks is one we shall dwell more on since it is quite unique in performance and choice. The neural networks are used for predicting what structural family (or actually fold class) a certain protein sequence belongs to. The methodology involves a special type of neural networks, the cascade-correlation network, that achieved a remarkably high performance for this purpose. Using a hierarchical scheme of fold classification, a cascade-correlation network was trained to construct features that characterize the membership of a fold class. At the highest level, a 4-class scheme was used and the network performed with a high accuracy of about 90%. In the case of fold classes defined in the so-called 3D-ALI (3-dimensional alignment) scheme in the presence of similar substructures or a certain percentage (30% - 60%) of sequence identity, the network determined the correct fold class for a set of about 100 novel proteins (out of a total of 42 classes) to an accuracy of 81.6%. Even for those test proteins with a sequence identity of less than 25%, compared to the training set of proteins the prediction accuracy is well above 70%. Such a scheme is very useful for assessing protein structural topology from sequence information alone and serves as a basis for further detailed homology modeling. On the basis of this prediction scheme a data base has been constructed for fold class and domain predictions of all available sequences in the comprehensive SWISS-PROD data base for protein sequences.

4.1   Introduction to Protein Fold Classes

In most definitions of fold classes, each member would have more than 50% sequence identity to each other although domains with far less sequence similarity could belong to the same class. It is important that each protein within a class would have a structure with a large topological similarity and a similar packing pattern to other members of the class.

The notion of fold classes is important for predicting new protein structures using homology modeling. In homology modeling an unknown 3-dimensional protein structure is inferred from other known 3-dimensional protein structures whose amino acid sequences are similar to the sequence of the protein in question. It has been shown (Goldstein et al. 1992; Reczko and Bohr 1994a) that one can predict or model protein structures to high accuracy by using structural information from proteins belonging to the same fold class or family. Several techniques have been developed for inferring homology at the structural level from fold class membership. Some of these incorporate a combination of secondary-structure prediction schemes, functional similarity, recognition of key structural motifs, and use of machine-learning methods for sequence/structure mapping (Reczko et al. 1994; Holm and Sander 1996; Sippl 1990; Jones et al. 1992). The proposed scheme, that consists of two steps, rests on the result that neural networks can be effectively trained to deduce features from a system that characterizes it. In a first step, a feed-forward neural network is used to determine the fold class of a protein from its sequence data. In a second step, the predicted fold class with its characteristic domains is used as input into a large recurrent neural network to predict the distance matrix for the protein (Reczko and Bohr 1994a). Such a distance-matrix prediction should be accurate enough for constructing the

3-dimensional backbone structure for the protein, which can then be subsequently refined by side-chain placement and molecular mechanics methods.

In the following section 4.2 the neural network methodology of cascade-correlation type for predicting the fold class of a protein will be discussed. In subsequent sections some results from neural network studies are presented. A hierarchy of fold classifications is used in our scheme and this is shown to yield the best prediction of fold classes.

4.2   Methodology of the Cascade-Correlator

Large success in the present application is obtained with a training and construction procedure, called Cascade-Correlation (Fahlman and Lebiere 1990). This algorithm optimizes both the weights in a feed-forward network and the number of hidden units by adding units during the training process (see Fig. 1). The initial network contains only input and output units and is first trained using the normal delta-rule that is a special case of the back-propagation algorithm without hidden units. Thus, the first phase of the training leads to the same solution that would be obtained by a perceptron and maps only those input patterns that may be linearly separated onto different output patterns. This linear part of the mapping may cover already a lot of input/output pattern pairs in the training set. To further reduce the error, one hidden unit, that is initially not connected, is added to the output layer. The weights leading into this unit are adapted by maximizing the correlation between the activity of this unit and the residual error occurring at each output unit. After this adaptation, all weights into this unit are frozen and the new hidden unit is connected to the output layer with the new weights set to 0. All weights connected to the output units are trained again to minimize the error function. The process of adding new hidden units that maximizes the correlation between their activity and the remaining error at the output layer is repeated until the mapping has the desired accuracy. Since each new hidden unit is also connected to all existing hidden units, the network contains as many hidden layers as hidden units.

4.3   Representation of the Data

The actual neural networks for predicting fold classes are constructed from the SNNS (Stuttgart Neural Network Simulator) environment (Zell et al. 1991) and are of the feed-forward type. The networks are trained on a selection of proteins from each of 42 fold classes containing domain segments of proteins or often the whole proteins. The input representation for each protein domain is a 20×20 matrix containing the relative frequencies of dipeptides occurring in neighbouring positions in the primary sequence of the domain. To calculate these frequencies, the number of occurrences of a dipeptide is counted in the protein sequence and divided by the total number of residues in that sequence. All protein domains are transformed in this way

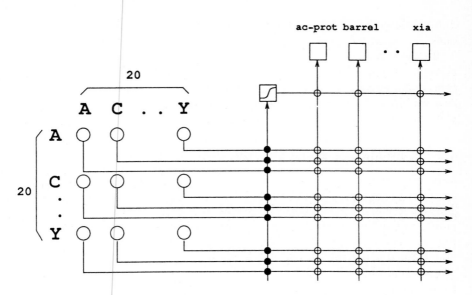

Fig. 1. The Cascade-Correlation architecture for predicting fold classes from dipeptide frequencies. On the left hand side of the figure is pictured the input matrix of 20×20 dipeptide frequencies while the right hand side top of the figure shows the output represented by two standard fold classes, the ac-proteases and the barrel fold class.

into one input pattern of fixed size. Small insertions and deletions from the protein sequence cause only small changes in the dipeptide frequencies, unlike the case of the traditional coding used for the previous applications. The same holds true for rearrangements of larger elements in the sequence that do not change the local sequences. There are many cases where members of the same fold class differ mostly by permutations of sequence elements. Such permutations of the primary sequence lead to very similar dipeptide matrices which supports similar classification results. Each fold class is represented by one output unit which should have an activation close to 1.0 if the domain coded in the input layer is a member of that fold class. In all other cases the activity should be close to 0. If an unknown sequence is classified, the fold class corresponding to the largest activation at the output unit is assigned to the sequence. This is the usual "winner-takes-all" evaluation of the output of a classifier. In order to facilitate the interpretation of misclassifications all the fold classes were grouped into larger super-fold classes that have a natural one-dimensional order inferred from physical properties of the folds. The super-fold class prediction and the fine-grained classification should then assign classes that are close in this order.

4.4   Fold Class Prediction Results

The main results in this chapter give the prediction of fold classes from sequence alone. The training set and the testing set are both constructed from
the data set of the 42 classes of domains, listed in Table 1. Roughly half of
each fold class domains are used for training. The rationale for choosing 42
classes from the Pascarella and Argos definition of folds, was to make certain
that there are enough members in each class in order to perform a valid test.
The fold class predictions are performed at three different levels of detail.
The first classification uses the four super-fold classes based entirely on the
secondary-structure composition and arrangement in the proteins. The classifications are based on proteins containing the following secondary structures
(notation in brackets): only α-helices (α), only β-sheets (β), alpha and beta
domains separated ($\alpha + \beta$) and one containing a combination of alpha and
beta secondary-structure elements intertwined ($\alpha + \beta$). In the second scheme,
the full set of 42 classes is used for a fine-grained classification.

For the first case of four super-fold classes a network trained up to 97.2%
accuracy had a test score of 90.4% with an average Mathews coefficient of
0.81. This is a very high performance compared to other secondary-structure
content predictors usually having a Mathews coefficient of around 0.4. The
third case, based on much better distributed classification, yields a remarkable performance of 100% on the training set and with a test score of 78% in
predicting a fold class correctly on the basis of the sequence. Furthermore,
adding the output of the four super-fold classes network to the input of the
42 class based network enhanced its performance to 81.6% on the test with
an average Mathews coefficient of 0.7. The results are presented in the *permutation matrix* of Table 1 where the number in row i and column j counts
all cases where a test protein that is predicted to be in class j in fact belongs
to class i. Optimally, all test cases should be counted on the main diagonal of
the permutation matrix. For the case of the 42 fold class prediction, the relation between maximal sequence identity of a test sequence to the sequences
used in the training set and prediction accuracy is given in Fig. 2. The four
points at 25%, 50%, 75% and 100% sequence identity defining the solid line
give the average prediction accuracy for those test cases that have a maximal
sequence similarity between 0% and 25% , 25% and 50%, 50% and 75%, 75%
and 100% to the training set. The fold class prediction is still more than
71% correct for those test sequences with 0% to 25% sequence identity to the
training set.

4.5   Fold Class Database

As a spin-off of the rather successful fold class predictions a database has
been constructed for public domain usage (Reczko and Bohr 1994b). The
DEF (Database for Expected Fold-classes) is made for protein fold-class predictions from sequences in the SWISS-PROT protein-sequence database and

Table 1. The permutation matrix for the test predictions of the 42 fold classes if the output of the secondary structure content and arrangement predictor is used as additional input.

Permutation matrix for 42 fold classes using predictions from secondary structure content network as additional input

predicted foldclass

true foldclass		α (01–11)	α/β (12–17)	$\alpha+\beta$ (18–27)	β (28–42)	C_M
1	helix.bndl					0.000
2	cyte					0.718
3	hmr					1.000
4	wrp					0.000
5	globin					0.787
6	lzm					0.704
7	cyp					0.397
8	ca.bind					0.701
9	tln					-0.008
10	cts					1.000
11	pep					0.000
12	crn					1.000
13	cpp					0.704
14	wga					1.000
15	ma					0.704
16	plipase					1.000
17	gap					0.000
18	inhibit					1.000
19	xia					1.000
20	kinase					0.704
21	binding					0.813
22	barrel					0.501
23	eglin					1.000
24	pgk					0.000
25	dfr					1.000
26	abl					1.000
27	s.prot					0.874
28	cpa					1.000
29	ferredox					0.704
30	fxc					0.704
31	pti					1.000
32	rdx					0.866
33	virus					1.000
34	virus.prot					1.000
35	gcr					0.862
36	igb					0.870
37	il					0.658
38	ac.prot					0.922
39	tox					1.000
40	plasto					1.000
41	lin					0.813
42	hoe					0.573

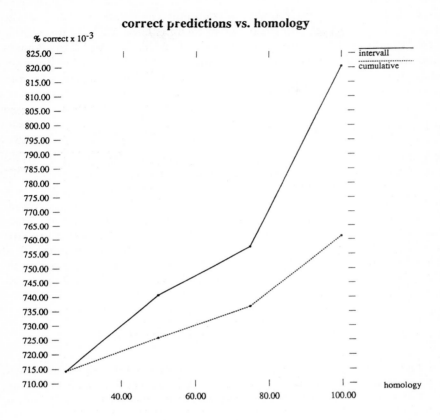

Fig. 2. The test prediction accuracy vs. homology with the training set.

is used for making predictions of fold-classes for any new sequence. In the DEF database a sequence of amino acids is assigned a specific overall fold-class, a super fold-class with respect to secondary structure content and a profile of possible fold-classes along the sequence. The assignment of a fold-class is one out of 45 well-known folds derived from the 3-dimensional protein structures in the Brookhaven Protein Data Bank, PDB. Most of these 45 fold-classes are contained in the set given by Pascarella and Argos (Pascarella and Argos 1992) and are roughly in accordance with similar selections of folds (Holm and Sander 1996; Jones et al. 1992). Apart from the fold-classes contained in the previous studies, some extra fold-classes, for example folds found in membrane-bound structures, were added in order to cover a wider range of structures. The 45 protein fold-classes are given in Table 2 which shows the format of the DEF database. The table actually represents a fold class prediction of Papain. One can also make use of the fold-class predictor to produce a profile of fold predictions if a window containing 100 residues scans over the sequence and produces an assessment in each instance. Such a procedure is interesting in the (most often occuring) case if one has no idea of any domain boundaries in the protein structure or whether it is an oligomer. However, with such a profile one has a chance of detecting boundaries between two structural domains as is clearly seen in Table 2.

4.6   Conclusions of the Fold Class Prediction Problem

In summary, we have used a hierarchical fold-classification scheme for training neural networks in two stages for protein fold class predictions. While this approach does not provide a complete solution to the protein structure prediction problem, it enables us to obtain a topological description of a protein whose structure is unknown. The prediction method has reached the performance with a score of correctness of about 82% even for the cases of proteins with a low sequence similarity to the training set. The present method is shaped to be added to the larger scheme of prediction of distance-matrices at a second stage and provides also an initialization to different homology-based approaches.

5   Prediction of the 3-D Structure of Proteins

During the last years there has been much progress in predicting 3-dimensional protein structures from the sequence although one is still far from achieving accurate tertiary structures of proteins with little homology to already known proteins. It is important to stress the last statement of little homology or, more precisely, sequence similarity since the whole field of structure construction from homology-modeling is a highly used technology that is based on some sequence similarity to sequences of known protein structures. One usually talks about "the twilight" zone (around 25% sequence similarity) above which homology modeling is quite successful. Therefore, other methodologies, such as that of neural networks, must be able to extend the tertiary-structure prediction of proteins below the twilight zone with respect to sequence similarity.

In the following we shall finally describe a neural network scheme for prediction of protein backbone structures in the distance-matrix approach. We concentrate here on the backbone structure or actually the C-α atomic positions although the full tertiary structure of proteins also involves the side-chain atomic coordinates. Their prediction is dealt with in a similar study (Bohr et al. 1993a). The prediction of the three-dimensional structure of protein backbones in this distance-based approach consists of two elements (1) prediction of the distance matrix of C-α atomic distances, (2) subsequent

Table 2. An example of an output from the DEF mail-server and the database. It shows a successful prediction of Papain (9pap). Below, a graphical representation of the profile is given where a window containing 100 residues glides over the sequence providing a prediction in each instance.

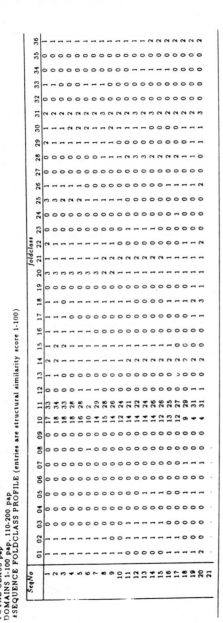

generation of the three-dimensional backbone structure from the distance matrix (Bohr et al. 1993b).

5.1 Prediction of Distance Matrices from Fold Classification

As mentioned above a general prediction of the 3-dimensional structure of a novel protein on the basis of its sequence of amino acids is only likely to be successful by computer techniques and especially neural networks, if the fold class to which the protein belongs can be determined first. This could, for example, be done by employing the fold-class predicting network mentioned in the previous chapter. If that is successfully done a subsequent determination of the 3-dimensional structure of the protein can be obtained in some cases through a prediction of the distance-matrix that represents the 3-dimensional backbone structure. The distance-matrix prediction can be carried out by a neural network trained on the protein folds from the same fold-class.

The neural networks for distance-matrix prediction have been described in detail elsewhere (Bohr et al. 1988b; Reczko et al. 1994) but one promising network was recurrent. Sequences are read into a symmetrical window of the size of about 40 residues and the correlated output would be a binary (or real-valued) distance between the middle residue and the 20 residues to the left. The C_α atomic position was used as coordinate for each residue. Thus having an input window scanning over the sequence of a protein domain makes the network produce a band of distance-values along the diagonal of a distance matrix. The band width is determined by the "horizon" of the correlation of the outputs. It is possible to have a larger output "window" than an input window in a recurrent network due to time delay. These distance-matrix predicting networks are trained on known proteins or protein domains from the fold class that had been determined from the previous network. The training set corresponding to each fold class will on average contain 10 proteins or protein segments each about 100 residues long.

We took two cases of proteins, the globular protein, Myoglobin 1MBD and Rubredoxin 1RDG predicted by a network to belong to the correct fold class and then trained networks on these fold classes, Globulin and RNX, respectively, to produce the corresponding distance matrices shown in Fig. 3 (RDX) where the predicted ones are compared to the real matrices derived from the crystallographical PDB data. Finally, these predicted distance-matrices were used in a minimization procedure (Bohr et al. 1993b) to construct the 3-dimensional backbone structure for the RDX proteins that is presented in Fig. 4. We obtained an accuracy of about 3.9 Å in RMS in the case of 1RDG that was 64% homologous to the closest related protein in the training set.

In the case of the globin class the distance matrix could be predicted about 90% correctly while in a test case with immunoglobulines the score was considerably lower, perhaps because only the variable chains (fab) were used. The resulting minimization procedure gave a 3-dimensional backbone

Fig. 3. Binary band of 30 residues along the diagonal of the native (left) and predicted (right) distance matrices of Rubredoxin (1RDG). A "1" indicates a C_α distance smaller than 16. The protein sequence is imagined to be represented along each axis and the distance matrix is rotated 45 degrees.

structure for myoglobin, that like rubredoxin was about 4 Å in RMS compared to the native structures while in the case of the immunoglobulin the RMS was larger than 8 Å. In the case of rubredoxin side-chains could be fairly correctly positioned resulting in an RMS of about 5 Å. A short molecular-dynamics test with CHARMM could verify that the full structure would be energetically stable (i.e., kept together) but without adding a solvent the structure would end up after a temperature cycle in a conformation with a higher RMS value.

6   Conclusion on the 3-Dimensional Structure Prediction

With the help of fold class prediction it is possible to improve distance-matrix prediction and hence generate the 3-dimensional structure of proteins

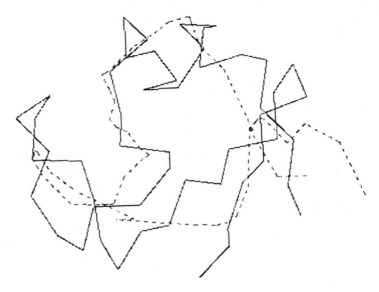

Fig. 4. Picture of the minimized structure of 1RDG on the basis of the best (30 residues long) band-matrix predicted by the network. The native backbone structure is superimposed on the predicted structure (dashed line).

to higher accuracy by employing a tiered, modular neural network system that can produce distances from sequence input. Research on such a scheme is on-going but the major problem is still that of predicting distance matrices from any sequence since it requires a certain amount of sequence similarity among the protein members of a fold class (Reczko et al. 1994). Some fold classes have that required similarity whereas others have not, such as the class of neuro-toxins where the secondary structures have are fairly regular throughout the class, except for the loop regions, and all the members have a very low sequence similarity to each other.

It is always a difficult task to use many cooperative neural networks if each net does not give perfect predictions. On the other hand, experiments with various modular systems of networks did prove that networks, such as the ones for predicting distance matrices, can be improved if the prediction task can be subdivided into several limited subtasks, i.e., first predicting, for example, what superclass a protein is a member of and then go for finer classifications. This was very clearly demonstrated in the case of fold class prediction (Reczko et al. 1994) where an intermediate step of secondary-structure content prediction helped the overall performance when it was included in the input of the larger net. Such intermediate networks could be considered as kinds of extra hidden layers of neurons but by dividing the task into separate networks the number of free parameters (the synaptic weights and thresholds) over the number of constraints is diminished.

References

Bohr, H., Bohr, J., Brunak, S., Cotterill, R. M. J., Lautrup, B., Nørskov, L., Olsen, O. H., Petersen, S. B. (1988a): Protein Secondary Structure and Homology by Neural Networks. FEBS Lett. **241**, 223–228.

Bohr, H., Bohr, J., Brunak, S., Cotterill, R. M. J., Lautrup, B., Fredholm, H., Petersen, S. B. (1988b): A novel Approach to Prediction of the 3-dimensional Structure of Protein Backbones by Neural Networks. FEBS Lett. **261**, 43–46.

Bohr, H., Goldstein, R. A., Wolynes, P. G. (1992): Predicting Surface Structures of proteins by Neural Networks. AMSE Periodicals, Modelling, Measurements and Control C **31**, 53–58.

Bohr, H., Irwin, J., Mochizuki, K., Wolynes, P. G. (1993a): Classification and Prediction of Protein Side-chains by Neural Network Techniques. Int. Journal of Neural Systems, (Supplementary Issue), 177–182.

Bohr, J., Bohr, H., Brunak, S. Cotterill, R. M. J., Lautrup, B., Fredholm, H., Petersen, S. B. (1993b): Protein Structure from Distance Inequalities. J. Mol. Biol. **231**, 861–869.

Brunak, S., Engelbrecht, J., Knudsen, S. (1990): Cleaning up Gene Databases. Nature **343**, 123–124.

Chothia, C. (1992): One Thousand Families for the Molecular Biologist. Nature **357**, 543–544.

Chou, P. Y., Fasman, G. D. (1974): Prediction of the Secondary Structure of Proteins from their Amino Acid Sequence. Advan. Enzymol. **47**, 45–148.

Crippen, G. M. (1991): Prediction of Protein Folding from Amino Acid Sequence over discrete Conformational Space. Biochemistry **30**, 4232–4237.

Fahlman, S. E., Lebiere, C. (1990): *Advances in Neural Information Processing systems II*. Ed. D.S. Touretzky, (Morgan Kaufmann, Los Altos, CA) 524–532.

Goldstein, R. A., Luthey-Schulten, Z., Wolynes, P. G. (1992): Protein tertiary Structure Recognition using optimized Hamiltonians with local Interaction. PNAS USA **89**, 9029–9033.

Holley, H. L., Karplus, M. (1989): Protein Secondary Structure Prediction with Neural Networks. Proc. Natl. Acad. Sci. USA **86**, 152–156.

Holm, L., Sander, C. (1996): Mapping the Protein Universe. Science **273**, 595–602.

Jones, J. T., Taylor, W. R., Thornton, J. M. (1992): A new Approach to Protein Fold Recognition. Nature **358**, 86–89.

Kabsch, W., Sander, C. (1983): Dictionary of Protein Secondary Structure: Pattern Recognition of Hydrogen-Bonded and Geometrical Features. Biopolymers **22**, 2577–2637.

Lesk, A. (1991): *Protein Architecture* (Oxford Press, Oxford).

Mathews, B. W. (1975): Comparison of the predicted and observed Secondary Structure of T4 Phage Lusozyme. Biochem. Biophys. Acta **405**, 442–451.

Pascarella, S., Argos, P. (1992): A Data Bank merging related Protein Structures and Sequences. Protein Engineering **5**, 121–137.

Qian, N., Sejnowski, T. J. (1988): Predicting the Secondary Structure of globular Proteins using Neural Networks Models. J. Mol. Biol. **202**, 865–884.

Reczko, M., Bohr, H., Subramaniam, S., Pamidighantam, S. V., Hatzigeorgiou, A. (1994): *Protein Structure by Distance Analysis* Ed. H. Bohr and S. Brunak, IOS Press, Amsterdam), 277–286.

Reczko, M., Bohr, H. (1994a): *Protein Structure by Distance Analysis* Eds. H. Bohr and S. Brunak (IOS Press, Amsterdam), 87–97.

Reczko, M., Bohr, H. (1994b): The DEF data base of sequence based Protein Fold Class Predictions. Nucleic Acid Research **22**, 3616–3619.

Reczko, M., Martin, A. C. R., Bohr, H., Suhai, S. (1995): Prediction of hypervariable CDR-H3 Loop Structure in Antibodies. Protein Engineering **8**, 389–395.

Rost, B., Sander, C. (1993): Prediction of Protein Secondary Structure at better than 70 % accuracy. J. Mol. Biol. **232**, 584–599.

Rumelhart, D. E., McClelland, J. L. and the PDP Research Group (1986): *Parallel Distributed Processing: Explorations in the Microstructure of Cognition*, Vols. 1 and 2 (MIT Press, Cambridge, MA).

Sander, C., Schneider, R. (1991): Database of Homology derived Protein Structures and the structural Meaning of Sequence Alignment. Proteins: Str., Func. and Gen. **9**, 56–64.

Sippl, M. J. (1990): Calculation of Conformational Ensembles from Potentials of Mean Force. J. Mol. Biol. **213**, 859–883.

Wade, R., Bohr, H., Wolynes, P. G. (1992): Prediction of Water Binding Sites on Proteins by Neural Networks. JACS **114**, 8284–8286.

Williams, R. J., Zipser, D. (1989): Neural Computation, **1**, 270–280.

Zell, A., Mache, N., Sommer, T., Korb, T. (1991): *Applications of Neural Networks Conf.*. SPIE, Aerospace Sensing Intl. Symposium, (Orlando Florida), 708–719.

Evolution Teaches Neural Networks to Predict Protein Structure

Burkhard Rost

LION Bioscience AG, Im Neuenheimer Feld 517, 69120 Heidelberg, Germany,
Columbia, Dep. of Biochem. and Molecular Biophys., 630 West 168th Str, New
York, N.Y. 10032, USA,
EMBL, 69 012 Heidelberg, Germany; rost@embl-heidelberg.de.
http://www.embl-heidelberg.de/~rost/

Abstract. In the wake of the genome data flow, we need – more urgently than
ever – accurate tools to predict protein structure. The problem of predicting protein
structure from sequence remains fundamentally unsolved despite more than three
decades of intensive research effort. However, the wealth of evolutionary information
deposited in current databases enabled a significant improvement for methods predicting protein structure in 1D: secondary structure, transmembrane helices, and
solvent accessibility. In particular, the combination of evolutionary information with
neural networks has proved extremely successful. The new generation of prediction
methods proved to be accurate and reliable enough to be useful in genome analysis, and in experimental structure determination. Moreover, the new generation
of theoretical methods is increasingly influencing experiments in molecular biology.
Neural networks have been applied to many pattern classification problems. Here,
I review applications to the problem of predicting protein structure from protein
sequence. Initially, methods were designed as a 'quick and dirty' demonstration that
artificial intelligence-based machines could solve real-life problems. At that stage,
biologists typically reached higher levels of accuracy when using their expertise than
computer scientists when using their machines. However, more thorough investigations introduced the information used by experts into neural network-based tools.
Now, some tools are – on average – as accurate as the best experts, and experts
using such tools often arrive at even more accurate predictions. Thus, several neural
network-based methods have eventually contributed significantly to advancing the
field of bio-informatics, and some are clearly influencing molecular biology.
Key words: protein structure prediction, evolution, neural networks.

1   Introduction

Proteins constitute life's machinery. The first bacterial genome was sequenced
in 1995 [1]; the first eukaryote (yeast) followed in 1996 [2]. Meanwhile, more
than ten other genomes have been published [3], and the human genome
(200 times larger than yeast) is expected to be sequenced as one of the first
milestones in the next millennium. Why bother? Because genomes contain
the blueprint for all parts of life's machinery. The machinery itself consists
of proteins that perform all important tasks in organisms (catalysis of bio-
chemical reactions, transport of nutrients, recognition, and transmission of

signals). Proteins are formed by joining 20 different amino acids (dubbed residues, when joined in proteins) into a stretched chain. In water, the chain folds into a unique three-dimensional (3D) structure (Fig. 1; introduction to protein structure: [4]).

Fig. 1. Representation of the 3D structure of Leishmania mexicana triose phosphate isomerase (TIM, Protein Data Bank [58] code 1amk). The trace of the protein chain in 3D is plotted schematically as a ribbon. Strands are indicated by arrows, helices by open coiled-tubes. Graph made with RASMOL (Roger Sayle, ras@32425@ggr.co.uk), and Molscript (Peer Kraulis, http://www.avatar.se/molscript, [60]). The TIM-barrel is named after the barrel formed by the strands in the centre of the molecule. The enzyme is found with a similar structure in most of the known life-forms, and thus represents a billion-year's old perspective at the complexity of the shapes of life.

Sequence determines structure determines function. The world of proteins is governed by shape: interactions between proteins are mediated by the 'key-hole' principle, i.e., two proteins interact when they fit to one another like a key into a hole. Thus, protein structure determines protein function. What determines structure? All information about the native structure of a protein is coded in the amino acid sequence, plus its native solution environment [5]. Can we decipher the code, i.e., can we predict 3D structure from sequence? In principle, we could; in practice, such approaches are frustrated by the difficulty of the task resulting from the high complexity of protein structure formation [6]. For over 40 years, there has been an ardent search for methods predicting protein structure from sequence (reviews: [7,8,6]; books: [9]). Many methods were found which looked initially very promising – but always the hope has been dashed [10]. The most successful predictions are achieved by experts starting combining machine-based predictions with their intuition and expertise [11].

How can neural networks predict protein structure? In practice, the most successful structure predictions extract patterns from data bases of known protein structures. Neural networks comprise a particular tool for pattern recognition and classification [12,13]. To which extent do neural networks contribute to predicting protein structure, in practice? Initially, researchers applied black boxes, and searched improvements through optimising the internal free parameters (training speed, network architecture). Later, researchers have opened the black boxes by extracting, or implementing rules, by carving specific knowledge into the networks, and by using networks to detect errors or outliers in data bases. More recently, the full potential of the tool has been explored by combining neural networks with evolutionary information. Now, applications of neural networks are amongst the most widely used methods in everyday's bioinformatics.

Here, I sketch neural network based methods (PHD series) for the prediction of 1D aspects (secondary structure, transmembrane helices, solvent accessibility) of protein structure. The methods illustrate that (1) neural networks as black boxes failed to improve prediction accuracy, (2) neural networks were sufficiently flexible to carve expertise from biology into the tool, (3) the quantum leap in prediction accuracy achieved in the 90's unearthed from implementing evolutionary information into neural networks, and (4) that the new generation of prediction methods is extremely useful in assisting, facilitating, and speeding-up experiments in molecular biology.

2 Carving Biology into Neural Networks

2.1 Conventional Prediction of Secondary Structure

Simplifying the structure prediction problem: The rapidly growing sequence-structure gap (number of known protein structures vs. number of known protein sequences) has enticed theoreticians to solve simplified prediction problems [8]. An extreme simplification is the prediction of protein structure in one dimension (1D), as represented by strings of, e.g., secondary structure, and residue solvent accessibility. Theoreticians are lucky not only because the 1D prediction problem is not only the task they can accomplish best, but in that even partially correct predictions of 1D structure are useful, e.g., for predicting protein function, or functional sites.

Basic idea of secondary structure prediction: The usual goal of secondary structure prediction methods is to classify a pattern of adjacent residues as either H (a-helix), E (for extended b-strand), or L (for loop, i.e., no regular structure). The principal idea underlying most secondary structure prediction methods is the fact that segments of consecutive residues have preferences for certain secondary structure states [4,14]. Thus, the prediction problem becomes a pattern-classification problem tractable by pattern recognition algorithms. The goal is to predict whether the residue at the centre of a segment of typically 13-21 adjacent residues is in a helix, a strand, or in none of the two regular structures.

First and second generation prediction methods: The first generation of 1D prediction methods was based on physico-chemical principles, expert rules, and statistics of single residues [15-17,4]. The second generation incorporated the influence of residues adjacent to the residue for which 1D structure was predicted (local information). These secondary structure prediction methods shared three major shortcomings: (1) prediction accuracy was limited to about 60% accuracy (percentage of residues predicted correctly in either of the three states H, E, L), (2) strands were predicted at typically < 40% accuracy, (3) predicted secondary structure segments were, on average, only half as long as observed segments. Methods were tailored to overcome one of these problems (long-range information: [18,19]; strand accuracy: [20]; length: [21]). However, the basic assumption was that these problems originated from using only local information (13-21 adjacent residues). It was assumed that, in general, 65% of the secondary structure formation is determined by local interactions, and that strands are dominantly determined by long-range interactions [22].

2.2   Improving Secondary Prediction by Neural Networks

No improvement by simple network: A simple tool that classified sequence stretches into three secondary structure states was a neural network (more precisely a multi-layered feed-forward network) [23-25]. Input was the sequence vector composed of 13-21 residues; output the secondary structure state of the central residue (Fig. 2). However, this simple device was not better than any other good prediction method. In particular, none of the three problems (prediction accuracy limited to 60%, strand accuracy around 40%, short segments) of conventional methods could be solved by such a device [26]. (However, due to inappropriate choices of the test sets this was not revealed by the first publications [27].)

Better prediction of strand by balanced training: Prediction accuracy for each of the three secondary structure classes approximately mirrored the observed occurrence of these classes in the training set [28,14]. In particular, only 21% of the correctly predicted residues belonged to the class E. Looking

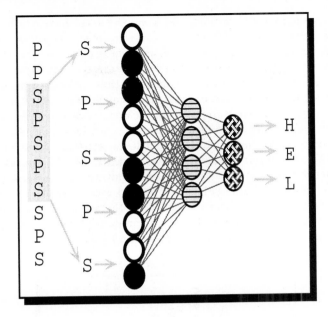

Fig. 2. Simple neural network for secondary structure prediction. For simplification the protein sequence given consists of two amino acid types (S and P). The protein sequence is translated into patterns by shifting a window of w adjacent residues (shown $w = 5$; typical values in practice are $w = 13$-21) through the protein. The output of the network is uniquely determined. Suppose the output would be: 0.2, 0.4, 0.5 for the three output states (H, E, L). For known examples the desired output is also known (1, 0, 0 if the central residue is in a helix). Consequently, the network error is given by the difference between actual network output and desired output. The only free variables are the connections. Training or learning means changing the connections such that the error decreases for the given examples. A training set typically comprises some 30,000 examples. If training is successful, the patterns are correctly classified.

at the training dynamics of the network revealed that the network learned H, and L ten times faster than E. Consequently, the idea was to improve the prediction for strand residues by simply increasing the frequency in presenting strand residues during training. Thus, instead of presenting in 1000 iteration time steps 220 examples for E, 310 for H and 470 for L (according to database distribution, dubbed unbalanced training), now at each time step one example for each class was used for training (balanced training). (1) All three classes were predicted almost equally well [28]. (2) Overall accuracy decreased, as the loop residues that were predicted more accurately by the unbalanced network comprised almost 50% of all residues. However, a balanced network proved that the inferior prediction of strand did NOT result primarily from long-range interactions, but from a technical problem.

Better prediction of segment length by 2nd level network: The average length of a helix is about 10 residues. However, helices predicted by the network were, on average, four residues long. The reason was that the network failed to learn correlating the secondary structure state of adjacent residues. The fact that, e.g., helices span over, at least, three residues was obscured by the particular training dynamics necessary to avoid unwanted database bias: examples presented in time steps $t1$ and $t2$ were chosen at random from the training set (and, thus, were usually not adjacent in sequence). This problem was corrected by introducing a 2nd level (structure-to-structure) network [28,14] . The input of this 2nd level network was the output of the 1st level (sequence-to-structure) network; the output was the secondary structure state of the central residue (Fig. 3). The 2nd level network had almost no effect in terms of overall accuracy. However, the average predicted helix extended over more than seven residues, i.e., predictions appeared considerably more protein-like than for the 1st level network [28,14].

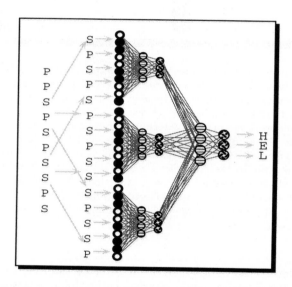

Fig. 3. Second level neural network [14] . (1) The window of w adjacent residues is shifted through the protein (here $w = 5$). For each window secondary structure is predicted for the central residue (shown three windows with central residues S, P, S). (2) The prediction of this first level network is fed into a second level network. This is again realised by shifting a window of w adjacent predictions through the protein (for the second level $w = 3$). The final prediction of secondary structure is valid for the central residue of the second window (here a P).

Better overall accuracy by averaging over many networks: Networks classify patterns separating them by lines. A particular training run results in a particular classification associated with a particular error. Part of this error usually is random noise. Furthermore, unbalanced, and balanced training occasionally yielded quite different predictions, in detail. Which to choose? The answer was to average over both networks, and to attempt reducing the random noise by generating even more differently trained networks (over $2 \times 2 = 4$ networks: 1st level: balanced, unbalanced, 2nd level: balanced, unbalanced). This 3rd level average over different networks improved prediction accuracy by 1-2 percentage points, and elegantly combined differently focused specialists [28-30].

Several problems solved, but accuracy still rather low: Incorporating facts about protein structures into the specific choice of the training dynamics and the combination of many independent neural networks solved two of the problems of conventional prediction methods (inaccurate prediction of strand, short segments). However, the overall prediction accuracy was still limited to about 65% [29,26]. Long-range information was not incorporated into the method. Increasing the window size (number of adjacent residues in protein fragment fed into the network) failed, as the signal-to-noise ratio increased considerably for longer windows. This problem was also reflected by that the networks did hardly use higher-order correlations in the input information: networks with and without hidden layers performed almost equally well [25,26].

2.3 The Issue of Appropriate Cross-Validation

Publishing optimistic results? The fact that the early applications of neural networks did not use higher-order information was obfuscated by the inappropriate choices of the test sets. Interestingly, the problem did not arise from 'computer geeks' intruding into the unknown space of biology. Rather, in the early 90's protein structure prediction experts just started to become aware of the importance of the issue of appropriately testing. Part of the problem is 'social' (better results, better journals, more grants). For example, the history of secondary structure prediction has partly been a hunt for highest-accuracy scores, with over-optimistic claims by predictors seeding the scepticism of potential users. However, the CASP experiments (protein structures are predicted before the structures are known, then predictors meet every second year and assess how well they did [31,11]) have illustrated: exaggerated claims are more damaging than genuine errors. Even a prediction method of limited accuracy can be useful if the user knows what to expect. For the editors of scientific journals this implies that no protein structure prediction method should be published that has not been sufficiently cross-validated. This raises a difficult question: how to evaluate prediction methods?

Full jack-knife test: The prediction of protein structure is an excellent example to illustrate the traps of 'positive thinking'. Given a data set of N proteins of known structure. The ideal testing is by jack-knifing: take $N - 1$ proteins for training, one for testing, and repeat N times. This recipe is simple, but does it suffice? Not, without ascertaining additional constraints. (1) No information of the performance on the one protein used for testing each time ought to steer the training. That is, developers do not want to adjust free parameters (network architecture, training speed, start conditions, training stop) such that the performance on the test protein is optimal. (2) The $N-1$ training proteins and the one test proteins ought to be distinct. For most protein structure prediction methods the definition of 'distinct' is simple: the percentage of sequence identity between the one test protein and any of the $N-1$ training proteins ought to be below 30%. If the two proteins have a higher level of pairwise sequence identity, we know from experience that they will adopt similar structure. Thus, we can predict structure by simply aligning the two proteins (much better and simpler than by any fancy device).

Complete cross-validation: The full jack-knife (N repeats of the experiment with splitting the N proteins of known structure into $N-1/1$) is often prohibited by limited computer resources. A simple alternative to full jack-knifing is the complete cross-validation: the N proteins are split into $N-N/F$ training and N/F test proteins, this is repeated F times. For say $F=10$, we refer to this as a complete ten-fold cross-validation. Again, the same constraints apply to each of the separations: (1) no information from F test proteins used for training, and (2) no overlap between training and testing set. In practice, we first have to play around with the neural networks to find a combination of network parameters for which we believe the method would be optimal. How can we thus avoid to look at what we pretend to be unknown (the structure for the testing proteins)? The solution brings forward a third data set, the validation set. Now the N proteins are split into $(N-N/F-V)$ training proteins, V validation proteins, and N/F test proteins. The V validation proteins are used to find the optimal parameters for the neural networks (and the conditions between training and testing set now apply to all three sets).

Number of cross-validation experiments not important: Are higher values of F (number of cross-validation experiments) better than lower ones? We often find an affirmative answer to this question in the literature. However, the exact number of F is not important provided the test set is representative, comprehensive and the cross-validation results are NOT miss-used to again change parameters. Developers usually have an interest to choose F as high as possible as that permits maximal use of the available information. However, the choice of F is of no meaning for the user provided the cross-validation is done appropriately.

Comprehensive data sets: All available unique proteins should be used for testing prediction methods (currently about 1000). The reason for taking as many proteins as possible is simply that proteins vary considerably in structural complexity; certain features are easy to predict, others harder. Thus, we can easily select a large subset from the database of known structures for which any prediction method looks much more accurate than any other.

This problem also raises the issue of comparing apples and oranges: no matter which data sets are used for a particular evaluation, a standard set for which results are published by others should also be included. Finally, the field of protein structure prediction offers an additional benefit: every month we witness the publication of about 50 novel protein structures. Thus, after having finalised the manuscript describing prediction methods, developers can simply apply the tool to all the new structures that were unknown by the time they started to invent their method.

3   Profiting from the Experiment Evolution

3.1   The Wealth of Evolutionary Information

Variation in sequence space: The exchange of a few residues can already destabilise a protein [32] . This implies that the majority of the $20N$ possible sequences of length N form different structures. But, has evolution created such an immense variety? Random errors in the DNA sequence lead to a different translation of protein sequences. These 'errors' are the basis for evolution. Mutations resulting in a structural change are not likely to be accepted, since the protein can no longer function appropriately. Furthermore, the universe of stable structures is not continuous: minor changes on the level of the 3D structure may destabilise the structure (due to high complexity). Thus, residue exchanges conserving structure are statistically unlikely. However, the evolutionary pressure to conserve function has led to a record of this unlikely event: structure is more conserved than sequence [33-35]. Indeed, all naturally evolved protein pairs that have 35 of 100 pairwise identical residues have similar structures [36,37]. But, the attractors of protein structures are larger, even: the majority of protein pairs of similar structures has levels of below 15% pairwise sequence identity [38,39].

Long-range information in multiple sequence alignments: The residue substitution patterns observed between proteins of a particular family, i.e., changes that conserved structure, are highly specific for the structure of that family. Furthermore, the substitutions realised by evolution, implicitly also carry information about long-range interactions: suppose residues i and $i+100$ are close in 3D, then the types of amino acids that can be exchanged (without changing structure) at position i are constrained by that their physico-chemical characteristics have to fit the amino acid types at position $i + 100$. Indeed, correlated mutations permit to predict inter-residue contacts [40].

Feeding profiles of residue exchanges into the networks: The simplest way to use evolutionary information was as following [14]. (1) A sequence of unknown structure (U) was aligned against the database of known sequences (i.e. no information of structure required!). (2) Proteins that had significant sequence identity to U to assure structural similarity [36,37] were extracted and re-aligned by the multiple alignment algorithm MaxHom [41]. (3) For each position the profile of residue exchanges in the final multiple alignment was compiled, and was used as input to the 1st-level sequence-to-structure network.

3.2   Secondary Structure Prediction (PHDsec)

Significant improvement in overall accuracy: Using evolutionary information in the simple way described improved prediction accuracy already from about 65% to over 70%. Further incorporation of specific information compiled from the multiple alignments [14] yielded a further improvement to levels of about 72% accuracy (system dubbed PHDsec). This number represented an average over a distribution (some proteins were predicted more accurately than others), with an approximate Gaussian form, and a standard deviation of about 10% [14]. The neural network system described, here, was the first to surpass the magic line of 70% accuracy [26], and proved four years after its implementation still to be the most accurate method at the Asilomar prediction contest in 1996 [42].

Predicting prediction accuracy: I failed to distinguish proteins predicted well from those predicted poorly based on their sequence characteristics. However, the strength of the prediction (measured as the normalised difference between the output unit with the highest and the one with the next highest value) provided an extremely useful index for the reliability of the prediction for each residue [14], and for the likelihood that the prediction for the entire protein was below, or above the average of 72% [14,42]. This allows in practice to focus on regions predicted with higher reliability.

3.3   Transmembrane Helix Prediction (PHDhtm)

Important class problematic for determining 3D: Even in the optimistic scenario that in the near future most protein structures will be either experimentally determined [39], one class of proteins will still represent a challenge for experimental determination of 3D structure: transmembrane proteins. The major obstacle with these proteins is that they do not crystallise, and are hardly tractable by NMR spectroscopy. Consequently, structure prediction methods are even more needed for this protein class than for globular water-soluble proteins. Fortunately, the prediction task is simplified by strong environmental constraints on transmembrane proteins: the lipid bilayer of the membrane reduces the degrees of freedom to such an extent that 3D structure formation becomes almost a 2D problem. Two major classes of membrane proteins are known: proteins that insert helices into the lipid bilayer, and proteins that form pores by a barrel of b-strands.

Failure of PHDsec to predict transmembrane helices: The neural network system designed to predict secondary structure for globular proteins failed in predicting transmembrane helices. Hence, the networks were trained again on proteins with transmembrane helices. Largely, the resulting prediction system (PHDhtm) was similar to the one used for predicting secondary structure

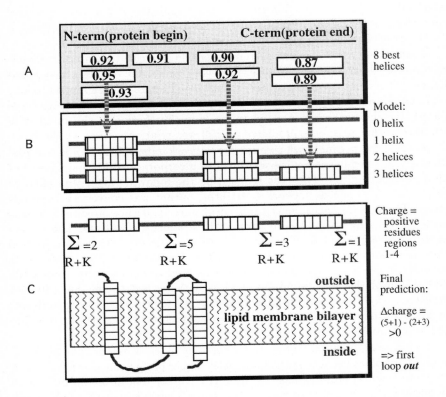

Fig. 4. Post-processing neural network output. A system of neural networks was trained to predict the location of transmembrane helices. The network output was treated as an 'energy-landscape' through which the best path was chosen given the constraint that transmembrane helices have a minimal (span of lipid bilayer) and a maximal length (longer energetically unfavourable). Thus, the normalised network outputs were used as input to a dynamic programming algorithm that found the model (number and locations of HTM's) representing the best path through all possible models consisting of HTM's between 18 and 25 residues by optimising the compatibility of the model with the neural network outputs. The final refined model output from the dynamic programming was used to apply the positive-inside rule: positively charged amino acids (Arginine: R, and Lysine: K) are more often observed inside [59]. **A**: pool of all possible membrane helices; **B**: successively building the prediction that is most compatible with the neural network output, given the assumption that the protein contains 0, 1, 2, 3 helices; **C**: final assignment of the helix orientation (topology) by the charge difference of the non-membrane regions.

for globular proteins [14]. One difference was the number of output units. PHDhtm distinguished two states: T (transmembrane helix), and non-T (i.e. a globular region). Again information from multiple alignments improved prediction accuracy significantly [14]. The final prediction system was, at least, as accurate as the best alternative prediction schemes [43,8,6].

Problems of PHDhtm: The system described so far had a major drawback: the 2nd level structure-to-structure network predicted too long membrane helices. This was corrected by introducing cut-off filters that chopped too long segments into several shorter ones. This procedure was relatively sensitive to parameter choices (when and where to cut). Furthermore, the number of transmembrane segments predicted overall was relatively often wrong.

Finding the optimal path through the network output: The problem of predicting transmembrane helices was ideal to incorporate additional aspects of globular information. This was realised by the following algorithm (Fig. 4) [44,45]. (1) The neural network (PHDhtm) output was converted to preferences. These preferences constituted an energy landscape predicted by the network. (2) The optimal path through this landscape was searched by dynamic programming. (3) The space of all possible predictions was limited by a minimal (18), and a maximal (25) length of transmembrane segments considered. (4) The final refined model was used to additionally predict the orientation of the transmembrane helices with respect to the cell (dubbed topology). (5) The average PHDhtm preference for the best transmembrane region possible was used to distinguish proteins bound to a membrane, and globular proteins.

Significant improvements by post-processing network output: The final system (PDtopology) achieved as significantly higher performance accuracy than the simple neural network-based system (PHDhtm) [44]: for about 89% of all membrane proteins all segments were predicted correctly, and for 86% of all proteins all segments, and the topology were correctly predicted (compared to 82% for PHDhtm). Furthermore, the number of false positives (globular proteins predicted to contain membrane spanning regions) fall from above 4% to below 2%. (Note: this number is extremely important to analyse entire genomes [44].)

3.4   Solvent Accessibility Prediction (PHDacc)

Important step towards predicting 3D? If secondary structure segments could be predicted sufficiently accurately, they may be arranged in space as rigid bodies to yield a model for 3D prediction [46]. One criterion for assessing each arrangement could be to use predictions of residue solvent accessibility. The solvent accessibility of a residue embedded in a protein structure can be described in several ways [47-49]. The simplest is a two-state description distinguishing between residues that are buried (relative solvent accessibility < 16%) and exposed (relative solvent accessibility ≥ 16%). The classical method to predict accessibility is to assign either of the two states, buried or exposed, according to residue hydrophobicity [50-52].

Evolutionary information improves prediction accuracy: Solvent accessibility at each position of the protein structure is evolutionarily conserved within sequence families [53,54]. This fact was used to develop another neural network method for predicting accessibility from multiple alignment information (PHDacc) [53,14]. For this method, I skipped the 2nd-level network since accessibility was hardly correlated between adjacent residues. The network output comprised ten units. Unit n, for $n = 0, \ldots, 9$ coded for a relative accessibility A in the interval $n^2 \leq A < (n+1)^2$. This encoding reflected the observation that in protein structures residues flip more easily between 70% and 100% relative accessibility than between 0% and 5%. The final network system predicted about 75% of the residues correctly in either of the two states buried, or exposed. This was more than five percentage points higher than for methods not using alignment information.

4   Conclusions: Do Neural Networks Help Biology?

Structure prediction: work in progress... Native 3D structures of proteins are encoded by a linear sequence of amino acid residues. To predict 3D structure from sequence is a task challenging enough to have occupied a generation of researchers. Have we finally succeeded? The bad news is: no, we still cannot predict structure for any sequence. The good news are: we have come closer, and growing databases facilitate the task.

Predictions in 1D: significant improvement by larger databases: The rich information contained in the growing sequence and structure databases enables improving the accuracy of 1D predictions. Here I sketched, how evolutionary information input to neural network systems yielded better predictions of secondary structure, solvent accessibility, and transmembrane helices. These predictions of protein structure in 1D are significantly more accurate, and more useful than five years ago.

Conditions to become useful: In the field of structure prediction we have witnessed blooming over-optimism [10], as well as, more or less intended cheating. The Asilomar meetings [31] to some extent are succeeding in separating the wheat from the chaff. However, Asilomar does not change the basic formula: when you develop a prediction method you ought to spend more than 70% of the time on appropriate evaluation of the performance [55,8]. The sustained levels of prediction accuracy published for the PHD methods were, supposedly, one of the major reason for their success. Another important issue is that of making the method available. Molecular biologists do NOT have the time to become experts in running programs. Thus, methods should be easy-to-use, and available via the internet [56,57].

Learning from evolution to help studying evolution... Mastering protein structure prediction has several impacts on the advance of biology. Firstly, prediction methods have assisted the experimental determination of protein structures. Secondly, predictions helped unravelling unknown protein functions, and improving our understanding of the mechanisms of partially known functions. Thirdly, prediction methods allow to separate the wheat from the chaff: we can explore which biological information improves the performance of neural networks and which does not. This is a simple, yet effective, means of shedding light into the understanding protein structure formation, and protein function (since we fail to predict structure and function, we thoroughly do NOT understand the underlying principles!).

What next? Most breakthroughs in protein structure prediction were achieved over the last six years. Thus, although we still cannot solve the general prediction problem, progress has been made. In general, however, we could ask the question – is it worth persevering with structure prediction, given that it is clearly such a difficult task? The answer is: yes. The methods which have spun off from structure prediction have already given us considerable insight into the first four complete genomes. Perseverance with structure prediction will yield fruit in about five years time when the human genome will be known.

Acknowledgements

Thanks – in alphabetical order – to all those who contributed ideas, and helped with motivating discussions: Michael Braxenthaler (Hoffman-LaRoche, New Jersey), Søren Brunak (CBS, Copenhagen), Rita Casadio (Univ., Bologna), Sean O'Donoghue (EMBL, Heidelberg), Piero Fariselli (Univ. Bologna), Terry Gaasterland (Univ. Chicago), Gunnar von Heijne (Univ. Stockholm), Tim Hubbard (Sanger, Hinxton), Rainer Kühnen (Univ. Heidelberg), Chris Sander (Millenium, Boston), Michael Scharf (Take5, Heidelberg), Reinhard Schneider (LION, Heidelberg), Manfred Sippl (Univ. Salzburg), Sara Solla (Western Univ., Chicago), Anna Tramantano (IRBM, Rome), Alfonso Valencia (CNB, Madrid), Gerrit Vriend (EMBL, Heidelberg). Thanks to Friedrich von Bohlen (LION, Heidelberg) for financial support.

References

1. Fleischmann, R. D., et al.: Whole-genome random sequencing and assembly of Haemophilus influenzae Rd. Science **269** (1995) 496-512.
2. Goffeau, A., et al.: Life with 6000 genes. Science **274** (1996) 546-567.
3. Gaasterland, T.: Genome sequencing projects. WWW document (http://www.mcs.anl.gov/home/gaasterl/genomes.html), Univ. Chicago (1998).
4. Brändé n, C., Tooze, J.: Introduction to Protein Structure . New York, London: Garland Publ. (1991).
5. Anfinsen, C. B.: Principles that govern the folding of protein chains. Science **181** (1973) 223-230.

6. Rost, B., O'Donoghue, S. I.: Sisyphus and prediction of protein structure. CABIOS **13** (1997) 345-356.

7. Barton, G. J.: Protein secondary structure prediction. Curr. Opin. Str. Biol. **5** (1995) 372-376.

8. Rost, B., Sander, C.: Bridging the protein sequence-structure gap by structure predictions. Annu. Rev. Biophys. Biomol. Struct. **25** (1996) 113-136.

9. Doolittle, R. F.: Computer methods for macromolecular sequence analysis . San Diego: Academic Press (1996).

10. Honig, B., Cohen, F. E.: Adding backbone to protein folding: why proteins are polypeptides. Folding & Design **1** (1996) R17-R20.

11. Moult, J., Hubburad, T., Bryant, S. H., Fidelis, K., Pedersen, J. T.: Critical assessment of methods of protein structure prediction (CASP): Round II. Proteins Suppl **1** (1997) 2-6.

12. Arbib, M.: The handbook of brain theory and neural networks . Cambridge, MA: Bradford Books/The MIT Press (1995).

13. Fiesler, E., Beale, R.: Handbook of Neural Computation . New York: Oxford Univ. Press (1996).

14. Rost, B.: PHD: predicting one-dimensional protein structure by profile based neural networks. Meth. Enzymol. **266** (1996) 525-539.

15. Schulz, G. E., Schirmer, R. H.: Principles of Protein Structure . Heidelberg: Springer (1979).

16. Kabsch, W., Sander, C.: How good are predictions of protein secondary structure? FEBS Lett. **155** (1983) 179-182.

17. Fasman, G. D.: Prediction of protein structure and the principles of protein conformation . New York, London: Plenum (1989).

18. Maxfield, F. R., Scheraga, H. A.: Improvements in the Prediction of Protein Topography by Reduction of Statistical Errors. Biochem. **18** (1979) 697-704.

19. Zvelebil, M. J., Barton, G. J., Taylor, W. R., Sternberg, M. J. E.: Prediction of protein secondary structure and active sites using alignment of homologous sequences. J. Mol. Biol. **195** (1987) 957-961.

20. Gascuel, O., Golmard, J. L.: A simple method for predicting the secondary structure of globular proteins: implications and accuracy. CABIOS 4 (1988) 357-365.

21. Kabsch, W., Sander, C.: Segment83. unpublished (1983).

22. Garnier, J., Levin, J. M.: The protein structure code: what is its present status? CABIOS **7** (1991) 133-142.

23. Bohr, H., Bohr, J., Brunak, S., Cotterill, R. M. J., Lautrup, B., Nørskov, L., Olsen, O. H., Petersen, S. B.: Protein secondary structure and homology by neural networks. FEBS Lett. **241** (1988) 223-228.

24. Qian, N., Sejnowski, T. J.: Predicting the secondary structure of globular proteins using neural network models. J. Mol. Biol. **202** (1988) 865-884.

25. Holley, H. L., Karplus, M.: Protein secondary structure prediction with a neural network. Proc. Natl. Acad. Sc. U.S.A. **86** (1989) 152-156.

26. Rost, B., Sander, C.: Secondary structure prediction of all-helical proteins in two states. Prot. Engin. **6** (1993) 831-836.

27. Rost, B., Sander, C., Schneider, R.: Progress in protein structure prediction? TIBS **18** (1993) 120-123.

28. Rost, B., Sander, C.: Prediction of protein secondary structure at better than 70accuracy. J. Mol. Biol. **232** (1993) 584-599.
29. Rost, B., Sander, C.: Improved prediction of protein secondary structure by use of sequence profiles and neural networks. Proc. Natl. Acad. Sc. U.S.A. **90** (1993) 7558-7562.
30. Rost, B., Sander, C.: Combining evolutionary information and neural networks to predict protein secondary structure. Proteins **19** (1994) 55-72.
31. Moult, J., Pedersen, J. T., Judson, R., Fidelis, K.: A large-scale experiment to assess protein structure prediction methods. Proteins **23** (1995) ii-iv
32. Dao-pin, S., Söderlind, E., Baase, W. A., Wozniak, J. A., Sauer, U., Matthews, B. W.: Cumulative site-directed charge-change replacements in bacteriophage T4 lysozyme suggest that long-range electrostatic interactions contribute little to protein stability. J. Mol. Biol. **221** (1991) 873-887.
33. Chothia, C., Lesk, A. M.: The relation between the divergence of sequence and structure in proteins. EMBO J. **5** (1986) 823-826.
34. Doolittle, R. F.: Of URFs and ORFs: a primer on how to analyze derived amino acid sequences. Mill Valley California: University Science Books (1986).
35. Lesk, A. M.: Protein Architecture - A Practical Approach . Oxford, New York, Tokyo: Oxford University Press (1991).
36. Sander, C., Schneider, R.: Database of homology-derived structures and the structural meaning of sequence alignment. Proteins **9** (1991) 56-68.
37. Rost, B.: Twilight zone of protein sequence alignments. J. Mol. Biol. (1998).
38. Rost, B.: Protein structures sustain evolutionary drift. Folding & Design **2** (1997) S19-S24.
39. Rost, B.: Marrying structure and genomics. Structure **6** (1998) 259-263.
40. Goebel, U., Sander, C., Schneider, R., Valencia, A.: Correlated mutations and residue contacts in proteins. Proteins **18** (1994) 309-317.
41. Schneider, R.: Sequenz und Sequenz-Struktur Vergleiche und deren Anwendung fr die Struktur- und Funktionsvorhersage von Proteinen. Ph.D. thesis, Univ. of Heidelberg (1994).
42. Rost, B.: Better 1D predictions by experts with machines. Proteins Suppl. **1** (1997) 192-197.
43. von Heijne, G.: Membrane proteins: from sequence to structure. Annu. Rev. Biophys. Biomol. Struct. **23** (1994) 167-192.
44. Rost, B., Casadio, R., Fariselli, P.: Topology prediction for helical transmembrane proteins at 86% accuracy. Prot. Sci. **5** (1996) 1704-1718.
45. Rost, B., Casadio, R., Fariselli, P.: Refining neural network predictions for helical transmembrane proteins by dynamic programming. In States, D., et al. eds. Fourth International Conference on Intelligent Systems for Molecular Biology . St. Louis, M.O., U.S.A.: Menlo Park, CA: AAAI Press (1996) 192-200.
46. Cohen, F. E., Presnell, S. R.: The combinatorial approach. In Sternberg, M. J. E. eds. Protein structure prediction . Oxford: Oxford Univ. Press (1996) 207-228.
47. Lee, B. K., Richards, F. M.: The interpretation of protein structures: estimation of static accessibility. J. Mol. Biol. **55** (1971) 379-400.
48. Chothia, C.: The nature of the accessible and buried surfaces in proteins. J. Mol. Biol. **105** (1976) 1-12.
49. Connolly, M. L.: Solvent-accessible surfaces of proteins and nucleic acids. Science **221** (1983) 709-713.

50. Tanford, C.: The hydrophobic effect: formation of micelles and biological membranes . New York: John Wiley & Sons (1980).
51. Kyte, J., Doolittle, R. F.: A simple method for displaying the hydrophathic character of a protein. J. Mol. Biol. **157** (1982) 105-132.
52. Eisenberg, D., Weiss, R. M., Terwilliger, T. C.: The hydrophobic moment detects periodicity in protein hydrophobicity. Proc. Natl. Acad. Sc. U.S.A. **81** (1984) 140-144.
53. Rost, B., Sander, C.: Conservation and prediction of solvent accessibility in protein families. Proteins **20** (1994) 216-226.
54. Rost, B.: Average conservation of 1D structure between remote homologues. WWW document (http://www.embl-heidelberg.de/~rost/Res/96E-ConservationOf1D.html), EMBL Heidelberg, Germany (1996).
55. Rost, B., Sander, C.: Progress of 1D protein structure prediction at last. Proteins **23** (1995) 295-300.
56. Rost, B.: PredictProtein - internet prediction service. WWW document (http://www.embl-heidelberg.de/predictprotein), EMBL (1997).
57. Rost, B., Schneider, R.: Pedestrian guide to analysing sequence databases. In Ashman, K. eds. Core techniques in biochemistry . Heidelberg: Springer (1998) (in press).
58. Bernstein, F. C., Koetzle, T. F., Williams, G. J. B., Meyer, E. F., Brice, M. D., Rodgers, J. R., Kennard, O., Shimanouchi, T., Tasumi, M.: The Protein Data Bank: a computer based archival file for macromolecular structures. J. Mol. Biol. **112** (1977) 535-542.
59. von Heijne, G.: Membrane protein structure prediction. J. Mol. Biol. **225** (1992) 487-494.
60. Kraulis, P. J.: J. Appl. Crystallography **24** (1991), 946-950.

An Application of Artificial Neural Networks in Linguistics

Jure Zupan

National Institute of Chemistry, Hajdrihova 19, SI-1000, Ljubljana, Slovenia

Abstract. Several problems associated with the role of statistics in the field of linguistics are outlined. Descriptors of sentences which allow classification of written material according to the style (not to the content!) are described, first in a more general view and next with respect to the Slovenian language. The computer requirements, algorithms and data bases necessary for such studies are briefly commented upon. Next, the artificial neural network (ANN) methodology used in these studies, i.e. the Kohonen ANNs, is described briefly and the role of weight maps obtained by the Kohonen ANNs is stressed. Finally, several results of two comparative linguistic studies on the styles of authors are given. The first study concerns the *clustering* of 37 articles of five contemporary Slovenian journalist-columnists, each of whom is writing a weekly column of approximately the same length in various journals, while the second example is used to show the possibility of *classification* of poems of "unknown" authors using the Kohonen mapping. In this example, 33 epic poems of three Slovenian poets who wrote their works in three different time domains (1820–1850, 1890–1910, and 1950–1980) are used.

1 Introduction

In order to extract general information about the flexibility of authors in writing, linguistic studies were always involved in various statistical methods. Studies of the frequencies of most common words and searching for concordances of specific phrases have been known since medieval times. For example, one of the first known concordances (i.e. appearances of selected words in the context of several neighboring words) of the *Vulgate* (Latin translation of the Bible by St. Jerome in the fourth century) was made by cardinal Hugo de Sancoto Caro as early as 1262.

Unfortunately, until computers were commonly available for text handling and sophisticated language-oriented software was developed, such tasks were carried out manually, which is an extremely time-consuming process. With more powerful computers, especially where the storage for various dictionaries and supporting software is concerned, the task of text handling or pre-processing of text for further linguistic studies has become more attractive. However, in spite of quite remarkable progress of software technology, text pre-processing of different languages still requires different types of dictionaries, inverted and/or pointer files, etc. For example, the classification of a given word according to its type (i.e., whether it is used as a noun, verb, or something else) is not only difficult because such classification depends on context, but also because the *form* in which the type of word is *"disguised"* in a given part of the text differs very much from language to language. In some languages, the most important factor for the decision of a given word type is the sequence of neighboring words. In other languages, the prevailing features are prefixes and suffices. In still others, the decision factor for the word type might be the presence and type of prepositions (or a lack of them).

2 Handling of Texts by Computers

It must always be kept in mind that we are talking about the automatic (i.e. computer supported) classification of encountered words in the text, and not about the recognition performed by humans. The computer does not know the meaning of the word. What a computer "sees" as a "word" is only a set of characters (separated from each other either by blanks and/or by punctuation marks) and it tries to match this set, its subsets, and/or additional neighboring sets to various patterns in accordance with procedures specified by the programmers. Of course, the standard or the *type setting patterns* for such decisions are different for each language.

For the determination of the frequency distributions of the word types in a given text, the most important piece of software is a so-called "dictionary-entry machine". Its purpose is to determine the lexicographic (dictionary) entry for each word encountered in the text in whatever form it may be written. For example, from any textual form of the word "water" (whether as a verb, a noun, etc.), which can be *water, water's, waters', waters, watering,* or *watered,* the computer must be able to deduce the correct entry as it stands in the dictionary, i.e. the form "water". Once the correct entry is found, another routine should decide whether the form in the text is representing a verb, a noun, an adjective, or if there are various possibilities.

The described task seems to be an easy one, but in real applications it turns out to be almost impossible to carry out automatically. Hence, in most applications a lot of checking still has to be done by hand. In the English language, at least the first part of the described procedure (i.e. finding the lexicographic entry) seems to be a relatively easy task. Unfortunately, this is not the case in other languages (regarding only the Indo-European ones), notably the ones belonging to the Slavic group of languages, and especially in the Slovenian language.

In declination of nouns (be it in the masculine, feminine or neuter forms) Slovenian language distinguishes six falls and three forms: singular, dual, and plural. This means that each noun can be found in a text in any of the 18 different forms possible. Table 1 gives the declination of the Slovenian word *konj (Engl.: the horse)* together with the translation to English. Although Slovenian language distinguishes 18 forms in the declination of the word *konj,* only nine of them are actually different. Further determination of the correct case out of the 18 possible ones has to be deduced from the context. The mentioned set of nine endings is only one of 52 different sets for declination of masculine nouns, and there are still 40 and 32 different ending sets for feminine and neuter nouns. In order to find the proper lexicographic entry for nouns the computer program should know 124 sets where each contains 18 endings (clearly, in each set several endings are equal—however, not always at the same positions). Compared to only four different endings for nouns in English (-, 's, s', and s), which are almost always the same and always at the same positions, the software handling a variety of endings in Slovenian language must be more complex.

Table 1. Declination of the word *konj* (the horse) in Slovenian language compared to the same forms of the English grammar in which the dual and the fifth and the sixth fall are not known.

Fall	Singular		Dual		Plural	
1	konj	the (a) horse	konj-a	two horse-s	konj-i	the horse-s
2	konj-a	the horse-'s	konj-(ev)	two horse-s'	konj-(ev)	the horse-s'
3	konj-u	to the horse	konj-ema	to two horse-s	konj-em	to the horse-s
4	konj-a	the horse	konj-a	two horse-s	konj-e	the horse-s
5	pri konj-u	at the horse	pri konj-ih	at two horse-s	pri konj-ih	at the horse-s
6	s konj-em	with the horse	s konj-em	with two horse-s	s konj-i	with the horse-s

The declination of nouns is even more complex than the declination of adjectives, because the same adjective has to match the nouns of any of the three genders in any of the 18 forms. There are, therefore, $18 \times 3 = 54$ different possibilities. It is true that not all of the 54 possible endings of the adjective declination are different, but there are still 12 different ones. Fortunately, for adjectives there are only 20 different sets and each of them has exactly 12 different endings (compared to 54 theoretically possible ones) which are always at exactly the same positions. To complete the story, the Slovenian language altogether has 256 different sets with various numbers of endings (ranging from 6 to 21) and all of them have to be, at least in principle, considered at every single word encountered during the text processing. Clearly, with sophisticated programming techniques such as hash-coding of roots in random access files[1,2] and checking the variable root lengths, the performance of such tests can be increased considerably.

In order to use such sets of endings efficiently, a complete dictionary of all possible roots for a given language associated with these ending sets must be prepared in advance. Fortunately, such a task, requiring many man-years of effort, has to be made only once. For the Slovenian language, several computer readable vocabularies of different lengths exist[3,4] The largest one now available[5] contains 155,000 entries ("roots"), each associated with one of the 256 ending sets.

After associating each word in the text with all possible roots and corresponding endings that yield the identical text query, the possible choices can be determined by some linguistic rules[6]. For example, the Slovenian word *stavka (Engl.: the strike)* can be composed:

- from the root "stav" and the ending "ka" specifying the first or the fourth fall of the dual form or the second fall of the singular form of the noun *stavek (Engl.: the sentence)*,
- from the root "stavk" and the ending "a" specifying the first fall of the singular form of the noun *stavka (Engl.: the strike)*, or
- from the root "stavka" without ending, specifying the act of a third person in the present time expressed by the verb *stavkati (Engl.: to strike)*.

Sometimes, the decision as to which possibility is the correct one can be made easily by implementing several grammatical rules that are language-specific [7]. For example, the presence or absence of a preceding adjective and/or a pronoun can determine whether a word is a noun or a verb. Similarly, the rule: "two consecutive nouns are not allowed" can exclude many wrong conclusions. However, no matter how complex a particular grammar is, and consequently how sophisticated the programs performing the rules and tables are, there always remain cases where one must decide upon the context. This means that a unique classification of some words in a specific context cannot be done automatically, but must be made by visual inspection.

Visual re-checking and corrections of word types are the most time-consuming tasks in text analysis. In spite of the fact that a large part of today's artificial intelligence (AI) research in automatic text recognition is focused on this issue the problem is far from being solved satisfactorily for any language. Nevertheless, specifications of words in all texts and poems in this study were checked manually, the sets of possible suggestions made by the computer were inspected, and the proper decisions implemented.

3   Representation of Texts

In the present study, each text or poem was represented as a string of 21 variables. These strings were obtained after each text was scanned by the author's program[5] and, for each word in the text, the unique classification of word types was made. The following variables were determined and used throughout the study:

1. percentage of nouns
2. percentage of adjectives
3. percentage of verbs
4. percentage of adverbs
5. percentage of auxiliary verbs
6. percentage of numerals
7. percentage of conjunctions
8. percentage of prepositions
9. percentage of pronouns
10. percentage of unknown words
11. percentage of different nouns
12. percentage of different adjectives
13. percentage of different verbs
14. percentage of different adverbs
15. percentage of different words
16. average number of words in a sentence
17. maximal number of words in a sentence
18. average number of words between two punctuation marks
19. average number of commas and semicolons per sentence
20. average number of all punctuation marks per sentence
21. percentage of different 3-word combinations
    (out of all possible 3-word combinations).

The percentage of interjections was omitted because of the extremely low frequencies in all texts. "Unknown" words also appear in all texts. These are words from different languages, slang words, or words "invented" by the authors. The variables eleven through fifteen describe the richness of the vocabulary used by each author. A separate study of the same authors has revealed that the maximal and minimal numbers of different words used by a specific author (2,320 and 1,850 different words per 6,000 used words, respectively) is significant regarding the estimated error at 2,000 words, $\sigma = \pm 50/2000 = 2.5\%$. However, this variable alone (variable 15) cannot distinguish between the authors within a 95% confidence interval ($\pm 1.96\sigma$) if they are separated by less than 3.92σ which is about 200 different-words per 6,000 used words.

Variables 16 through 20 were normalized to the largest number of words, or largest average number of words, or number of punctuation marks encountered for the specific variable in the entire study. The last (twenty-first) variable is an interesting one. It has led us to pinpoint some empirical grammar rules discussed in the above paragraphs. When scanning the variation and the distribution of 3-word sequences, the flexibility and ability to compose sentences in different ways was evident. An example for the 3-word sequence can be compounded from a *pronoun*, an *auxiliary verb* and a *verbal substantive*. This 3-word sequence is abbreviated as "PN-AV-VB", and can be found at the beginning of the sentence "I am writing this manuscript". For eleven different word types in the study, there are $11^3 = 1331$ different 3-word combinations. Although the same 3-word-combination *preposition-adjective-noun*, "PP-AJ-NN", is the most frequently used sequence by all authors in the study, the variation of different sequences used is quite scattered, ranging from 20 to 30% of all possible sequences.

4   Kohonen Neural Networks

In this short essay I will show a specific use of artificial neural networks (ANNs), Kohonen ANNs[8,9] in this case. However, these networks have a broader range of applicability[10] than that discussed here (namely, linguistic studies). ANNs have several advantages over statistical techniques, e.g. they have the ability to continuously adapt to new data through the use of less rigid assumptions about the underlying data distribution, they allow to built models without knowing the actual modeling functions, useful information about the relations between various variables can be extracted from Kohonen ANNs, etc.

However, it is important to stress that there is no "best" or "unique" method which would be optimal, either for all problems or merely for a specific problem. Therefore, complementary information can and should be obtained by application of several different techniques and methods, among which ANNs are as good the standard statistical methods. In this sense, examples of the analysis of several texts and poems using Kohonen ANN are explained here. The reader is of course challenged to adapt its use to his or her own problem accordingly. Due to its simplicity and additional information (top-map, weight-maps) that a Kohonen ANN offers[10], it is often a good choice as a supplementary or alternative method to statistical explorations.

The task we are trying to accomplish in this essay (clustering and classification) is mainly approached using one of the so-called *unsupervised* learning methods. Unsupervised learning requires only that the objects of the study are represented by input vectors X_s. Contrary to supervised learning methods, unsupervised ones do not require target vectors T_s associated with each input vector X_s to be known in advance. In unsupervised learning, knowing the associated target vector T_s or class q_s to which the input vector X_s belongs is required merely for checking the results after the learning is finished and not for learning itself. In a Kohonen ANN, which applies unsupervised learning, the results (either the object X_s belongs to the category q_s, or not) are implied by the position of vector X_s in the 2-d map of neurons. It can be said that the obtained Kohonen ANN serves as a pointer to the look-up table where the actual results are stored.

To generate, to train, or to teach the final ANN means to input all p objects X_s to the network a number of times. The correction of weights in the ANN p-times, i.e. once for each object in the study, is considered to be one cycle (one epoch) of the learning procedure. The input of all objects to the

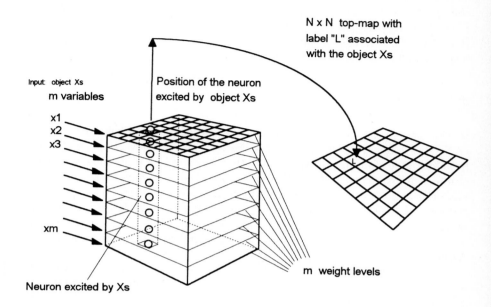

Fig. 1. Kohonen ANN. The neurons are represented as columns and are arranged in a box. The length of the columns m corresponds to the number of weights (circles) and hence to the dimension of the input vector X_s (shown at the left side of the Kohonen ANN). All the weights of the i-th weight level receive the i-th component x_{is} of the input vector X_s simultaneously. The excited neuron c having a label "L" is chosen on the basis of the "best match" criterion, eq. (1).

ANN is repeated until the Kohonen ANN correctly recognizes all p objects or until the number of pre-specified number of epochs is exceeded. In the sense of Kohonen learning, the phrase "correct recognition" of the training objects means that each and every object in the training set will *excite* exactly the same neuron in two consecutive epochs of the learning procedure. This in turn means that no further relocation of objects (in a topological sense) is possible by continuing the learning procedure. Of course, the weights in the Kohonen ANN can still be further adapted and slightly changed, but this is usually fully accomplished during the second identical epoch in the sequence.

The Kohonen ANN resembles biological NN probably most closely of all ANN architectures and learning schemes. A Kohonen-type ANN is based on a single layer of neurons arranged in a box exhibiting a 2-dimensional plane of responses on its top (Fig. 1).

Neurons (columns in the scheme shown in Fig. 1) have 8, 16, 24, etc. neighbors in its first, second, third, etc., neighborhood rings. The topology of the Kohonen ANN can be considerably improved if *toroidal* boundary conditions are fulfilled, i.e., if the same number of neighbors at all positions, including the borders of the Kohonen ANN are considered. Toroidal boundary conditions mean that the network at its right or top edge is continued (in a computational sense) at its left and lower edge, respectively, and *vice versa*.

Unfortunately, the toroidal conditions decrease the available mapping area by a factor of four compared to the same dimensional non-toroidal ANN. Namely, the maximal topological separation of two neurons in the $N_{net} \times N_{net}$-dimensional toroidal ANN is $N_{net}/2$ neuron positions long, while in the $N_{net} \times N_{net}$-non-toroidal ANN the maximal topological distance between two neurons is equal to N_{net} positions. In cases when a large area for mapping is needed, the possibility of a two times larger maximal topological distance between two neurons can be a predominant factor for the use of the non-toroidal ANN. A larger maximal topological distance between neurons offers a better possibility for the separation of clusters in a non-toroidal ANN compared to a toroidal ANN of the same size. However, it has to be pointed out that a maximal topological distance between two neurons excited by two different objects does not necessarily mean that these two objects are the most separated ones, i.e., that they are separated by the largest possible distance in the measurement space of the treated objects. The topological distance of excited neurons in the Kohonen ANNs depends on the frequency distribution of the objects as well.

At a given step of learning, each multi-dimensional vector X_s is passed to all neurons as input. At each neuron, a comparison between vector X_s and the neuron's weights is made. Despite this, after the input of one object in a Kohonen ANN, only one neuron is selected. This means that the neurons are competing among themselves which one will be selected as the *excited* or as *central* one. Therefore, such learning is usually referred to as *competitive learning*. It is evident that the selection of this neuron is crucial. The neuron selected from among the $N_{net} \times N_{net}$ neurons is called the *winning*, the most *excited*, or the *central* neuron and is labeled with the letter c (for *central*). The actual selection of the winning neuron is based on the comparison between all weight vectors $W_j = (w_{j1}, w_{j2}, \ldots, w_{jm})$ and the input signal $X_s = (x_{s1}, x_{s2}, \ldots, x_{sm})$:

$$\text{neuron } c \leftarrow \min \left\{ \sum_{i=1}^{m} (x_{si} - w_{ji})^2 \right\} , \quad j = 1, 2, \ldots, c, \ldots N_{net} \times N_{net} . \quad (1)$$

After the winning neuron c is found, the correction of weights (i.e., adaption of the ANN or the learning) starts. The most characteristic feature of a Kohonen ANN which makes it very similar to a biological NN is its implementation of the corrections. The corrections do not cover the entire network. They do not even cover the same number of neurons at different stages of the learning process—the number and the extent of the corrections changes during the learning. It can be said that the learning procedure considers the local feed-back only, with the emphasis on local. This means that the correction of weights does not affect all neurons in the ANN, but only a small number of them: the ones that are topologically close to the winning one. Such local feed-back of corrections causes the topologically close neurons to start acting similarly if similar input objects are presented to the network. The goal of learning is that two similar objects shall excite two topologically close neurons, and two very different objects shall excite (select) winning neurons topologically far away from each other.

The main requirement for learning is that the weights w_{ci} of the neuron c excited by object X_s should be corrected in such a way that the distance between neuron c and object X_s will be even smaller (according to (1)) the next time object X_s is presented. At any step of the learning procedure, not only the excited neuron c, but also its $8, 16, 24, \ldots, 8p$ neighbors in the first, second,..., p-th, neighborhoods are stimulated. However, to which p and to which extent the neurons are stimulated, depends on the parameters a_{max}, a_{min}, N_{net}, and i_{tot} of the learning strategy:

$$\Delta w_{ji} = [(a_{max} - a_{min})(p/N_{net} + a_{min})][1 - d/(p+1)](x_{si} - w_{ji}^{old}) ,$$

$$d = 0, 1, 2, \ldots, p . \tag{2}$$

The entire expression on the left of the term $(x_{si} - w_{ji}^{old})$ in (2) describes how the correction of the weight w_{ji} decreases with increasing learning time. The topological distance d refers to the distance between the neuron j which is to be corrected and the central neuron c. During learning the range p to which the neurons are still corrected decreases as:

$$p = (i_{tot} - i_{it}) N_{net} / (i_{tot} - 1) . \tag{3}$$

At the beginning of learning ($i_{it} = 1$) p covers the entire network ($p = N_{net}$) while at the end of the learning iteration steps ($i_{it} = i_{tot}$), p is limited only to the central neuron c ($p = 0$). Whether the difference $x_{si} - w_{ji}^{old}$ in (2) is positive or negative, i.e. whether x_{si} is greater or smaller than the weight w_{ji}^{old}, the weight w_{ji}^{new} will be closer to x_{si} than was w_{ji}^{old}.

At the end of the training, after all training vectors X_s have been presented to the ANN i_{tot} times (i_{tot} epochs), the complete set of vectors X_s is once more run through the ANN. In its last run, the labeling of the neurons excited by the input vectors is made into the table called the *top-map*. This top-map has exactly $N_{net} \times N_{net}$ entries, each of which corresponds to exactly one neuron in the Kohonen ANN (Fig. 1). The labels in the top-map serve for better visualization of how the final Kohonen ANN responds to all input objects. Usually, the labels in the top-map correspond to some property of the input object exciting the associated neuron (to the class number to which the object belongs, for example). Eventually, the labels are simply the ID numbers of objects. It may well happen that one neuron is excited by more than one object X_s. In such cases, called conflicts, either the labels of all objects exciting this neuron are stored (if the program allows such storage) or only the most representative one is retained and the others are discarded. In the latter case, the top-map with labels has to be regarded only as partial information, good for a quick visualization only. Nevertheless, the top-map is a useful tool for a number of applications.

The number of weights in each neuron is equal to the dimension m of the input vector $X_s = (x_{s1}, x_{s2}, \ldots, x_{sm})$ (Fig. 1), and therefore each ($N_{net} \times N_{net}$) Kohonen ANN consists of ($N_{net} \times N_{net}$) $\times m$ weights. Before learning starts, all weights in the ANN are randomized in the interval $[0, 1]$. Each neuron has the same number of weights, and to each weight w_{ji} at a fixed position i in neuron j the same variable x_i is always passed, e.g. the first weight w_{j1} handles only the first variable x_1, the second weight w_{j2} the second variable x_2, etc. Hence, in each level of weights, only data of one specific variable are handled. Consequently, at the end of learning, in each level a map showing the distribution of values of the particular variable is formed. It is important to keep in mind that at the end of learning each and every weight in the Kohonen ANN is adapted to the data even if certain neurons (and, of course, their weights) were never excited by an object.

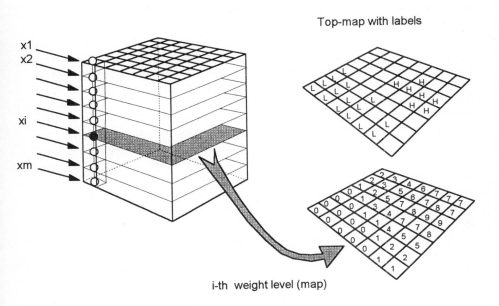

Fig. 2. Comparison of the i-th weight level with the top-map. All weight levels can be presented as contour maps of variable values. Depending on the software used, maps of different qualities can be drawn. Here the weight map is presented as a 10-step square look-up table. Although only 22 objects (22 letters in the top-map) were used in training all boxes in all weight maps contain numbers.

As can be seen from the top-map in Fig. 2, fifteen objects labeled as *"L"* and seven objects labeled as *"H"* have participated in learning and have been clustered into two groups: group *"L"* in the lower left part and group *"H"* in the middle right part of the top-map. By inspection of all levels and the corresponding maps, it can be found, for example, that the distribution of the values of the i-th variable (weight map of i-th weight level) correspond quite well to the grouping of objects found in the top-map of the Kohonen ANN. The conclusion is that a rule for separation of unknown objects based on the values of the variable i can be suggested. Of course, a first inspection of the weight maps obtained on a small number of objects can merely yield a suggestion. For a more reliable decision rule, more objects and more tests have to be performed, but the idea can be clearly seen how the rules should be formed based on the weight maps of Kohonen ANN.

5   Linguistic Examples

The first example concerns clustering of written texts. The aim of this experiment is to show whether the texts of various authors can form clearly separated clusters, i.e. whether the variables in question carry enough information for clustering. The objects in the study are 37 articles (mainly weekly columns) of five contemporary Slovenian journalists: Alenka Puhar *(A)*, five texts; Boris Jež *(B)*, six texts; Marko Crnkovič *(C)*, six texts; Danilo Slivnik *(D)*, eight texts; and Stane Sedlak *(S)*, six texts. As a standard, a short story *Martin Krpan* by the novelist Fran Levstik (1831–1887) *(F)*, the founding father of Slovenian journalistic writing, was divided into seven parts, each of which was treated as a standalone text. Due to the requirement that approximately the same number of words per author have to be considered, and because the texts of authors were of different lengths, different numbers of texts of each author were taken into account: each author was represented by texts containing altogether approximately 6,000 words.

All 37 texts were processed as described above, and for each of them 21 variables (percentages of words, percentages of unique words, sentence lengths, distributions of punctuation marks and percentage of different 3-word combinations) were obtained using home-developed software[5]. In a preliminary data exploration, the two most commonly used statistical clustering methods were employed. The first one is the standard hierarchical agglomerative clustering method[11,12], while the second one is a more complex 2-dimensional mapping method called *principal component analysis* (PCA)[13, 14]. The hierarchical clustering yields a dendrogram of groups and clusters linked together at various levels of the similarity measure defined in the measurement space. Depending on how the two distances (one distance between two objects and the other between two clusters) are defined, slightly different dendrograms can be obtained. Nevertheless, none of the dendrograms obtained for the 37 texts has led to the formation of clusters relevant to the journalists.

The other method, PCA[13,14], is a linear transformation $X^{new} = LX^{old}$ transforming the old 21-variable-coordinate system into a new one through the minimization of the correlation matrix of the objects $\{X_s\}$ in the study. The new 21 orthogonal axes x_i^{new} are ordered according to the percentage of the total variance v_i which a particular new axis x_i^{new} carries. The first axis x_1^{new} is found in such a way that the variance between all objects and the new axis is maximal, i.e. v_1 is associated with $\max\{v_i\}$. The second axis x_2^{new} (orthogonal to the x_1^{new}) defines the direction of the second-largest percentage of the total variance v_2, etc. Depending on the type of normalization of the representations of 37 texts, the first two (the principal) components carry between 54% and 88% of the total variance, $v^{total} = \sum_i v_i$. Again, regardless of the data pre-processing method, none of the PC1/PC2 plots was able to produce clusters in which the text of particular authors is well separated from each other. The PCA mapping has clearly shown that the 37 texts in the study are not linearly separable. Hence, some other technique has to be employed.

To demonstrate the ability of Kohonen mapping, several Kohonen ANNs (all composed of neurons having 21 weights) were tested. The seven tested ANNs were composed of as few as $36 = 6 \times 6$ neurons to $144 = 12 \times 12$ neurons. All Kohonen ANNs were trained with about 200 epochs, i.e. the entire set of 37 texts went through the particular ANN 200 times. As expected, the smallest ANN was not able to separate all 37 texts because it has only 36 neurons, however, already the next ANN, with $49 = 7 \times 7$ neurons was able to produce six clusters in each of which only the texts of one author were found. There was only one text (S_1) that was placed in the middle of another journalist's cluster. The particular outlier (S_1) turns out to be a persistent "trouble-maker" throughout the entire study, regardless of the size of the ANN; its style of writing could not be distinguished on the basis of the chosen variables. Due to the fact that the 7×7 Kohonen ANN did not clearly separate the clusters (clusters have common borders and/or they touch at several places) larger ANNs were tried as well. Figure 3 shows the final 12×12 top-map in which all clusters are well separated. Each text is marked by a two-letter label which is placed over the position of a neuron that was excited by this particular text. The labels mark journalists (A - Alenka Puhar, B - Boris Jež, C - Marko Crnkovič, D - Danilo Slivnik, L - Fran Levstik, and S - Stane Sedlak), while the labels' indices are assigned to the corresponding texts of each author.

Now that it has been shown that a Kohonen ANN is capable of doing the clustering of texts according to the selected variables, we can proceed with the second example. In the comparative studies of different novelists and poets, especially when studying older authors or some less known works, the question of authenticity of short stories or poems is raised. To offer a possi-

Uniform text representation as input

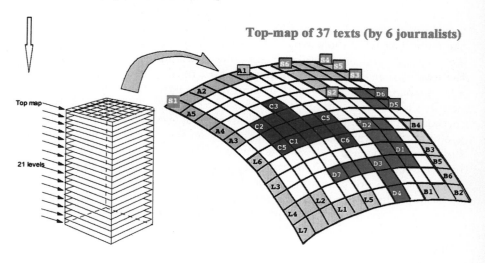

Fig. 3. Top-map of 144 neurons (12×12) showing the clusters of texts. All clusters are not only well formed, but they are also well separated from each other. The only exception is the text "S_1" (Stane Sedlak, text No. 1) which is placed in the middle of the texts A_i of the journalist Alenka Puhar. The more two texts are separated from each other, the more different of styles they have. A larger area for an author indicates a greater diversity in his or her writing.

ble tool to handle such cases, thirty epic poems of three different Slovenian poets (Fig. 4) were pre-processed exactly in the same manner as described previously. Due to the fact that in this experiment only 30 texts (epic poems) with much smaller numbers of words (on the average about 350 words per poem) are considered, the clusters are not so well separated as in the previous example. The left side of Fig. 5 shows the top-map of the 10×10 Kohonen ANN in which 30 poems are clustered. The top-map is obtained in exactly the same manner as the ANNs in the previous example. The poems of Prešeren were, in all ANN experiments (no matter how large or small the applied Kohonen ANN was), always separated into two clusters, both always containing the same poems. The small cluster of three Prešeren's poems touches two clusters of both other poets. It is interesting to note two poems P_1 and M_1 which are the translation of the same poem by the English poet Lord Byron, namely "Parisina." In spite of the fact that the content of both translations is exactly the same, the influence of the writing style of each translator is so personalized that both poems join the proper cluster of the corresponding poet.

Dr. France Prešeren   **Simon Gregorčič**   **Janez Menart**
1800 - 1849   **1844 - 1906**   **1929 -**

Fig. 4. Three Slovenian poets, poems of which were taken into consideration for the second experiment.

The essence of the second experiment was not grouping or clustering of the poems by itself—the clustering ability of linearly non-separable objects by Kohonen ANN was already shown in the first example. The actual goal of the second experiment was to show the ability of classification, i.e., the ability of correct identification of authors on the basis of their texts using an already existing top-map. To demonstrate the classification ability of the trained ANN, three more poems P_x, G_x, and M_x (one additional poem by each poet, Prešeren, Gregorčič, and Menart, respectively), *not* considered during the training of the ANN, were processed in the same way as the thirty poems used for training and were input to the trained 10×10 ANN. The three new poems of the "unknown" poets triggered the neurons indicated by circles on the 10×10 top-map shown in Fig. 5 (right). As it can be seen, the poems P_x and M_x are located clearly within the clusters associated with the correct poet, while the poem G_x, albeit close to Gregorčič's area, did not identify the author unambiguously.

To conclude the discussion of these experiments, a lesser-known feature of the Kohonen mapping will be discussed, namely, the formulation or extraction of logical rules with the help of the weight maps obtained during the Kohonen ANN training. As mentioned above, the correction of weights (eq. (2)) is applied not only to the central (or excited) neuron, but also to the weights of all neurons in the neighborhood close to the central one. It is true that during the learning procedure the extent of the neighborhood area is diminishing

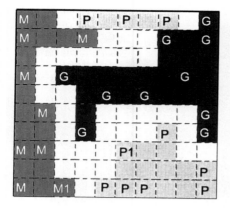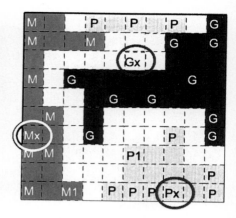

Fig. 5. The top-map of the 10 × 10 Kohonen ANN on the left side shows clusters of 30 epic poems written by Prešeren - *P*, Gregorčič - *G*, and Menart - *M*. The poems assigned with M_1 and P_1 are two different translations of the same poem "Parisina" written in *1815* by the English poet Lord Byron. The first translation P_1 was made around *1835*, while the second one M_1 was made around *1960*. Into the existing 10 × 10 ANN, three additional poems P_x, G_x, and M_x, of "unknown" poets (actually written by Prešeren, Gregorčič, and Menart, respectively) were input and all three were correctly classified within the areas assigned to the particular poet (right).

and at the end of the training only the weights of the central neuron are corrected. Nevertheless, the mere fact that at the beginning the corrections of neighboring weights cover the entire map on each weight level enables the formation of complete weight maps. The term *complete weight map* implies that each weight in each neuron is adapted to its neighborhood regardless of how many times a neuron was excited, i.e. even in the neurons that have never been selected as the central one, the values of their weights were influenced by the corrections of excited neurons in their neighborhoods. Because each level of weights is associated with only one variable and the weights in all levels are in exact superposition (in one-to-one correspondence), the resulting map at all weight levels reflects the topological distribution of the values of this variable exactly in the way the top-map describes the topology of the objects. The final topology of objects depends on the final position of the neurons that are excited by the objects of the training set (Fig. 2). Thus, if the top-map with labels of objects is overlapped by any of the weight maps the correlation between the particular variable and the resulting clusters of objects can be visually inspected.

In Fig. 6, the 10th and the 14th weight level maps are shown superimposed over the 10 × 10 top-map of 30 poems . The shaded area of the 10th weight

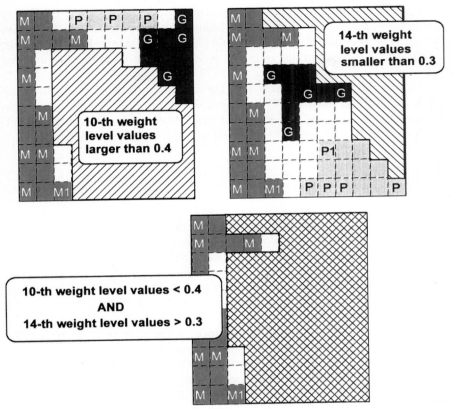

Fig. 6. The 10th and 14th weight level maps superimposed over the 10×10 top-map for 30 poems (upper two maps). In the 10th and 14th weight maps, the weights larger than 0.40 and smaller than 0.30, respectively, are shaded. The overlap of the top-map with both the 10th and the 14th weight level map defines an "if-then" rule for classifying Menart's poems. A low value of the 10th variable (percentage of unknown words), and relatively high 14th variable, (percentage of different adverbs, are thus characteristic for ten Menart's poems (lower map).

level map represents the area where the objects (poems) have values of the 10th variable, percentage of unknown words, larger than 0.40, i.e. the poems placed on the shaded area have more than 40% of the maximal absolute percentage of the unknown words in all poems in the study.

On the other hand, the shaded area of the 14th weight level map marks the positions where the values of the 14th variable, percentage of different adverbs, are smaller than 0.30, i.e. this area represents poems with relatively low percentage of different adverbs. The overlap of the top-map with both the 10th and the 14th weight level maps defines the "if-then" rule for classification of Menart's poems. The logical "if-then" rule obtained by such overlap simply states that the area in which all ten Menart's poems are located is defined by a low percentage of unknown words and by a relatively high percentage of different adverbs. Similar rules can be deduced for all areas shown in the top-map.

Due to the fact that correlations between 21 variables in this study are weak (mostly well bellow 0.7) the logical "if-then" rules can be quite complicated and might combine several "if-then" statements with several "and", "or", and "not" relations between them and involve many variables.

6   Conclusion

The aim of this short essay is to show that Kohonen ANNs can be used in linguistic studies for solving problems for which no simple or linear models exist. It was shown that Kohonen ANNs can do *unsupervised* clustering, classification of unknown objects, and that logical "if-then" rules can be formulated by the use of weight level maps. With the help of logical rules, a so called *formal knowledge* base can be created, which can be further implemented as a basis for various problem-solving options in dedicated linguistics oriented *expert systems*.

In any scientific study the importance of the quality of data and the choice of the proper representation should be mentioned. The quality of data depends on the quality of experiments, which involves proper experimental design (adequate covering of the measurement space of the problem with objects entering the study), the selection of the relevant variables and, above all, on the quality and reliability of the measurements itself. In linguistic studies in the past, the quality of data was traditionally dependent on the manually collected facts. Now, more and more computer supported software is used, but the checking of data and finally the responsibility for their validity is still with the human expert. The extraction of items (variables, features, factors, etc.) that enables a determination of the statistical parameters is a matter of personal choice which should be verified through statistical validation. As is well known, each language has its own specialities and, at the time being, no generally applicable common rules or common software exist that could be used for all languages or, for that matter, for at least for some of them. No matter how complex the applied software is, the linguistic studies still require a large amount of "manual" checks and corrections, which makes research on large textual material quite rare.

The second important issue in such studies is the representation of linguistic objects (text blocks, poems, chapters, sentences, concordances, etc.). The determination of proper representation involves the definitions of variables and subsequently the selection of the relevant ones. Even with the most sophisticated information extraction methods used, the study will not yield reliable results if the chosen representation, i.e. the selected descriptors of objects (variables, features), are not relevant to the problem. In this study, most of the 21 variables were shown to be relevant to the problem. However, for such a small amount of data, an extensive selection of a smaller sub-set would be desirable. Due to the limited space, the problem of selection of the most relevant variables was not addressed here. Nevertheless, the inspection of the overlap between various weight level maps, as explained towards the end of our study, can help to distinguish between the relevant and less important variables at least on a qualitative level.

At the end it should be said again that these examples were shown just for the purpose of demonstration of the ability of Kohonen ANNs. For an actual use in linguistic studies, or in order to come to some generally valid conclusions and results, many more texts of each author and preferably more different authors, together with more meticulous and time-consuming feature selection should be performed. It is evident that 37 articles of six authors or 33 poems of the three poets is far too small a data base for the generation of a solid benchmark or for establishing a generally valid writing or style standard. However, the data base was large enough to show the main principles and methods that can be used in such studies and for explanation of a general idea behind such studies. It is hoped that the reader will regard the essay in this context.

Acknowledgements

The author is highly indebted to the Organizing Committee of the 194th Heraeus Seminar in Bad Honnef for the invitation to talk about linguistics-oriented problems in connection with ANNs. The financial support of a part of this study concerning the application of ANNs through the Grant No. J1-8900-104 by the Ministry of Science and Technology of Slovenia (MZS) is gratefully acknowledged.

References

1. Knuth, D. E.: The Art of Computer Programming, Vol. 2. Seminumerical Algorithms, Addison-Wesley, Reading, Mass. 1975, Chap. 3.4.
2. Zupan, J., Algorithms for Chemists, Wiley, Chichester 1989, Chap. 3.3.
3. for example: BesAna, v 2.03, (Spell-checking Program for Slovenian Texts), Instruction Manual, Amebis, d.o.o., Kamnik, Slovenia 1993, and J. Zupan, SLON-CEK Spell-check program for Slovenian Language (in Slovenian), Informatica 15, (1993)(3), 21–32.
4. Dictionary of the Slovenian Literary Language, DZS, Ljubljana, 1994 (available on CD-ROM).
5. Zupan, J.: Extended version of the *Root-Dictionary of Slovenian Language* (mentioned in Zupan's article in the newspaper DELO, 5th February 1998, p.16).
6. Jakopin, P. and Bizjak, A.: Computer Supported Tagging of Slovenian Texts (in Slovenian, with extended English summary), SLR 45, (1997)(3–4), 421–580.

7. Partee, B. H., Muelen, A., and Wall, R. E.: Mathematical Methods in Linguistics, Studies in Linguistics and Philosophy, Vol. **30**, Kluwer Academic Publishers, Dordrecht, The Netherlands, (1990).

8. Kohonen, T.: Self-Organization and Associative Memory Third Edition, Springer Verlag, Berlin, Germany (1989).

9. Kohonen, T.: An Introduction to Neural Computing, Neural Networks, **1**, (1988), 3–16.

10. Zupan J., and Gasteiger, J.:Neural Networks for Chemists: An Introduction, VCH, Weinheim, (1993).

11. Spaeth, H.: Cluster Analysis for Data Reduction and Classification of Objects, Ellis Horwood, Chichester, (1980).

12. Zupan, J.: Clustering of Large Data Sets, Research Studies Press, (J. Wiley) Chichester, (1982).

13. Massart, D. L., Vandengiste, B. G. M., Deming, S. N. Michotte, Y., and Kaufman, L.: Chemometrics: a textbook, Elsevier, Amsterdam, 1988, p. 201, and 330 ff..

14. Varmuza K., and Lohninger, H.: Principal Component Analysis of Chemical Data, in PCs for Chemists, Ed. Zupan, J.: Elsevier, Amsterdam, 1990, pp. 43–64.

Optimization with Neural Networks

Bo Söderberg

Department of Theoretical Physics, Lund University, Sölvegatan 14A, S-22362
Lund, Sweden

Abstract. The recurrent neural network approach to combinatorial optimization
has during the last decade evolved into a competitive and versatile heuristic method
that can be used on a wide range of problem types. In the state-of-the-art neural
approach the discrete elementary decisions (not necessarily binary) are represented
by continuous Potts mean-field neurons, interpolating between the available discrete
states, with a dynamics based on iteration of a set of mean-field equations. Driven by
annealing in an artificial temperature, they will converge into a candidate solution.

1 Introduction

Combinatorial optimization problems often require a more or less exhaustive
state-space search to achieve exact solutions, with the computational effort
growing exponentially or faster with system size. Various kinds of heuristic
methods are therefore often used to provide good approximate solutions.

Artificial neural networks (ANN) provide a very versatile method for obtaining approximate solutions to combinatorial optimization problems. In
contrast to most other methods, it does not rely on a direct exploration of
the discrete state-space. Instead, it is based on using a set of continuous
variables, giving a probabilistic representation of the elementary decisions
involved in the problem. These variables evolve through a continuous, interpolating space, feeling their way towards a good solution driven by annealing
in an artificial temperature.

Early versions were entirely based on binary (Ising) variables for representing elementary decisions, but modern variants typically involve more
general multi-state (Potts) variables.

The ANN approach yields high-quality solutions to a variety of problem
types, spanning from simple scheduling and assignment problems to complex
communications routing. It also has the advantage that the dynamics can be
directly implemented in VLSI, allowing for an unusually tight bond between
algorithm and hardware. The pure ANN approach can also be extended in
various ways to yield methods for approaching more general problem types.

Key elements in the neural approach are the *mean-field* approximation
(Hopfield and Tank 1985; Peterson and Söderberg 1989), *annealing*, and
for many problems the *Potts* formulation (Peterson and Söderberg 1989).
Recently, also *propagator* methods have proven most valuable for handling
topological complications (Lagerholm et al. 1997; Häkkinen et al. 1998).

2   The Hopfield Model

Simple models for magnetic systems ("spin glasses") strongly resemble recurrent networks – with the direction of a microscopic elementary magnet seen as analogous to the firing-state of a neuron – and have been a strong source of inspiration for neural network studies.

Recurrent networks first attracted interest in the context of associative memory. The Hopfield model (Hopfield 1982) of an associative memory is based on the energy function

$$E(s_1 \ldots s_N) = -\frac{1}{2} \sum_{i,j} w_{ij} s_i s_j \ , \tag{1}$$

in terms of N binary variables (*Ising* neurons) $s_i = \pm 1$, and a set of symmetric weights w_{ij}, appropriately chosen depending on the patterns to be stored. The patterns will then appear as local minima of $E(s)$. With an asynchronous local optimization dynamics

$$s_i(t+1) = \mathrm{sgn}\left(\sum_{j \neq i} w_{ij} s_j(t)\right) \ , \tag{2}$$

the state of the system will be attracted to a nearby local minimum representing one of the patterns. Thus the system has the functionality of an associative memory.

3   Optimization with Hopfield Networks

In a similar way, a Hopfield-type energy function (1) can be used to represent an optimization problem (Hopfield and Tank 1985), with a dedicated choice of weights such that an optimal solution to the problem is represented by minimizing $E(s)$. This yields an archetype of the neural approach to optimization.

The performance is enhanced by using a softer dynamics, with the step function $\mathrm{sgn}(\cdot)$ in (2) replaced by the sigmoid $\tanh(\cdot/T)$, with T an artificial temperature. With annealing in T the resulting mean-field (MF) neurons will, as $T \searrow 0$, relax to a stable configuration representing a tentative solution to the problem. The key problem here is to reach the global minimum or at least a very low-lying local minimum.

Below we will discuss this approach in some detail, using the graph bisection (GBis) problem as an example. Systems with generic multi-state (Potts) spins instead of binary Ising spins can be treated in a similar way (Peterson and Söderberg 1989), and will be discussed later.

3.1   The Graph Bisection Problem

The ANN approach is particularly transparent for GBis because of its binary nature (Peterson and Anderson 1988). The problem is defined as follows. A graph of N nodes is to be partitioned into two halves, such that the cut-size (the number of connections between the two halves) is minimal (Fig. 1A). The graph defines a symmetric connection matrix J such that $J_{ij} = 1$ if nodes

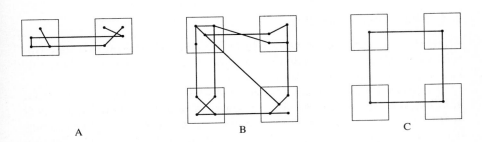

Fig. 1. A, a graph bisection problem. **B**, a $K = 4$ graph partition problem. **C**, an $N = 4$ TSP problem.

i and j are connected, and 0 if they are not (or if they are identical). The distribution of nodes between the two halves can be represented by binary spins as follows. To each node i a binary spin $s_i = \pm 1$ is assigned, representing whether the node winds up in the left or right partition, in terms of Fig. 1A. With this notation, $J_{ij} s_i s_j$ will be non-vanishing only if nodes i, j are connected, yielding 1 if they are in the same partition and -1 if not. Thus, apart from a constant, the cut-size is measured by $-1/2 \sum_{ij} J_{ij} s_i s_j$.

This is not enough for an energy function: the cut-size is trivially minimized by putting all the nodes in the same "half". We need to add a *global constraint term* that penalizes situations where the nodes are not evenly partitioned. Since $\sum s_i = 0$ only for a balanced partition, a term proportional to $(\sum s_i)^2$ will subsequently do the trick. Discarding the constant diagonal part $\sum_i s_i^2 = N$, we can thus represent the problem with a Hopfield energy function (1) with $w_{ij} = J_{ij} - \alpha (1 - \delta_{ij})$:

$$E = -\frac{1}{2} \sum_{ij} J_{ij} s_i s_j + \frac{\alpha}{2} \left(\left(\sum_i s_i \right)^2 - \sum_i s_i^2 \right) , \qquad (3)$$

where the constraint coefficient α sets the relative strength of the penalty term.

The generic form of (3) is

$$E = \text{Cost} + \text{Global constraint} , \qquad (4)$$

which is characteristic of combinatorial optimization problems. The source of the difficulty inherent in this kind of problem is very transparent here: the system is frustrated by the competition between the two terms, cost and global constraint. This frequently leads to the appearance of many local energy-minima.

The next step is to find an efficient procedure for minimizing E, such that suboptimal local minima are avoided as much as possible.

3.2   Simulated Annealing

Attempting to minimize E according to a local optimization rule (like (2)) will very likely give the result that the system ends up in some suboptimal local minimum close to the starting point.

A better strategy is then to employ a stochastic algorithm that allows for uphill moves. One such method is Simulated Annealing (SA) (Kirkpatrick et al. 1983), in which sequences of configurations are generated with neighborhood search methods, in a way designed to emulate the Boltzmann distribution

$$P[s] = \frac{1}{Z}e^{-E[s]/T} \ . \tag{5}$$

Here, Z is the *partition function*

$$Z = \sum_{[s]} e^{-E[s]/T} \ , \tag{6}$$

while $T > 0$ is a artificial temperature representing the noise level of the system. For $T \searrow 0$ the Boltzmann distribution becomes concentrated to the configuration minimizing E. If configurations are generated with a slowly decreasing T – annealing –, they are less likely to get stuck in local minima than if T is set to 0 from the start. The disadvantage of this method is that in order to be reasonably certain to hit the global minimum, one has to employ a very slow annealing schedule, which is very CPU consuming.

3.3   The Mean-Field Approximation

The MF approach aims at approximating the stochastic SA method with a deterministic dynamics, based on the MF approximation, which can be derived as follows. Introduce for each spin s_i a new variable v_i, living in a linear space (in this case $\mathbb{R}$) containing the compact state-space ($\{\pm 1\}$) of the spin, and set it equal to the spin by means of a Dirac delta function. This yields

$$Z = \sum_{[s]} \int d[v]e^{-E[v]/T} \prod_i \delta(s_i - v_i) \ . \tag{7}$$

Next, Fourier-expand the delta functions in terms of conjugate variables u_i, giving

$$Z \propto \sum_{[s]} \int d[v] \int d[u] e^{-E[v]/T} \prod_i e^{u_i(s_i - v_i)} \ . \tag{8}$$

Then carry out the (by now trivial) sum over the spins $\{s\}$, and write the resulting product of cosh factors as a sum of logarithms in the exponent:

$$Z \propto \int d[v] \int d[u] e^{-E[v]/T - \sum_i u_i v_i + \sum_i \log \cosh u_i} \ . \tag{9}$$

The partition function is now rewritten entirely in terms of the new, continuous variables $\{u, v\}$, with an effective energy $E_{\text{eff}}[v, u] = E[v] + T \sum_i (u_i v_i - \log \cosh u_i)$ in the exponent. So far no approximation has been made.

We next make a saddle-point approximation, assuming that the integral in (9) is dominated by an extremal value of E_{eff}. This occurs for $\partial E_{\text{eff}} / \partial v_i = \partial E_{\text{eff}} / \partial u_i = 0$, yielding

$$v_i = \tanh u_i \ , \tag{10}$$

$$u_i = -\frac{1}{T} \frac{\partial E[v]}{\partial v_i} \ . \tag{11}$$

Upon elimination of $\{u\}$, we obtain the *MF equations*

$$v_i = \tanh \left(-\frac{1}{T} \frac{\partial E[v]}{\partial v_i} \right) \ . \tag{12}$$

The solutions $\{v_i\}$, the *MF neurons*, represent approximations to the thermal averages $\langle s_i \rangle_T$ of the original binary spins.

For the Hopfield energy function (1) we obtain

$$v_i = \tanh \left(\frac{1}{T} \sum_j w_{ij} v_j \right) \ . \tag{13}$$

The result (13) is obviously a softer version of (2).

3.4   The Mean-Field Dynamics

The MF equations (12) or (13) are solved iteratively, either synchronously or asynchronously, under annealing in T. This yields a deterministic dynamics, characteristic of a recurrent ANN. The resulting neurons $\{v_i\}$ will not be confined to ± 1, but will populate the intervening interval $[-1, 1]$.

The MF dynamics typically exhibits two phases. At high temperatures the sigmoid $\tanh(\cdot/T)$ is very smooth, and the system typically relaxes into a trivial fixed point $v_i^0 = 0$. As the temperature is lowered the sigmoid becomes steeper, and at some critical temperature T_c a bifurcation occurs, where v_i^0 becomes unstable and non-trivial fixed points v_i^* emerge. As $T \to 0$, the dynamics of (2) is recovered, and v_i^* are forced to ± 1, representing a specific decision made as to the solution of the problem in question (Fig. 2).

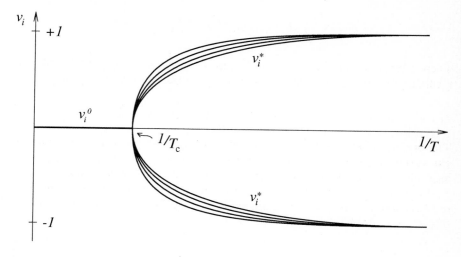

Fig. 2. Fixed points of the binary MF dynamics and the bifurcation, schematically depicted in terms of $\{v_i\}$ as a function of $1/T$.

The position of T_c can be estimated by linearizing (13) around the trivial fixed point $v_i^0 = 0$, i.e. by replacing the sigmoid function (tanh) by its argument. This leads to

$$v_i = \frac{1}{T} \sum_j w_{ij} v_j \tag{14}$$

valid for small $\{v\}$.

For *synchronous* updating it is clear that the trivial fixed point becomes unstable as soon as an eigenvalue λ of the matrix $\mathbf{w}/T$ is > 1 in absolute value. This happens if T equals the largest positive eigenvalue of $\mathbf{w}$ ($\Rightarrow \lambda = 1$), or minus the largest negative one ($\Rightarrow \lambda = -1$), whichever is larger. Prior estimation of the largest eigenvalues is straightforward; apart from determination of T_c, this is also important for avoiding oscillatory behavior, resulting for $\lambda = -1$.

In the preferred case of *serial* updating, the philosophy is the same but the analysis slightly more complicated; the main change (in the absence of diagonal terms) is that the fixed point can only loose stability with $\lambda = 1$, eliminating oscillatory behavior (Peterson and Söderberg 1989). This is why the diagonal terms where removed in (3); as a consequence all self-couplings are eliminated, in the sense that the updated value of v_i does not depend on its old value.

4   Optimization with Potts Neural Networks

For GBis and many other optimization problems, an encoding in terms of binary elementary variables is natural. However, for a generic problem, the natural elementary decisions are not always binary, but often of the type one-of-K with $K > 2$.

In early attempts to approach such problems by neural network methods, the problem was forced into a binary form by the use of *neuron multiplexing* (Hopfield and Tank 1985). For each elementary K-fold decision, a set of K binary 0/1-neurons was used, with the additional constraint that precisely one of them be on (=1). These *syntactic* constraints were implemented in a soft manner as penalty terms. As it turned out in the original work on the traveling salesman problem (Hopfield and Tank 1985), as well as in subsequent investigations for the graph partition problem (Peterson and Söderberg 1989), this approach does not generally yield high-quality solutions in a parameter-robust way.

An alternative and better encoding results from using *Potts neurons* with the syntactic constraint built in. In this way the dynamics is confined to the relevant parts of the solution space (Fig. 3), leading to a drastically improved performance.

4.1   Potts Spins

A K-state Potts spin should have K possible values (states). For our purposes, the best representation is in terms of a vector of 0/1-components. Thus, denoting a spin by $\mathbf{s} = (s_1, s_2, \ldots, s_K)$, the ath possible state is given by setting the ath component of $\mathbf{s}$ to 1, and the rest to zero. The state vectors point to the corners of a regular K-simplex (see Fig. 3 for the case of $K = 3$), and fulfill by definition the syntactic constraint $\sum_a s_a = 1$.

Potts Mean Field Equations. The MF equations for a system of Potts spins $\mathbf{s}_i$ with a given energy function $E(\mathbf{s})$ are derived following the same path as in the Ising neuron case: rewrite the partition function as an integral over $\mathbf{u}_i$ and $\mathbf{v}_i$ (which now live in $\mathbb{R}^K$), and make a saddle-point approximation. One obtains

$$u_{ia} = -\frac{\partial E(\mathbf{v})}{\partial v_{ia}}/T \tag{15}$$

$$v_{ia} = \frac{e^{u_{ia}}}{\sum_b e^{u_{ib}}} \tag{16}$$

to be solved by iteration, sequentially in i, with annealing in T.

From (16) it follows that the *Potts neurons* $\mathbf{v}_i$, approximating the thermal averages of $\mathbf{s}_i$, satisfy

$$v_{ia} > 0, \qquad \sum_a v_{ia} = 1 . \tag{17}$$

One can think of the neuron component v_{ia} as the *probability* for the ith Potts spin to be in state a. For $K = 2$ one recovers the formalism of the Ising case in a slightly disguised form.

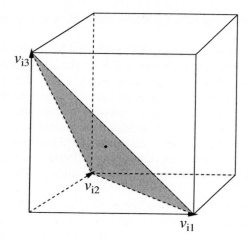

Fig. 3. The volume of solutions corresponding to the neuron multiplexing encoding for $K = 3$. The shaded plane represents the solution space of the corresponding Potts encoding; its corners correspond to the allowed Potts spin states, while the black dot at the center marks the trivial fixed point of no decision.

As in the binary case, there is at high T a trivial fixed point $v_{ia}^0 = 1/K$ corresponding to no decision, and a linear stability analysis yields a critical temperature T_c. For $T \to 0$, the Potts neurons are forced to a decision, with one component approaching unity, and the others zero (see Fig. 3).

4.2   Straightforward Applications

For a large class of combinatorial optimization problems, a straight-forward Potts MF ANN approach can be used, with the following basic steps:

- Map the problem onto a Potts neural network by a suitable encoding of the solution space and an appropriate choice of energy function.
- Compute a suitable starting temperature, e.g. by using a linear stability analysis of the MF dynamics to compute T_c.
- While annealing, solve the MF equations iteratively.
- When the system has settled, the solutions are checked with respect to constraint satisfaction. If needed, one may perform a simple corrective post-processing, or rerun the system (possibly with modified constraint coefficients).

Note that with a Potts encoding there is no difficulty in treating problems having a varying K, in the sense that distinct elementary decisions have different order.

Below, a selected sample of straightforward applications will be briefly discussed, with references to relevant original articles.

The Graph Partition Problem. A generalization of graph bisection is *graph partitioning* (GP): an N-node graph, defined by a symmetric connection matrix $J_{ij} = 0, 1$, $i \neq j = 1, \ldots, N$, is to be partitioned into K subsets of N/K nodes each, while minimizing the cut-size, i.e. the number of connected node pairs winding up in different subsets (see Fig. 1B).

This problem is naturally encoded in terms of Potts spins as follows: for each node $i = 1, \ldots, N$, a K-state Potts spin $\mathbf{s}_i = (s_{i1}, \ldots, s_{iK})$ is assigned, where the component s_{ia} being 1 represents the decision that node i should belong to subset a. A suitable energy function is then given by (cf. (3))

$$ E = -\frac{1}{2} \sum_{i,j=1}^{N} J_{ij} \mathbf{s}_i \cdot \mathbf{s}_j + \frac{\alpha}{2} \left(\left(\sum_{i=1}^{N} \mathbf{s}_i \right)^2 - \sum_{i=1}^{N} \mathbf{s}_i^2 \right) , \qquad (18) $$

where the first term is a cost term (cut-size), while the second is a penalty term with a minimum when the nodes are equally partitioned into the K subsets. Note that the (trivial) diagonal contributions are eliminated in the second term; this stabilizes the dynamics for sequential updating, in analogy to the case for binary neurons (Peterson and Söderberg 1989).

The Traveling Salesman Problem. In the traveling salesman problem (TSP) the coordinates $\mathbf{x}_i \in \mathcal{R}^2$ of a set of N cities are given. A closed tour of minimal total length is to be chosen such that each city is visited exactly once. This problem is somewhat reminiscent of (the trivial) $K = N$ graph partition (see Fig. 1C). We define an N-state Potts neuron $\mathbf{s}_i$ for each city $i = 1, \ldots, N$, such that the component s_{ia} $(a = 1, \ldots, N)$ is 1 if city i has the tour number a, and 0 otherwise. Let d_{ij} be the distance between cities i and j. Then a suitable energy function is given by

$$ E = \sum_{i,j=1}^{N} d_{ij} \sum_{a=1}^{N} s_{ia} s_{j(a+1)} + \frac{\alpha}{2} \left(\left(\sum_{i=1}^{N} \mathbf{s}_i \right)^2 - \sum_{i=1}^{N} \mathbf{s}_i^2 \right) , \qquad (19) $$

where the first term is a cost term, and the second a soft constraint term penalizing configurations where two cities are assigned the same tour number (Peterson and Söderberg 1989; Peterson 1990).

Scheduling Problems. Scheduling problems have a natural formulation in terms of Potts neurons. In its purest form, a scheduling problem has the simple structure: for a given set of events, a time-slot and a location are to be chosen, each from a set of allowed possibilities, such that no clashes occur. Such a problem consists entirely in fulfilling a set of basic no-clash constraints, each of which can be encoded as a *penalty term* that will vanish when the constraint is obeyed (Gislén et al. 1989).

In many realistic scheduling applications, however, there exist additional preferences within the set of legal schedules, that lead to the appearance also of *cost terms*. In Gislén et al. 1992b, a set of real-world scheduling problems was successfully dealt with.

5 Refinements of the Potts MF Approach

For many optimization problems, complications arise that need special treatment, and the straight-forward method of the previous section will have to be modified or supplemented. In this section we will address a set of such common complications.

5.1 Non-Quadratic Energy Functions

Not all optimization problems can be formulated in terms of a quadratic energy function, even though the state-space can be encoded in terms of a set of Potts neurons. This presents no principal difficulty, and one can still use (15). However, a possible practical problem arises from the induced self-couplings in the energy function (terms with non-linearities in a single spin), that might affect performance (Peterson and Söderberg 1989).

With a quadratic E, self-couplings can be avoided by removing all diagonal terms, $s_{ia}s_{ib} \to \delta_{ab}s_{ia}$. Such a procedure can be generalized to any polynomial E. Although in principle any energy-function of a finite number N of Potts spins can be rewritten as a polynomial of at most degree N, this can be difficult in practice for large N.

An efficient and general method for avoiding self-couplings altogether is to replace the derivative in (15) by a difference:

$$u_{ia} = -\frac{1}{T}\left[E(\mathbf{v})|_{\mathbf{v}_i=\mathbf{e}_a} - E(\mathbf{v})|_{\mathbf{v}_i=\mathbf{0}}\right] \ , \tag{20}$$

where $\mathbf{v}_i = (v_{i1}, \ldots, v_{iK})$, and $\mathbf{e}_a$ is the principal unit-vector in the a-direction. Whenever E is free of self-couplings, (15) and (20) are equivalent.

5.2 Inequality Constraints

In the problems mentioned in the previous section, the constraints considered were all of the *equality* type, $f(s) = 0$, that could be implemented with quadratic penalty terms $\propto f(s)^2$. However, in many optimization problems, in particular those of resource allocation type, one has to deal with *inequalities*. An inequality constraint, $g(s) \leq 0$, can be implemented with a penalty term, e.g. proportional to

$$\Phi(g) = g\Theta(g) \ , \tag{21}$$

with Θ the Heaviside step function: $\Theta(x) = 1$ if $x > 0$ and 0 otherwise. Of course, such a non-polynomial term in the energy function must be handled using (20).

A problem category with inequality constraints is the *knapsack problem*, where one has a set of N *items* i with associated *utilities* c_i and *loads* a_{ki}. The goal is to fill a "knapsack" with a subset of the items such that their total utility,

$$U = \sum_{i=1}^{N} c_i s_i \; , \tag{22}$$

is maximized, subject to a set of M load constraints,

$$\sum_{i=1}^{N} a_{ki} s_i \leq b_k, \qquad k = 1, \ldots, M \; , \tag{23}$$

defined by load *capacities* $b_k > 0$. In (Ohlsson et al. 1993), a set of difficult random knapsack problems were approached with a neural method, based on the energy function

$$E = -\sum_{i=1}^{N} c_i s_i + \alpha \sum_{k=1}^{M} \Phi \left(\sum_{i=1}^{N} a_{ki} s_i - b_k \right) \tag{24}$$

in terms of binary spins $s_i \in \{1, 0\}$, representing whether or not item i goes into the knapsack. In (Ohlsson and Pi 1997) this approach was extended to multiple knapsacks and generalized assignment problems, using Potts MF neurons.

5.3   Routing Problems

Many *network-routing* problems can be conveniently handled in a Potts MF approach. We will briefly illustrate the method on the simple shortest-path problem: given a set of N cities connected by a network of roads, find the shortest path from city a to city b.

Regarding the cities as nodes in a network, and denoting by d_{ij} the length of a direct path, an *arc*, between cities i and j, the problem is equivalent to finding the shortest sequence of arcs leading from node a to b.

This problem can be solved in polynomial time using, e.g. the Bellman-Ford (BF) algorithm (Bellman 1958), where every node i estimates its distance D_{ib} to b, minimized with respect to the choice of a continuation node j among its neighbors (nodes directly connected to i via an arc):

$$D_{ib} = \min_j \left(d_{ij} + D_{jb} \right), \quad i \neq b \; , \tag{25}$$

while $D_{bb} = 0$. Iteration of (25) gives convergence in less than N steps, and D_{ab} as well as the path can be directly read off.

By introducing a Potts MF neuron to handle the neighbor choice for each node, we obtain a softer version of the BF algorithm,

$$D_{ib} = \sum_j v_{ij} \left(d_{ij} + D_{jb} \right), \qquad \text{with} \qquad v_{ij} = e^{-(d_{ij}+D_{jb})/T} \ . \qquad (26)$$

In the limit of $T \to 0$, a strict minimum is obtained, and BF is recovered. Note that this is an example of a problem where the natural Potts encoding lead to a varying K, since different nodes typically have a different number of neighbors.

In a similar way also more complex routing problems can be formulated in terms of an optimal local neighbor choice, that naturally lends itself to a MF recast. An appealing feature of such an approach is the *locality* inherited from BF: the information needed for the neighbor choice is local to the node and its neighbors.

Note that the basic philosophy here, borrowed from the BF algorithm, is slightly unconventional. Instead of attempting to minimize a global energy function, each node strives to minimize its own, local objective function.

In (Häkkinen et al. 1998) a set of communication routing problems, based on various combinations of unicast and multicast in a network with arcs of finite capacity, were approached along these lines.

The Propagator. Typical of the type of approach sketched above, is the use of *node-node* MF Potts neurons v_{ij} for the local direction of a path from a node to its neighbors.

As a convenient tool for monitoring the global topological properties of fuzzy paths (occurring for $T > 0$), a propagator formalism has been devised (Häkkinen et al. 1998), defined in terms of a propagator matrix $\mathbf{P} = (\mathbf{1} - \mathbf{v})^{-1}$, or

$$P_{ij} = \delta_{ij} + v_{ij} + \sum_k v_{ik}v_{kj} + \sum_{kl} v_{ik}v_{kl}v_{lj} + \dots \ . \qquad (27)$$

In the absence of loops, P_{ij} will be 1 if there exists a path from i to j, and 0 if not (sharp path). For the case of fuzzy paths, the propagator has an obvious probabilistic interpretation. In addition, the propagator will diverge in the presence of loops, and can therefore be used to avoid loop formation.

By using suitable methods for doing the matrix-inversion, required for the computation of the propagator elements, in a clever way, the increase in computational demand can be kept manageable.

As an example of the use of an analogous formalism in a somewhat different setting, see (Lagerholm et al. 1997), where a set of semi-realistic airline-crew scheduling problems was treated, with v_{ij} in (27) controlling the transfer of a crew from a flight i to a connecting flight j.

5.4   Hybrid Approaches

A large class of optimization problems can be viewed as *parametric assignment* problems, containing elements of both discrete assignment and parametric fitting to given data, e.g. using templates with a known structure. The assignment part can then be encoded in terms of Potts neurons, while the template part may be formulated in terms of a set of continuous, adjustable parameters. This leads to a kind of hybrid approach, *deformable templates.*

Track finding in high energy physics, where circular or spiral-shaped tracks are to be extracted from a set of data points, is an example where this approach is suitable (Ohlsson et al. 1992; Gyulassy and Harlander 1991). In addition, certain pure assignment problems with a well-defined geometric structure can be recast in this form; an example is given by TSP (Durbin and Willshaw 1987).

6   Summary and Discussion

From the above-discussed applications of the Potts MF approach to optimization, the following general picture emerges: for a variety of problem types and sizes, the MF method consistently performs (without excessive fine-tuning) roughly in parity with available comparison methods, typically designed to perform well for the particular problem class. With a prior estimate of T_C, convergence is consistently achieved after a very modest number (typically 50–100) of iterations, independently of problem size.

Inequality constraints and other complications leading to non-polynomial energy functions, can be handled in a straight-forward manner.

For communication routing problems and other problems of a topological nature, a Potts approach can be devised, where a propagator formalism enables the monitoring of global topological features, while the dynamics is based on a strictly local information.

For parametric assignment problems, like track-finding, a deformable templates method can be used, utilizing Potts neurons in combination with analog variables; a similar hybrid approach can be applied also to a class of low-dimensional geometrical assignment problems, such as TSP.

Generalizations. A binary (Ising) spin can be considered as a vector living on a "sphere" in one dimension. The MF approach can be generalized to variables defined on spheres in higher dimensions (or indeed in any compact manifold). Such *rotor* neurons can be used in geometrical optimization problems involving angular variables (Gislén et al. 1992a).

The MF approximation can be viewed as a variational approach – the true energy $E(s_i)$ is approximated by a trial one $E_0(s_i; u_i)$ (in this case linear), which is optimized with respect to the variational parameters u_i (the coefficients). This general procedure can be used in a wide range of situations not necessarily confined to discrete optimization problems (see e.g. Jönsson et al. 1993, and references therein, for an application to polymers).

References

Bellman, R. (1958): On a routing problem. Quarterly of Appl. Math. **16**, 87–90.

Durbin, R., and Willshaw, D. (1987): An analog approach to the traveling salesman problem using an elastic net method. Nature **326**, 689–691.

Gislén, L., Peterson, C., and Söderberg, B. (1989): Teachers and Classes with Neural Networks. Int. J. Neural Syst. **1**, 167–176.

Gislén, L., Peterson, C., and Söderberg, B. (1992 a): Rotor neurons – Basic formalism and dynamics. Neural Computat. **4**, 737–745.

Gislén, L., Peterson, C., and Söderberg, B. (1992 b): Complex scheduling with Potts neural networks. Neural Computat. **4**, 805–831.

Gyulassy, M., and Harlander, H. (1991): Elastic tracking and neural network algorithms for complex pattern recognition. Comput. Phys. Commun. **66**, 31–46.

Häkkinen, J., Lagerholm, M., Peterson, C., and Söderberg, B. (1998): A Potts neuron approach to communication routing. Neur. Computat. **10**, 1587-1599.

Hopfield, J.J. (1982): Neural networks and physical systems with emergent collective computational abilities. Proc. Natl. Acad. Sci. USA **79**, 2554–2558.

Hopfield, J.J., and Tank, D.W. (1985): Neural computation of decisions in optimization problems. Biol. Cybern. **52**, 141–152.

Jönsson, B., Peterson, C., and Söderberg, B. (1993): A variational approach to correlations in polymers. Phys. Rev. Lett. **71**, 376–379.

Kirkpatrick, S., Gelatt, C.D., and Vecchi, M.P. (1983): Optimization by simulated annealing. Science **220**, 671–680.

Lagerholm, M., Peterson, C., and Söderberg, B. (1997): Airline crew scheduling with Potts neurons. Neur. Computat. **9**, 1589–1599.

Ohlsson, M., Peterson, C., and Söderberg, B. (1993): Neural networks for optimization problems with inequality constraints – The knapsack problem. Neural Computat. **5**, 331–339.

Ohlsson, M., Peterson, C., and Yuille, A. (1992): Track finding with deformable templates – The elastic arms approach. Comput. Phys. Commun. **71**, 77–98.

Ohlsson, M., and Pi, H. (1997): A study of the Mean Field Approach to Knapsack Problems. Neur. Netw. **10**, 263–271.

C. Peterson (1990): Parallel distributed approaches to combinatorial optimization problems – Benchmark studies on TSP. Neural Computat. **2**, 261–269.

Peterson, C., and Anderson, J. R. (1988): Neural Networks and NP-complete Optimization Problems – A Performance Study on the Graph Partition Problem. Compl. Syst. **2**, 59–89.

Peterson, C., and Söderberg, B. (1989): A new method for mapping optimization problems onto neural networks. Int. J. Neural Syst. **1**, 3–22.

Dynamics of Networks and Applications

R. Vilela Mendes

Grupo de Física-Matemática
Complexo Interdisciplinar, Universidade de Lisboa
Av. Gama Pinto, 2, 1699 Lisboa Codex Portugal
e-mail: vilela@alf4.cii.fc.ul.pt

Abstract. A survey is made of several aspects of the dynamics of networks, with special emphasis on unsupervised learning processes, non-Gaussian data analysis and pattern recognition in networks with complex nodes.

1 Recurrent Networks, Dynamics, and Applications

There are three large classes of neural networks. One is the class of *multilayered feedforward networks* which, through supervised learning, are used to approximate nonlinear functions. The second is the class of *relaxation networks* with symmetric synaptic connections, like the Hopfield network. Under time evolution the relaxation networks evolve to fixed points which, in successful applications, are identified with memorized patterns. In the last class one includes all networks with arbitrary connections which are neither completely feedforward nor symmetric. They are called *recurrent networks*.

Recurrent networks exhibit a complex variety of temporal behavior and are being increasingly proposed for use in many engineering applications (Sato, Murakami, and Joe 1990, Sato, Joe, and Hirahara 1990, Pearlmutter 1989, Lin and Horne 1996). In contrast to feedforward or relaxation networks, the rich dynamical structure of recurrent networks makes them a natural choice to process temporal information. Even for the learning of nonlinear functions the feedback structure improves the reproduction of discontinuities or large derivative regions (Jones 1993). Also, in some cases, the feedback structure is a way to enhance, through a choice of architecture, the sensitivity of the network to particular features to be detected (Amorim et al. 1998). The learning algorithms used for feedforward networks may be generalized to the recurrent case (Pineda 1987-9, Almeida 1987, Almeida 1988, Amorim et al. 1998).

Feedback loops and coexistence of quiescent, oscillatory, and chaotic behavior are also present in biological systems and only recurrent networks are appropriate models for these phenomena (Skarda and Freeman 1987, Amit and Brunel 1997).

The whole field of recurrent networks, both for engineering and biological applications, is developing in many directions. Their global analysis is rather

involved (Hertz 1995). Their characterization as dynamical systems is sharpened by a decomposition theorem. Continuous state - continuous time neural
networks, as well as many other systems (Grossberg 1988), may be written
in the Cohen-Grossberg (Cohen and Grossberg 1983) form

$$\frac{dx_i}{dt} = a_i(x_i) \left\{ b_i(x_i) - \sum_{j=1}^{n} w_{ij} d_j(x_j) \right\} . \tag{1}$$

Dynamical systems of this type have a decomposition property. Define

$$
\begin{aligned}
w_{ij} &= w_{ij}^{(S)} + w_{ij}^{(A)} , \\
w_{ij}^{(S)} &= \tfrac{1}{2} \left(w_{ij} + w_{ji} \right) , \\
w_{ij}^{(A)} &= \tfrac{1}{2} \left(w_{ij} - w_{ji} \right) , \\
V^{(S)} &= - \sum_{i=1}^{n} \int^{x^i} b_i(\xi_i) d_i'(\xi_i) d\xi^i + \tfrac{1}{2} \sum_{j,k=1}^{n} w_{jk}^{(S)} d_j(x^j) d_k(x^k) , \\
H &= \sum_{i=1}^{n} \int^{x^i} \frac{d_i(\xi_i)}{a_i(\xi_i)} d\xi_i ,
\end{aligned}
\tag{2}
$$

such that we have the following

Theorem (Vilela Mendes and Duarte 1992). If $a_i(x_i)/d_i'(x_i) > 0 \ \forall x, i$
and the matrix $w_{ij}^{(A)}$ has an inverse then the vector field $\overset{\bullet}{x}_i$ in Eq.(1) de-
composes into one gradient component and one Hamiltonian component,
$\overset{\bullet}{x}_i = \overset{\bullet}{x}_i^{(G)} + \overset{\bullet}{x}_i^{(H)}$, where

$$
\begin{aligned}
\overset{\bullet}{x}_i^{(G)} &= -\frac{a_i(x_i)}{d_i'(x_i)} \frac{\partial V^{(S)}}{\partial x_i} = - \sum_j g^{ij}(x) \frac{\partial V^{(S)}}{\partial x_j} , \\
\overset{\bullet}{x}_i^{(H)} &= - \sum_j a_i(x_i) w_{ij}^{(A)}(x) a_j(x_j) \frac{\partial H}{\partial x_j} = \sum_j I^{ij}(x) \frac{\partial H}{\partial x_j} ,
\end{aligned}
\tag{3}
$$

and

$$
\begin{aligned}
g^{ij}(x) &= \frac{a_i(x_i)}{d_i'(x_i)} \delta^{ij} , \\
\omega_{ij}(x) &= -a_i(x_i)^{-1} \left(w^{(A)-1} \right)_{ij}(x) a_j(x_j)^{-1} ,
\end{aligned}
\tag{4}
$$

$\left(\omega_{ij} I^{jk} = \delta_i^k \right)$. $g^{ij}(x)$ and $\omega_{ij}(x)$ are the components of a Riemannian metric
and a symplectic form.

Proof: The decomposition follows by direct calculation from (1) and (2).
The conditions on $a_i(x_i)$, $d_i'(x_i)$ and $w^{(A)}$ insure that g is a well defined metric
and ω is non-degenerate. Indeed let v be a vector such that $\sum_i v^i \omega_{ij} = 0$.
Then

$$0 = \sum_{ij} v^i \omega_{ij} a_j(x_j) w_{jk}^{(A)}(x) = - \frac{v^k}{a_k(x_k)}$$

would imply $v^k = 0 \ \forall k$. That ω is a closed form follows from the fact that
ω_{ij} depends only on x_i and x_j.

The identification, in the system (1), of just one gradient and one Hamiltonian component with explicitly known potential and Hamiltonian functions,

is a considerable simplification as compared to a generic dynamical system. Recall that in the general case, although such a decomposition is possible locally (Vilela Mendes and Duarte 1981), explicit functions are not easy to obtain unless one allows for one gradient and $n - 1$ Hamiltonian components. Notice that the decomposition of the vector field does not decouple the dynamical evolution of the components. In fact it is the interplay of the dissipative (gradient) and the Hamiltonian component that leads, for example, to limit-cycle behavior.

For the case of symmetric connections $w_{ij} = w_{ji}$ one recovers the Cohen-Grossberg result (Cohen and Grossberg 1983) that states that a symmetric system of the type (1) has a Lyapunov function $V^{(S)}$ of which Hopfield's (Hopfield 1984) "energy" function is a particular case. For the symmetric case the existence of a Lyapunov function guarantees global asymptotic stability of the dynamics. However not all vector fields with a Lyapunov function are differentially equivalent to a gradient field. Therefore the fact that a gradient vector is actually obtained gives additional information, namely about the structural stability of the model.

A necessary condition for the structural stability of the gradient vector field is the non-degeneracy of the critical points of $V^{(S)}$, namely $\det \left\| \frac{\partial^2 V^{(S)}}{\partial x_i \partial x_j} \right\| \neq 0$ at the points where $\frac{\partial V^{(S)}}{\partial x_i} = 0$. In a gradient flow all orbits approach the critical points as $t \to \infty$. If the critical points are non-degenerate then the gradient flow satisfies the conditions defining a Morse-Smale field, except perhaps the transversality conditions for stable and unstable manifolds of the critical points. However because Morse-Smale fields are open and dense in the set of gradient vector fields, any gradient flow with non-degenerate critical points may always be C^1-approximated by a (structurally stable) Morse-Smale gradient field. Therefore given a symmetric model of the type (1), the identification of its gradient nature provides an easy way to check its robustness as a physical model.

As an example of the decomposition applied to a biological model consider the Wilson-Cowan model of a neural oscillator without refractory periods, in the antisymmetric coupling case considered by most authors,

$$
\begin{aligned}
\dot{x}_1 &= -x_1 + S\left(\rho_1 + w_{11}x_1 + w_{12}x_2\right), \\
\dot{x}_2 &= -x_2 + S\left(\rho_2 + w_{21}x_1 + w_{22}x_2\right),
\end{aligned}
\tag{5}
$$

with $w_{12} = -w_{21}$ and S is the sigmoid function $(1 + e^{-x})^{-1}$. Changing variables to

$$
z_i = \rho_i + \sum_{i=1}^{2} w_{ij} x_j
$$

one obtains

$$
\begin{aligned}
\dot{z}_1 &= -\frac{\partial V}{\partial z_1} + w_{12} \frac{\partial H}{\partial z_2}, \\
\dot{z}_2 &= -\frac{\partial V}{\partial z_2} - w_{12} \frac{\partial H}{\partial z_1}
\end{aligned}
\tag{6}
$$

with
$$V = \frac{1}{2}\sum_i \left\{ z_i^2 - \rho_i z_i + w_{ii} \log\left(1 - S(z_i)\right)\right\} ,$$
$$H = \sum_i \log\left(1 - S(z_i)\right) .$$

The model is completely described by these functions, the bifurcation sets (Hoppensteadt and Izhikevich 1997), for example, being characterized by $\triangle V = 0$ for Andronov-Hopf bifurcations and by

$$\frac{\partial^2 V}{\partial z_1^2}\frac{\partial^2 V}{\partial z_2^2} + w_{12}^2 \frac{\partial^2 H}{\partial z_1^2}\frac{\partial^2 H}{\partial z_1^2} = 0$$

for saddle-node bifurcations.

2   Unsupervised Learning in Generalized Networks and the Processing of non-Gaussian Signals

2.1   Unsupervised Learning in General Networks

Doyne Farmer (Doyne Farmer 1990) has shown that there is a common mathematical framework where neural networks, classifier systems, immune networks and autocatalytic reaction networks may be treated in a unified way. The general model in which all these models may be mapped looks like a neural network where, in addition to the node state variables (x_i) and the connection strengths (W_{ij}), there is also a node parameter (θ_i) with learning capabilities (Fig.1). The node parameter represents the possibility of changing, through learning, the nature of the linear or non-linear function $f_i(\sum_j W_{ij}x_j)$ at each node. In the simplest case θ_i will be simply an intensity parameter. Therefore the degree to which the activity at node i influences the activity at other nodes depends not only on the connection strengths (W_{ij}) but also on an adaptive node parameter θ_i. In some cases, as in the B-cell immune network, the node parameter is the only means to control the relative influence of a node on others, the connection strengths being fixed chemical reaction rates.

Hebbian - type learning with a node parameter. We will denote by x_i the output of node i. Hebbian learning (Hebb 1949) is a type of unsupervised learning where a connection strength W_{ij} is reinforced whenever the product $x_i x_j$ is large. As shown by several authors, Hebbian learning extracts the eigenvectors of the correlation matrix Q of the input data

$$Q_{ij} = \langle x_i x_j \rangle \tag{7}$$

where $\langle ... \rangle$ means the sample average. If the learning law is local, the lines of the connection matrix W_{ij} all tend to the eigenvector associated to the largest eigenvalue of the correlation matrix. To obtain the other eigenvector directions one needs non-local laws (Sanger 1989, Oja 1989, Oja 1992). Sanger's

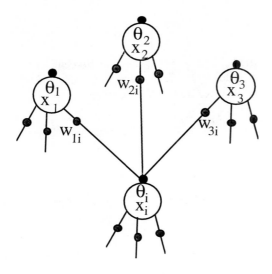

Fig. 1. A general connectionist network with state variables (x_i), connection strengths (W_{ij}) and node parameters (θ_i).

approach has the advantage of organizing the connection matrix in such a way that the rows are the eigenvectors associated with the eigenvalues in decreasing order. It suffers however from slow convergence rates for the lowest eigenvalues. The methods that have been proposed may, with small modifications, be used both for linear and non-linear networks. However, because the maximum information about a signal $\{x_i\}$, that may be coded directly in the connection matrix W_{ij}, is the principal components decomposition and this may already be obtained with linear units, we will discuss only this case.

The learning rules proposed below (Dente and Vilela Mendes 1996a) are a generalization of Sanger's scheme including a node parameter θ_i. We consider a one-layer feedforward network with as many inputs as outputs (Fig.2) and the updating rules proposed for W_{ij} and θ_i are

$$W_{ij}(t+1) = W_{ij}(t) + \gamma_w y_i(t) \left\{ x_j(t) - \sum_{k=1}^{i} \theta_k^{-1} y_k(t) W_{kj}(t) \right\}, \qquad (8)$$
$$\theta_i(t+1) = \theta_i(t) + \gamma_\theta y_i(t) \left\{ 1 - y_i(t) \right\},$$

where y_i is the output of node i

$$y_i = \theta_i \sum_j W_{ij} x_j \qquad (9)$$

and γ_w and γ_θ are positive constants that control the learning rate. As will be shown below the learning dynamics of Eqs. (8) has the capability to acceler-

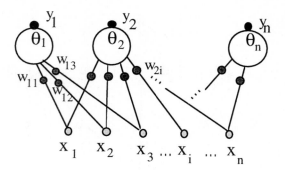

Fig. 2. One-layer neural feedforward network for Hebbian learning with node parameters.

ate the convergence rate for the small eigenvalues of Q. To avoid undesirable fixed points of the dynamics, where this acceleration effect is not obtained, the learning rules (8) are supplemented by the following prescription: "*The starting θ_i's are all positive and are not allowed to decrease below a value $(\theta_i)_{min}$. If, at a certain point of the learning process, θ_i hits the lower bound one makes the replacement $W_{ij} \rightarrow -W_{ij}$ in the line i of the connection matrix*" (see remark 3).

Define the N-dimensional vectors

$$
\begin{aligned}
(\mathbf{x})_i &= x_i , \\
(\mathbf{W}_i)_j &= W_{ij}
\end{aligned}
\tag{10}
$$

such that the following result holds (Dente and Vilela Mendes 1996a):

If the time scale of the system (8) is much slower than the averaging time of the input signal $\{x_i\}$ then:

a) the system (8) has a stable fixed point such that

$$
\begin{aligned}
Q\mathbf{W}_i &= \lambda_i \mathbf{W}_i , \\
\theta_i &= \tfrac{1}{\lambda_i} (\mathbf{W} \cdot \langle \mathbf{x} \rangle) ,
\end{aligned}
\tag{11}
$$

and $|\mathbf{W}_i| = 1$. $\lambda_i > 0$ (i=1...,N) are the eigenvalues of the correlation matrix,

b) convergence to the fixed point is sequential in the sense that $\mathbf{W}_i$ is only attracted to the eigenvector described in (11) if all vectors $\mathbf{W}_i$ for $i < j$ are already close to their corresponding eigenvector values.

Remarks:

(1) The matrix $\mathbf{W}$ of the connection strengths extracts the principal components of the correlation matrix Q. The node parameters θ_i at their fixed point (11) extract additional information on the mean value of the data vector

x and the eigenvalues of Q. To deal with data with zero mean it is convenient to change the θ_i−updating law to

$$\theta_i(t+1) = \theta_i(t) + \gamma_\theta \left\{ \theta_i(t) \sum_k W_{ij}(x_k + r_k) - \left(\theta_i(t) \sum_k W_{ik}x_k \right)^2 \right\} \quad (12)$$

where **r** is a fixed vector. The stable fixed point is now

$$\theta_i = \frac{1}{\lambda_i} \left(\mathbf{W}_i \cdot (\langle \mathbf{x} \rangle + \mathbf{r}) \right) . \quad (13)$$

If one wishes to separate the eigenvalues from the information on the average data $\langle \mathbf{x} \rangle$ one may add another parameter μ_i to each node with a learning law

$$\mu_i(t+1) = \mu_i(t) + \gamma_\mu \left\{ 1 - \mu_i^\alpha(t) \left(\sum_k W_{ik}x_k \right)^2 \right\} \quad (14)$$

which, with the same assumptions about time scales as before, converges to the stable fixed point

$$\mu_i = \left(\frac{1}{\lambda_i} \right)^{\frac{1}{\alpha}} . \quad (15)$$

The convergence rate near the fixed point is $\gamma_\mu \alpha \lambda_i^{\frac{1}{\alpha}}$. Therefore choosing a large value α accelerates the convergence for the small eigenvalues.

(2) In addition to its role in extracting additional information on the input signal, the node parameters θ_i also play a role in accelerating the convergence to the stable fixed point. For fixed $\gamma_w\theta_i$, the rate of convergence to the fixed point is very slow for the components associated to the smallest eigenvalues. This is the reason for the convergence problems in Sanger's method and one also finds sometimes that the results for the small components are quite misleading. On the other hand, increasing γ_w does not help because the time scale of the W_{ij} learning law becomes of the order of the time scale of the data and one obtains large fluctuations in the principal components. With a node parameter and the learning law (8) the situation is more favorable because for small eigenvalues the effective control parameter $(\gamma_w\theta_i)$ is dynamically amplified. This accelerates the convergence of the minor components without inducing fluctuations on the principal components.

(3) Here one examines the effect of the prescription to avoid the fixed points where some $\theta_i = 0$. Consider the average evolution of θ_i, assuming all the other variables fixed

$$\theta_i(t+1) = \theta_i(t) \left(1 + \gamma_\theta \mathbf{W}_i \cdot \langle \mathbf{x} \rangle \right) - \gamma_\theta \theta_i^2(t) \left(\mathbf{W}_i \cdot \mathbf{Q} \mathbf{W}_i \right) .$$

If $\mathbf{W}_i \cdot \langle \mathbf{x} \rangle > 0$ the stable fixed point is at $\mathbf{W}_i \cdot \langle \mathbf{x} \rangle / (\mathbf{W}_i \cdot \mathbf{Q} \mathbf{W}_i)$ and if $\mathbf{W}_i \cdot \langle \mathbf{x} \rangle < 0$ it is at zero (see Fig.3). If $\mathbf{W}_i \cdot \langle \mathbf{x} \rangle$ is < 0, θ_i moves towards the fixed point F0 at zero but, when it reaches $(\theta_i)_{\min}$, the change of the sign of

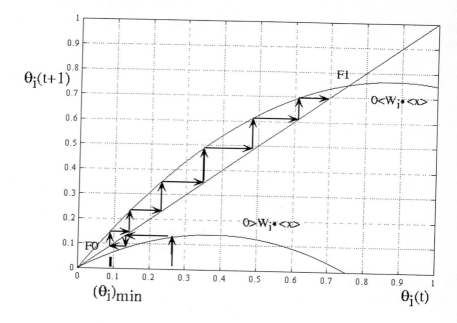

Fig. 3. Effect of changing the sign of W_i approaching the stable fixed points.

the corresponding row $\mathbf{W}_i$ in the connection matrix changes the dynamics and F1 is now the attracting fixed point.

To illustrate the effect of node parameters in principal component analysis (PCA), consider the following two-dimensional x signal, where t_i are Gaussian distributed variables with zero mean and unit variance:

$$x_1 = t_1 ; \quad x_2 = t_1 + 0.08t_2 .$$

The principal components of the signal and the eigenvalues are given in the following table:

λ_i	W_{i1}	W_{i2}
2.0032	0.7060	0.7082
0.0032	0.7082	−0.7060 .

A one-layer network with node parameters as in Fig.2 is used to perform PCA. Fig.4 shows the data and the principal directions that are obtained using the learning laws (8) and (12). The parameter values used are γ_w =0.015, γ_θ =0.005 and $\mathbf{r}$=(0.002,0.002). Fig.5 shows the convergence of the $\mathbf{W}$'s to their final values in the learning process. Fig.6 shows the variation of the node parameters.

Sanger's original algorithm is recovered by fixing the node parameters to unit values. In this case the principal components are also extracted, but

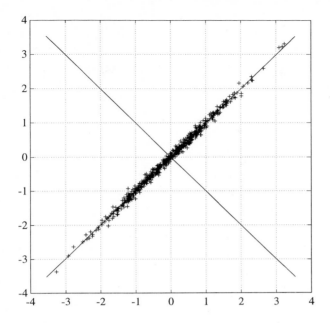

Fig. 4. Signal distribution and its principal directions.

the convergence of the process is much slower, as shown in Fig.7. With node parameters, improved convergence of the small eigenvalues is to be expected under generic conditions because the rates of convergence are controlled by $\gamma_w \theta_i$ and θ_i is proportional to $1/\lambda_i$. In the example the effect of θ_2 is further amplified by the fact that, before $\mathbf{W}_2$ starts to converge, $\mathbf{W}_2 \cdot \mathbf{r}$ is large. Hence, in this case at least, the node parameter acts like a variable learning rate for the minor component. Notice that the method does not induce oscillations in the major components as would happen with large γ_w values. Besides speeding up the learning process the node parameters also contain information about first moments and the eigenvalues.

Node parameters also have a beneficial effect on competitive learning algorithms (Dente and Vilela Mendes 1996a).

2.2 Non-Gaussian Data and the Neural Computation of the Characteristic Function

The aim of the principal component analysis (PCA) is to extract the eigenvectors of the correlation matrix from the data. There are standard neural network algorithms for this purpose (Sanger 1989, Oja 1989, Oja 1992, Dente and Vilela Mendes 1996a). However if the process is non-Gaussian, PCA

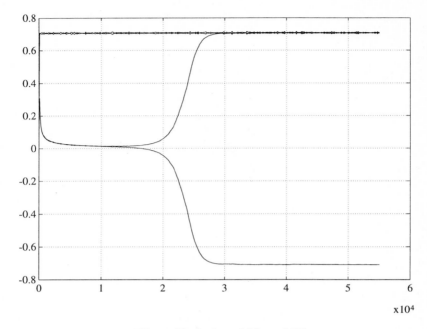

Fig. 5. Evolution of W_1 and W_2.

algorithms or their higher-order generalizations provide only incomplete or misleading information on the statistical properties of the data.

Let x_i denote the output of node i in a neural network. Hebbian learning (Hebb 1949) is a type of unsupervised learning where the neural network connection strengths W_{ij} are reinforced whenever the products $x_j x_i$ are large. The simplest form is

$$\Delta W_{ij} = \eta x_i x_j \ . \tag{16}$$

Hebbian learning extracts the eigenvectors of the correlation matrix Q

$$Q_{ij} = \langle x_i x_j \rangle \ . \tag{17}$$

However, if the learning law is local as in Eq.(16), all the lines of the connection matrix W_{ij} converge to the eigenvector with the largest eigenvalue of the correlation matrix. To obtain other eigenvector directions requires non-local laws. These principal component analysis algorithms find the characteristic directions of the correlation matrix $(Q)_{ij} = \langle x_i x_j \rangle$. If the data has zero mean ($\langle x_i \rangle = 0$) they are the orthogonal directions along which the data has maximum variance. If the data is Gaussian in each channel, it is distributed as a

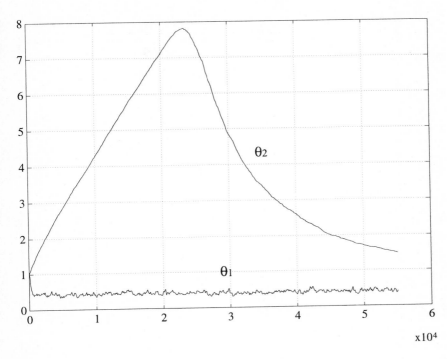

Fig. 6. Evolution of θ_1 and θ_2.

hyperellipsoid and the correlation matrix Q already contains all the information about the statistical properties. This is because higher-order moments
of the data may be obtained from the second-order moments. However, if
the data is non-Gaussian, the PCA analysis is not complete and higher-order
correlations are needed to characterize the statistical properties. This led
some authors(Softky and Kammen 1991, Taylor and Coombes 1993) to pro-
pose networks with higher-order neurons to obtain the higher-order statistical
correlations of the data. An higher-order neuron is one that is capable of ac-
cepting, in each of its input lines, data from two or more channels at once.
There is then a set of adjustable strengths $W_{ij_1}, W_{ij_1 j_2}, \ldots, W_{ij_1 \ldots j_n}$, n being
the order of the neuron. Networks with higher-order neurons have interesting
applications, for example in fitting data to a high-dimensional hypersurface.
However there is a basic weakness in the characterization of the statistical
properties of non-Gaussian data by higher-order moments. Existence of the
moments of a distribution function depends on the behavior of this function
at infinity and it frequently happens that a distribution has moments up to

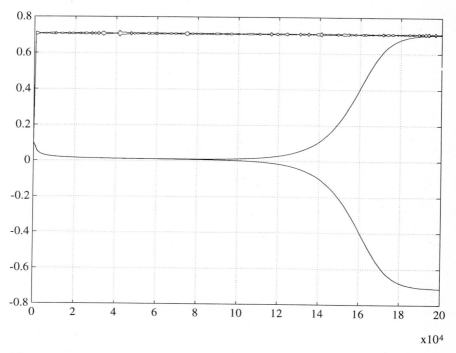

Fig. 7. Evolution of W_1 and W_2 using Sanger's algorithm, without node parameters.

a certain order but no higher. A well-behaved probability distribution might even have no moments of order higher than one (the mean). In addition a sequence of moments does not necessarily determine a probability distribution function uniquely (Lukacs 1970). Two different distributions may have the same set of moments. Therefore, for non-Gaussian data, the PCA algorithms or higher-order generalizations may give misleading results.

As an example consider the two-dimensional signal shown in Fig.8. Fig.9 shows the evolution of the connection strengths W_{11} and W_{12} when this signal is passed through a typical PCA algorithm. Large oscillations appear and finally the algorithm overflows. Smaller learning rates do not introduce qualitative modifications in this evolution. The values may at times appear to stabilize but large spikes do occur. The reason is that the seemingly harmless data in Fig.8 is generated by a linear combination of a Gaussian with the following distribution

$$p(x) = k(2 + x^2)^{-\frac{3}{2}} , \tag{18}$$

which has a first moment, but no moments of higher-order.

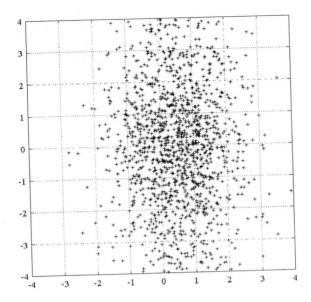

Fig. 8. A two-dimensional test signal.

Interest in non-Gaussian processes is not a pure academic exercise because in many applications adequate tools are needed to analyze such processes. For example, processes without higher-order moments, in particular those associated with Lévy statistics, are prominent in complex processes such as relaxation in glassy materials, chaotic phase diffusion in Josephson junctions and turbulent diffusion (Shlesinger et al. 1993, Zumofen and Klafter 1993, Zumofen and Klafter 1994). Moments of an arbitrary probability distribution may not exist. However, because every bounded and measurable function is integrable with respect to any distribution, the existence of the characteristic function f(α) is always assured (Lukacs 1970),

$$f(\alpha) = \int e^{i\alpha.x} dF(x) = \langle e^{i\alpha.x} \rangle \ , \tag{19}$$

α and x are N-dimensional vectors, x is the data vector, and $F(x)$ its distribution function. The characteristic function is a compact and complete characterization of the probability distribution of the signal. In addition, if one wishes to describe the time correlations of the stochastic process $x(t)$, the corresponding quantity is the characteristic functional (Hida 1980)

$$F(\xi) = \int e^{i(x,\xi)} d\mu(x) \tag{20}$$

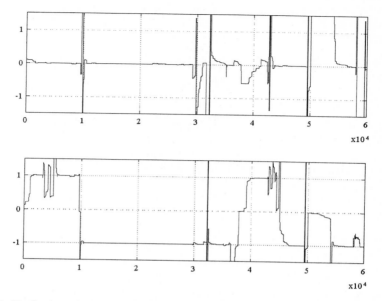

Fig. 9. Evolution of the connection strengths W_{11} and W_{12} in a PCA network for the data presented in Fig.8.

where $\xi(t)$ is a smooth function and the scalar product is

$$(x, \xi) = \int dt x(t)\xi(t) , \qquad (21)$$

$\mu(x)$ being the probability measure over the sample paths of the process.

Next, I describe an algorithm to compute the characteristic function from the data, by a learning process (Dente and Vilela Mendes 1997). The main idea is that, in the end of the learning process, one has a neural network which is a representation of the characteristic function. This network is then available to provide all the required information on the probability distribution of the data being analyzed.

Suppose we want to learn the characteristic function $f(\alpha)$ of a one-dimensional signal $x(t)$ in a domain $\alpha \in [\alpha_0, \alpha_N]$. The α-domain is divided into N intervals by a sequence of values $\alpha_0, \alpha_1, \alpha_2, ..., \alpha_N$ and a network is constructed with $N + 1$ intermediate layer nodes and one output node (Fig. 10).

The learning parameters in the network are the connection strengths W_{0i} and the node parameters θ_i. The existence of the node parameter means that

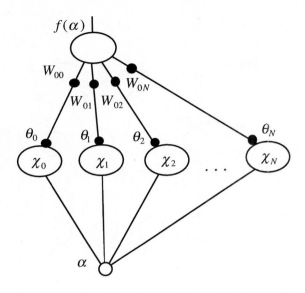

Fig. 10. Network to learn the characteristic function of a scalar process.

the output of a node in the intermediate layer is $\theta_i \chi_i(\alpha)$, χ_i being a non-linear function. The use of both connection strengths and node parameters in neural networks makes them equivalent to a wide range of other connectionist systems (Doyne Farmer 1990) and improves their performance in standard applications (Dente and Vilela Mendes 1996a). The learning laws for the network in Fig. 10 are:

$$\theta_i(t+1) = \theta_i(t) + \gamma(\cos \alpha_i x(t) - \theta_i(t)) \,,$$
$$W_{0i}(t+1) = W_{0i}(t) + \eta \sum_j \left(\theta_j(t) - \sum_k W_{0k}(t)\chi_k(\alpha_j)\theta_k(t)\right) \theta_i(t)\chi_i(\alpha_j) \,,$$

$$(22)$$

$\gamma, \eta > 0$. The intermediate layer nodes are equipped with a radial basis function

$$\chi_i(\alpha) = \frac{e^{-(\alpha-\alpha_i)^2/2\sigma_i^2}}{\sum_{k=0}^{N} e^{-(\alpha-\alpha_k)^2/2\sigma_k^2}} \tag{23}$$

where in general one uses $\sigma_i = \sigma$ for all i. The output is a simple additive node. The learning constant γ should be sufficiently small to ensure that the learning time is much smaller than the characteristic times of the data $x(t)$. If this condition is satisfied each node parameter θ_i tends to $\langle \cos \alpha_i x \rangle$, the real part of the characteristic function $f(\alpha)$ for $\alpha = \alpha_i$. The W_{oi} learning law was chosen to minimize the error function

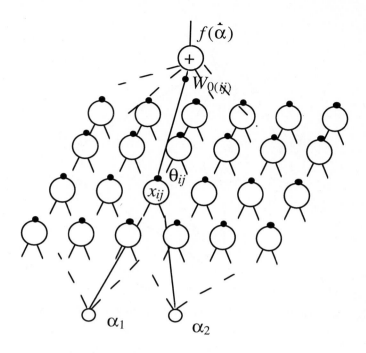

Fig. 11. Network to learn the characteristic function of a 2-dimensional signal $\vec{x}(t)$.

$$f(W) = \frac{1}{2} \sum_j \left(\theta_j - \sum_k W_{0k}(t)\chi_k(\alpha_j)\theta_k \right)^2 . \qquad (24)$$

One sees that the learning scheme is a hybrid one, in the sense that the node parameter θ_i learns, in an unsupervised way, (the real part of) the characteristic function $f(\alpha_i)$ and then, by a supervised learning scheme, the W_{0i}'s are adjusted to reproduce the θ_i value in the output whenever the input is α_i. Through the learning law (22) each node parameter θ_i converges to $\langle \cos \alpha_i x \rangle$ and the interpolating nature of the radial basis functions guarantees that, after training, the network will approximate the real part of the characteristic function for any α in the domain $[\alpha_0, \alpha_N]$. A similar network is constructed for the imaginary part of the characteristic function, where now

$$\theta_i'(t+1) = \theta_i'(t) + \gamma(\sin \alpha_i x(t) - \theta_i'(t)) . \qquad (25)$$

For higher-dimensional data the scheme is similar. The number of required nodes is N^d for a d-dimensional data vector $\vec{x}(t)$. For example for the 2-dimensional data of Fig.8 a set of N^2 nodes was used (Fig.11).

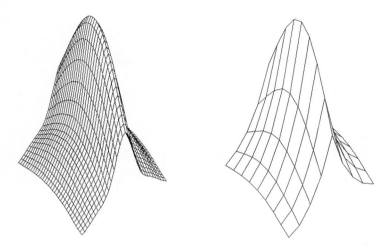

Fig. 12. Real part of the characteristic function for the data in Fig.8 (left) and the mesh of θ_i values (right) obtained by the network.

Each node in the square lattice has two inputs for the two components α_1 and α_2 of the vector argument of $f(\overrightarrow{\alpha})$. The learning laws are, as before,

$$\theta_{(ij)}(t+1) = \theta_{(ij)}(t) + \gamma(\cos \overrightarrow{\alpha}_{(ij)} \cdot \overrightarrow{x}(t) - \theta_{(ij)}(t)) ,$$
$$W_{0(ij)}(t+1) = W_{0(ij)}(t) + \eta \sum_{(kl)} \Big(\theta_{(kl)}(t)$$
$$- \sum_{(mn)} W_{0(mn)}(t)\chi_{(mn)}(\overrightarrow{\alpha}_{(kl)})\theta_{(mn)}(t) \Big) \theta_{(ij)}(t)\chi_{(ij)}(\overrightarrow{\alpha}_{(kl)}) . \tag{26}$$

The pair (ij) denotes the position of the node in the square lattice and the radial basis function is

$$\chi_{(ij)}(\overrightarrow{\alpha}) = \frac{e^{-|\overrightarrow{\alpha} - \overrightarrow{\alpha}_{(ij)}|^2 / 2\sigma_{(ij)}^2}}{\sum_{(kl)} e^{-|\overrightarrow{\alpha} - \overrightarrow{\alpha}_{(kl)}|^2 / 2\sigma_{(kl)}^2}} . \tag{27}$$

Two networks are used, one for the real part of the characteristic function, and another for the imaginary part with $\cos(\overrightarrow{\alpha}_{(ij)} \cdot \overrightarrow{x}(t))$, in Eqs.(27), replaced by $\sin(\overrightarrow{\alpha}_{(ij)} \cdot \overrightarrow{x}(t))$. Figs.12 and 13 show the values computed by the algorithm for the real and imaginary parts of the characteristic function corresponding to the two-dimensional signal shown in Fig.8.

On the left is a plot of the exact characteristic function and on the right the values learned by the network. In this case we show only the mesh corresponding to the θ_i values. One obtains a 2.0% accuracy for the real part and 4.5% accuracy for the imaginary part. The convergence of the learning

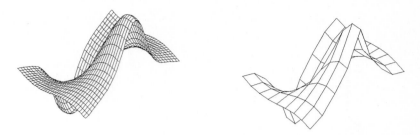

Fig. 13. Imaginary part of the characteristic function for the data of Fig.8 (left) and the mesh of θ_i-values (right) obtained by the network.

process is fast and the approximation is reasonably good. Notice in particular the discontinuous slope at the origin which reveals the non-existence of a second moment.

For a second example the data is generated by a Weierstrass random walk with a probability distribution

$$p(x) = \frac{1}{6} \sum_{j=0}^{\infty} \left(\frac{2}{3}\right)^j \left(\delta_{x,b^j} + \delta_{x,-b^j}\right) \tag{28}$$

and b=1.31 for a process of the Lévy flight type. The characteristic function obtained by the network is shown in Fig. 14.

Taking the $\log(-\log)$ of the network output one obtains the scaling exponent 1.49 near $\alpha = 0$, close to the expected fractal dimension of the random walk path (1.5).

These examples test the algorithm as a process identifier, in the sense that, after the learning process, the network is a dynamical representation of the characteristic function and may be used to perform all kinds of analyses of the statistics of the data.

3   Chaotic Networks for Information Processing

Freeman and collaborators (Skarda and Freeman 1987, Freeman and Baird 1987, Freeman 1987, Freeman, Yao, and Burke 1988, Yao and Freeman 1990, Yao et al. 1991) have extensively studied and modeled the neural activity in the mammalian olfactory system. Their conclusions challenge the idea that pattern recognition in the brain is accomplished as in an attractor neural network (Amit 1989). Pattern recognition in the brain is the process by which external signals arriving at the sense organs are converted into internal meaningful states. The studies of the excitation patterns in the olfactory

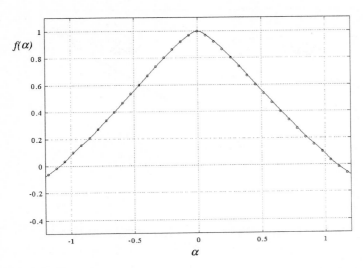

Fig. 14. Characteristic function for the Weierstrass random walk (b=1.31).

bulb of the rabbit lead to the conclusion that, at least in this biological pattern recognition system, there is no evolution towards an equilibrium fixed point nor does it seem to be minimizing an energy function. Other interesting conclusions of these biological studies are:

(1) the main component of the neural activity in the olfactory system is chaotic. This is also true in other parts of the brain, periodic behavior occurring only in abnormal situations like deep anesthesia, coma, epileptic seizures, or in areas of the cortex that have been isolated from the rest of the brain;

(2) the low-level chaos that exists in the absence of an external stimulus is, in the presence of a signal, replaced by bursts lasting for about 100 ms which have different intensities in different regions of the olfactory bulb. Olfactory pattern recognition manifests itself as a spatially coherent pattern of intensity;

(3) the recognition time is very fast, in the sense that the transition between different patterns occurs in times as short as 6 ms. Given the neuron characteristic response-time this is clearly incompatible with the global approach to equilibrium of an attractor neural network;

(4) the biological measurements that have been performed do not record the action potential of individual neurons, but the local effect of the currents coming out of thousands of cells. Therefore the very exis-

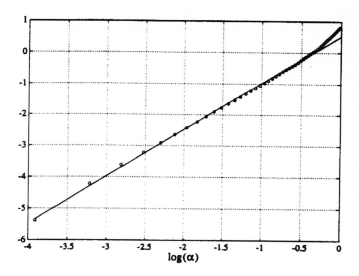

Fig. 15. log(-log) of the characteristic function for the Weierstrass random walk (b=1.31).

tence of measurable activity bursts implies a synchronization of local assemblies of many neurons.

Freeman, Yao and Burke (Freeman, Yao, and Burke 1988, Yao and Freeman 1990) model the olfactory system with a set of non-linear coupled differential equations, the coupling being adjusted by means of an input correlation learning scheme. Each variable in the coupled system is assumed to represent the dynamical state of a local assembly of many neurons. Based on numerical simulations they conjecture that olfactory pattern recognition is realized through a multilobe strange attractor. The system would be, most of the time, in a basal (low-activity) state, being excited to one of the higher lobes by the external stimulus.

To compute or even prove the existence of chaotic measures in a system of coupled differential equations is an awesome task. Therefore, even if it may be biologically accurate, the analytical model proposed by these authors is difficult to deal with and unsuitable for a wide application in technological pattern recognition tasks, although one such application has indeed been attempted by the authors (Yao et al. 1991). However the idea that efficient pattern recognition may be achieved by a chaotic system, that selects distinct invariant measures according to the class of external stimuli, is quite interesting and deserves further exploration. Inspired by the biological evidence a model has been developed (Dente and Vilela Mendes 1996b), that behaves roughly like an olfactory system (in Freeman's sense) and, at the same time, is easier to describe and control by analytical means. To play the role of the local chaotic assembly of neurons a Bernoulli unit is chosen. The connection between the units is realized by linear synapses with an input correlation learning law and the external inputs have adjustable gains, changing as in a biological potentiation mechanism. This last feature turns out to be useful to enhance the novelty-filter qualities of the system.

3.1   A Network of Bernoulli Units

The network is the fully connected system shown in Fig.16. The output of

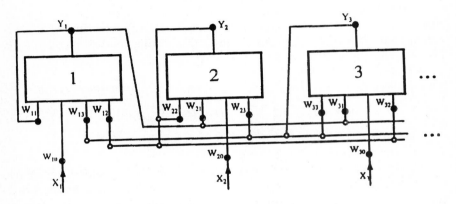

Fig. 16. The Bernoulli network.

the nodes is denoted by y_i and the x_i's are the external inputs. W_{ij} with $i, j \in \{1, 2, \cdots, N\}$ are the connection strengths and W_{i0} are the input gains. Both W_{ij} and W_{i0} are real numbers in the interval $[0,1]$. The input patterns are zero-one sequences ($x_i \in \{0, 1\}$). The learning laws for the connection strengths and the input gains are the following:

Let $S_{ij}(t) = W_{ij}(t) + \eta x_i(t) x_j(t)$ such that for $i \neq j$

$$W_{ij}(t + 1) = S_{ij}(t) , \quad \text{if } \sum_{k \neq i} S_{ik}(t) \leq C , \tag{29}$$

$$W_{ij}(t + 1) = C \frac{S_{ij}(t)}{\sum_{k \neq i} S_{ik}(t)} , \quad \text{if } \sum_{k \neq i} S_{ik}(t) > C , \tag{30}$$

for $i = j$,

$$W_{ii}(t+1) = 1 - \sum_{j \neq i} W_{ij}(t+1) \,, \tag{31}$$

and for the input gains

$$W_{i0}(t+1) = \alpha \frac{N_i(1)_t}{N_i(1)_t + N_i(0)_t} \,. \tag{32}$$

If according to Eqs.(29-32) an input pattern has a one in both the position i and position j, the correlation of the units i and j becomes stronger. $C < 1$ is a constant related to the node dynamics, that the sum of the off-diagonal connections is not allowed to exceed. η is a small parameter that controls the learning speed. Finally, the diagonal element W_{ii} is chosen in such a way that the sum of all connections entering each unit adds to unity.

In the input gain learning law, $N_i(1)_t$ (or $N_i(0)_t$) is the number of times that a one (or a zero) has appeared at the input i, up to time t. Eq.(32) means that if an input is excited many times during the learning phase, it becomes more sensitive.

The node dynamics is

$$y_i(t+1) = f\left(\sum_j W_{ij}(t)y_j(t) + W_{i0}x_i(t) \right) \,, \tag{33}$$

f being the function depicted in Fig.17.

The learning process starts with $W_{ii} = 1$ and $W_{ij} = 0$ for $i \neq j$. Each unit has then an independent absolutely continuous invariant measure that is the Lebesgue measure in $[0, C]$ and is zero outside. When the W_{ij} $(i \neq j)$ become different from zero and the inputs x_i are still zero, all variables y_i stay in the interval $[0, C]$ because of the convex linear combination of inputs imposed by the normalization of the W_{ij}'s. If some inputs x_i are $\neq 0$, there is a finite probability for an irregular burst in the interval $[C, 1]$, of some of the node variables, with reinjection into $[0, C]$ whenever the iterate falls on the interval $[\frac{1}{2} + \frac{C}{2}, \frac{1}{2} + C]$.

The bursts in the interval $[C, 1]$ in response to some of the input patterns represent the recognition mechanism of the network. The basal chaotic dynamics ensuring uniform covering of the interval $[0, C]$, the timing of the onset of the bursts depends only on the correlation probability and on the clock time of the discrete dynamics. We understand therefore why a chaos-based network may have a recognition time faster than an attractor network.

Learning, invariant measures, and simulations. Both the connection strengths and the nature of the bursts, for a given set of W's and an applied input pattern, may be estimated in probability.

In Eqs.(29-30), either the node i is not correlated to any other node and all off-diagonal elements W_{ij} are zero or, as soon as the input patterns begin

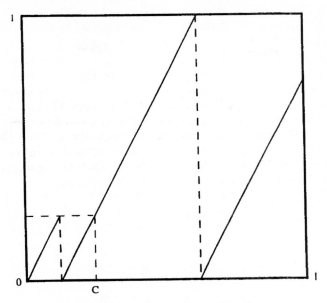

Fig. 17. The node-dynamics function.

to correlate the node i with any other node, the off-diagonal elements start to grow and only the second case (Eq.(30)) needs to be considered. Let the learning gain be small such that in first order in η we have

$$W_{ij}(t+1) = W_{ij}(t) + \eta x_i(t)x_j(t) - \frac{\eta}{C}W_{ij}(t)\sum_{k\neq i} x_i(t)x_k(t) \ .$$

For N learning steps, in first order in η,

$$W_{ij}(t+N) = W_{ij}(t) + \eta \sum_{n=0}^{N-1} x_i(t+n)x_j(t+n)$$

$$- \frac{\eta}{C}W_{ij}(t)\sum_{k\neq i}\sum_{n=0}^{N-1} x_i(t+n)x_k(t+n) \ .$$

Denote $p_{ij}(1)$ to be the probability for the occurrence, in the input patterns, of a one in the positions i and j, the above equation has the stationary solution

$$W_{ij} = C\frac{p_{ij}(1)}{\sum_{k\neq i} p_{ik}(1)} \ .$$

Now we establish an equation for the burst probabilities. Consider the case where C is much smaller than one, that is, the basal chaos is of low intensity. In this case, because of the normalization chosen for W_{ii}, the dynamics inside the interval $[0, C]$ is dominated by $y \to 2y(\mathrm{mod}\ C)$ and in the interval $[C, 1]$ by $y \to 2y(\mathrm{mod}\ 1)$. Hence, to a good approximation, we may assume uniform probability measures for the motion inside each one of the intervals. Denoting the interval $[0, C]$ as the state 1 and the interval $[C, 1]$ as the state 2, the dynamics of each node is a two-state Markov process with transition probabilities between the states corresponding to the probabilities of falling into some subintervals of the intervals $[0, C]$ and $[C, 1]$. Namely, the probability $p(2 \to 1)$ equals the probability of falling into the reinjection interval $[\frac{1}{2} + \frac{C}{2}, \frac{1}{2} + C]$ and the probability $p(1 \to 2)$ that of falling near the point C at a distance smaller than the off-diagonal excitation,

$$p_i(2 \to 1)_t = \frac{C}{2(1 - C)} \ , \tag{34}$$

$$p_i(2 \to 2)_t = 1 - p_i(2 \to 1)_t \ , \tag{35}$$

$$p_i(1 \to 2)_t = \left\{ \frac{1}{W_{ii}C} \left(W_{i0}x_i(t) - C(1 - W_{ii}) + \sum_{j \neq i} W_{ij}y_j(t) \right) \right\}^{\#} \ , \tag{36}$$

$$p_i(1 \to 1)_t = 1 - p_i(1 \to 2)_t \ , \tag{37}$$

where we have used the notation

$$f^{\#} = (f \vee 0) \wedge 1$$

for functions truncated to the range $[0, 1]$, i.e., $f^{\#} = 0$ if $f < 0$, $f^{\#} = 1$ if $f > 1$ and $f^{\#} = f$ if $1 \geq f \geq 0$.

The sum in the right-hand side of Eq.(36) is approximated in probability by

$$\sum_{j \neq i} W_{ij} \left(\frac{1}{2}p_j(2)_t + \frac{C}{2}(1 - p_j(2)_t) \right) \tag{38}$$

where $p_j(2)_t$ denotes the probability of finding the node j in the state 2 at time t. The probability estimate (38), for the outputs y_j, assumes statistical independence of the units. This hypothesis fails if there are synchronization effects, which are mainly to be expected if a small group of units is strongly correlated.

With the probability estimate for the y_j's and the detailed balance principle it is now possible to write a self-consistent equation for the probability $p_i(2)_t$ to find an arbitrary node i in the state 2 at time t,

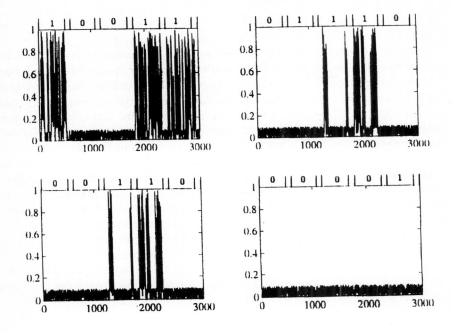

Fig. 18. Response of a 4-nodes network after being exposed to the patterns 1000 and 0110 ($C = 0.1, \alpha = 0.01$).

$$
\begin{aligned}
p_i(2)_t = \Bigg\{ &\frac{1}{W_{ii}C} \left(W_{i0}x_i(t) - C(1 - W_{ii}) \right. \\
&\left. + \sum_{j \neq i} W_{ij} \left(\frac{1}{2}p_j(2)_t + \frac{C}{2}(1 - p_j(2)_t) \right) \right) \Bigg\}^{\#} \\
\times \Bigg\{ &\Bigg\{ \frac{1}{W_{ii}C} \left(W_{i0}x_i(t) - C(1 - W_{ii}) \right. \\
&\left. + \sum_{j \neq i} W_{ij} \left(\frac{1}{2}p_j(2)_t + \frac{C}{2}(1 - p_j(2)_t) \right) \right) \Bigg\}^{\#} + \frac{C}{2(1 - C)} \Bigg\}^{-1} .
\end{aligned}
\tag{39}
$$

For each input pattern $x_i(t)$, one obtains an estimate for $p_i(2)_t$ by solving Eq.(39) iteratively. We find that the solution that is obtained is qualitatively similar to the numerically determined invariant measures, although it tends to overestimate the burst excitation probabilities when they are small. This may be understood from the synchronization effects between groups of units.

If one unit is not excited (not in state 2) the others also tend not to be excited, hence (38) overestimates the sum $\sum_{j \neq i} W_{ij} y_j(t)$.

We now illustrate how the network behaves as an associator and pattern recognizer. For simplicity, consider a network of four nodes that is exposed during many iterations to the patterns 1000 and 0110 where the first pattern appears twice as much as the second. After this learning period we expose the network to all possible zero-one input patterns for 500 time steps each and observe the network reaction. During the recall experiment no further adjustment of the W_{ij}'s is made. The result is shown in Fig.18.

The conclusion from this and other simulations is that, according to the nature of the learning patterns, the network acts, for the recall input patterns, as a mixture of memory, associator and novelty-filter. For example, in Fig.18 we see that after having learned the sequences 1000 and 0110, the network reproduces these patterns as a memory. The pattern 1001 is associated to the pattern 1000 and the pattern 1110 associated to a mixture of the two learned patterns. By contrast, the pattern 0100 is not recognized by the network that acts then as a novelty-filter. Fig.19 shows the invariant measures of the system (expressed in probability per bin) when the input patterns are 1100 and 0110.

In conclusion, the network based on Bernoulli units, with node dynamics as in Fig.17 and correlation learning described by Eqs.(29-32)

(1) is a chaos-based pattern recognizer with the capability of operating on distinct invariant measures that are selected by the input patterns;

(2) has a response time of the selection that is controlled by the magnitude of the invariant measures and the clock time of the basal chaotic dynamics;

(3) is a pattern recognizer with a mixture of memory, associator and novelty-filter. This, however, is sensitive to the learning algorithm that is chosen and, for the algorithm that is discussed here, is sensitive to the values of the parameters C and η.

3.2   Feigenbaum Networks

In this part we study dynamical systems composed of a set of coupled quadratic maps

$$x_i(t+1) = 1 - \mu_* \left\{ \sum_j W_{ij} x_j(t) \right\}^2 \tag{40}$$

with $x \in [-1, 1]$, $\sum_j W_{ij} = 1 \; \forall i$ and $W_{ij} > 0 \; \forall i, j$ and $\mu_* = 1.401155$. The value chosen for μ_* implies that, in the uncoupled limit ($W_{ii} = 1, W_{ij} = 0$ $i \neq j$), each unit transforms as a one-dimensional quadratic map in the accumulation point of the Feigenbaum period-doubling bifurcation cascade. This system will be called a Feigenbaum network.

The quadratic map at the Feigenbaum accumulation point is not in the class of chaotic systems (in the sense of having positive Lyapunov exponents).

However, it shares with them the property of having an infinite number of unstable periodic orbits. Therefore, before the interaction sets in, each elementary map possesses an infinite diversity of potential dynamical behavior. As we will show later, the interaction between the individual units is able to stabilize selectively some of the previously unstable periodic orbits. The selection of the periodic orbits that are stabilized depends both on the initial conditions and on the intensity of the interaction coefficients W_{ij}. As a result Feigenbaum networks appear as systems with potential applications in the fields of control of chaos, information processing, and as models of self-organization.

Control of chaos or of the transition to chaos has been, in recent years, a very active field (see for example Shinbrot 1995, and references therein). Several methods were developed to control the unstable periodic orbits that are embedded within a chaotic attractor. Having a way to select and stabilize at will these orbits we would have a device with infinite storage capacity (or infinite pattern discrimination capacity). However, an even better control might be achieved if, instead of an infinite number of unstable periodic orbits, the system possesses an infinite number of periodic attractors. The basins of attraction would evidently be small but the situation is in principle more favorable because the control need not be as sharp as before. As long as the system is kept in a neighborhood of an attractor the uncontrolled dynamics itself stabilizes the orbit.

The creation of systems with infinitely many sinks near an homoclinic tangency was discovered by Newhouse (Newhouse 1974-79) and later studied by several other authors (Robinson 1983, Gambaudo and Tresser 1983, Tedeschini-Lalli and Yorke 1986, Wang 1990, Nusse and Tedeschini-Lalli 1992). In the Newhouse phenomenon infinitely many attractors may coexist but only for special parameter values, namely for a residual subset of an interval. Another system, different from the Newhouse phenomenon, is a rotor map with a small amount of dissipation (Feudel et al. 1996) that also displays many coexisting periodic attractors.

Here one shows that for a Feigenbaum system with only two units and symmetrical couplings one obtains a system that has an infinite number of sinks for an open set of coupling parameters. Then one also analyzes the behavior of a Feigenbaum network in the limit of a very large number of units. A mean-field analysis shows how the interaction between the units may generate distinct periodic-orbit patterns throughout the network.

A simple system with an infinite number of sinks. Consider two units with symmetric positive couplings ($W_{12} = W_{21} = c > 0$)

$$
\begin{aligned}
x_1(t+1) &= 1 - \mu_* \left((1-c)x_1(t) + cx_2(t)\right)^2 , \\
x_2(t+1) &= 1 - \mu_* \left(cx_1(t) + (1-c)x_2(t)\right)^2 .
\end{aligned}
\tag{41}
$$

The mechanism leading to the emergence of periodic attractors from a system that, without coupling, has no stable finite-period orbits is the permanence of unstabilized orbits in a flip bifurcation and the contraction effect introduced by the coupling. The structure of the basins of attraction is also understood from the same mechanism. The result is (Carvalho et al. 1988):

For sufficiently small c there is an N such that the system (41) has stable periodic orbits of all periods 2^n for $n > N$.

Feigenbaum networks with many units, mean-field analysis. For the calculations below it is convenient to use the net input to the units, $y_i = \sum_j W_{ij} x_j$, as a variable such that Eq.(40) becomes

$$y_i(t+1) = 1 - \mu_* \sum_{j=1}^{N} W_{ij} y_j^2(t) . \tag{42}$$

For practical purposes some restrictions have to be put on the range of values that the connection strengths may take. For information processing (pattern storage and pattern recognition) it is important to preserve, as much as possible, the dynamical diversity of the system. That means, for example, that a state with all the units synchronized is undesirable insofar as the effective number of degrees of freedom is drastically reduced. From

$$\delta y_i(t+1) = -2\mu_* y(t) \left(W_{ii} \delta y_i(t) + \sum_{j \neq i} W_{ij} \delta y_j(t) \right) \tag{43}$$

one sees that an instability of the fully synchronized state implies $|2\mu_* y(t) W_{ii}| > 1$. Therefore, the interesting case is when the off-diagonal connections are sufficiently small to ensure that

$$W_{ii} > \frac{1}{\mu_*} . \tag{44}$$

For large N, provided there is no large scale synchronization effect, a mean-field analysis might be appropriate, at least to obtain qualitative estimates on the behavior of the network. For the unit i the average value $\langle 1 - \mu_* \sum_{j \neq i} W_{ij} y_j^2(t) \rangle$ acts like a constant and the mean-field dynamics is

$$z_i(t+1) = 1 - \mu_{i,eff} \, z_i^2(t) \tag{45}$$

where

$$z_i = \frac{y_i}{\langle 1 - \mu_* \sum_{j \neq i} W_{ij} y_j^2 \rangle} \tag{46}$$

and

$$\mu_{i,eff} = \mu_* W_{ii} \langle 1 - \mu_* \sum_{j \neq i} W_{ij} y_j^2 \rangle , \tag{47}$$

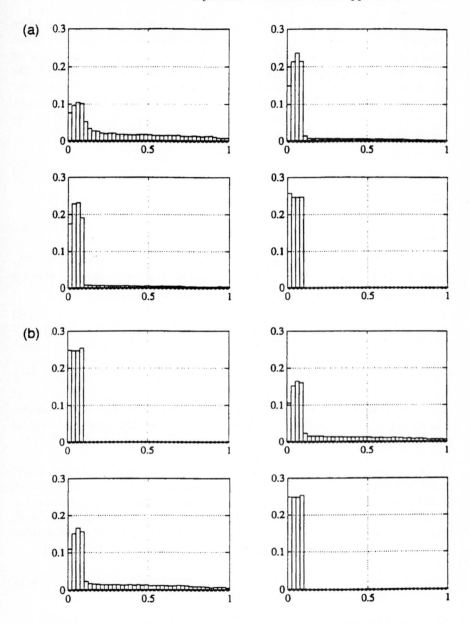

Fig. 19. Invariant measures for the input patterns (a)1000 and (b)0110.

$\mu_{i,eff}$ is the effective parameter for the mean-field dynamics of unit i. From (44) and (47) it follows $\mu_* W_{ii} > \mu_{i,eff} > \mu_* W_{ii}(2 - \mu_*)$. The conclusion is that the effective mean-field dynamics always corresponds to a parameter value below the Feigenbaum accumulation point. Therefore, one expects the interaction to stabilize the dynamics of each unit in one of the 2^n- periodic orbits. On the other hand, to keep the dynamics inside an interesting region we require $\mu_{i,eff} > \mu_2 = 1.3681$, the period-2 bifurcation point. With the estimate $\langle y^2 \rangle = \frac{1}{3}$ one obtains

$$\mu_* W_{ii} \left(1 - \frac{\mu_*}{3}(1 - W_{ii}) \right) > \mu_2 \tag{48}$$

that, together with (44), defines the interesting range of parameters for $W_{ii} = 1 - \sum_{j \neq i} W_{ij}$.

Feigenbaum networks as signal processors. Let, for example, the W_{ij} connections be constructed from an input signal x_i by a correlation learning process

$$\begin{aligned} W_{ij} \to W'_{ij} &= (W_{ij} + \eta x_i x_j)e^{-\gamma} \quad \text{for } i \neq j , \\ W_{ii} \to W'_{ii} &= 1 - \sum_{j \neq i} W_{ij} . \end{aligned} \tag{49}$$

The dynamical behavior of the network, at a particular time, will reflect the learning history, that is, the data regularities, in the sense that W_{ij} is being structured by the patterns that occur more frequently in the data. The decay term $e^{-\gamma}$ ensures that the off-diagonal terms remain small and that the network structure is determined by the most frequent recent patterns. Alternatively, instead of the decay term, we may use a normalization method and the connection structure would depend on the weighted effect of all the data.

In the operating mode described above the network acts as a *signal identifier*. For example if the signal patterns are random, there is little correlation established and all the units operate near the Feigenbaum point. Alternatively the learning process may be stopped at a certain time and the network is then used as a *pattern recognizer*. In this latter mode, whenever the pattern $\{x_i\}$ appears, one makes the replacement

$$\begin{aligned} W_{ij} \to W'_{ij} &= W_{ij} x_i x_j \quad \text{for } i \neq j , \\ W_{ii} \to W'_{ii} &= 1 - \sum_{j \neq i} W_{ij} . \end{aligned} \tag{50}$$

Therefore if W_{ij} was $\neq 0$ but either x_i or x_j is $= 0$ then $W'_{ij} = 0$. That is, the correlation between node i and j disappears and the effect of this connection on the lowering of the periods vanishes.

If both x_i and x_j are one, then $W'_{ij} = W_{ij}$ and the effect of this connection persists. Suppose however that for all the W_{ij}'s different from zero either x_i or x_j are equal to zero. Then the correlations are totally destroyed and the network comes back to the uncorrelated (nonperiodic) behavior. This case is what is called a *novelty-filter*. Conversely, by displaying periodic behavior, the network *recognizes* the patterns that are similar to those that, in the learning stage, determined its connection structure. Recognition and *association* of similar patterns is then performed (Carvalho et al. 1988).

References

Almeida L. B. (1987); IEEE First Int. Conf. on Neural Networks, M. Caudill and C. Butler (Eds.) p. 600.

Almeida L. B. (1988); in *Neural Computers*, R. Eckmiller and Ch. von der Malsburg (Eds.) p. 199, Springer, Berlin.

Amit D. J. (1989); *Modeling brain function. The world of attractor neural networks*, Cambridge Univ. Press, Cambridge.

Amit D. J. and Brunel N. (1997); Network: Comput. in Neural Systems **8**, 373.

Amorim A., Vilela Mendes R. and Seixas J. (1988); *Learning Energy Correction Factors by Neural Networks: A Case Study for the ATLAS Calorimeter*, CERN report.

Carvalho R., Vilela Mendes R. and Seixas J. (1998); chao-dyn/9712004, to appear in Physica D.

Cohen M. A. and Grossberg S. (1983); IEEE Transactions on Syst., Man and Cybern. **13**, 815.

Dente J. A. and Vilela Mendes R. (1996a); Network: Computation in Neural Systems **7**, 123.

Dente J. A. and Vilela Mendes R. (1996b); Phys. Lett. A**211**, 87.

Dente J. A. and Vilela Mendes R. (1997); Neural Networks **10**, 1465.

Doyne Farmer J. (1990); Physica D**42**, 153.

Feudel U., Grebogi C., Hunt B. R. and Yorke J. A. (1996); Phys. Rev. E**54**, 71.

Freeman W.J. (1987); Biological Cybernetics **56**, 139.

W. J. Freeman W. J. and Baird B. (1987); Behavioral Neuroscience **101**, 393.

Freeman W. J., Yao Y. and Burke B. (1988); Neural Networks **1**, 277.

Gambaudo J. M. and Tresser C. (1983); J. Stat. Phys. **32**, 455.

Grossberg S. (1988); Neural Networks **1**, 17.

Hebb D. O. (1949); *The organisation of behavior*, Wiley, New York.

Hertz A. V. M. (1995); in *Models of Neural Networks III*, E. Domany, J. L. van Hemmen and K. Schulten (Eds.), p. 1, Springer.

Hida T. (1980), *Brownian motion*, Springer, Berlin.

Hopfield J. J. (1984); Proc. Nat. Acad. Sciences USA **81**, 3088.

Hoppensteadt F. C. and Izhikevich E. M. (1997); *Weakly Connected Neural Networks*, Springer, Berlin.

Jones R. D. (1993); Los Alamos Science **21**, 195.

Lin T. and Horne B. G. (1996); IEEE Trans. on Neural Networks **7**, 1329.

Lukacs E. (1970); *Characteristic functions*, Griffin, London.

Newhouse S. E. (1974-79); Topology **13**, 9,1974; Publ. Math. IHES **50**, 102, 1979.

Nusse H. E. and Tedeschini-Lalli L. (1992); Commun. Math. Phys. **144**, 429.

Oja E. (1989); Int. J. of Neural Systems **1**, 61.

Oja E. (1992); Neural Networks **5**, 927.

Pearlmutter B. A. (1989); Neural Computation **1**, 263.

Pineda F. J. (1987-9); Phys. Rev. Lett. **59**, 2229, 1987; Neural Computation **1**, 161, 1989.

Robinson C: (1983); Commun. Math. Phys. **90**, 433.

Sanger T. D. (1989); Neural Networks **2**, 459.

Sato M., Joe K. and Hirahara T. (1990); Proc. Int. Joint Conf. on Neural Networks, San Diego, vol. 1, p. 581.

Sato M., Murakami Y. and Joe K. (1990); Proc. Int. Conf. on Fuzzy Logic and Neural Networks, 601.

Shinbrot T. (1995); Advances in Physics **44**, 73.

Shlesinger M. F., Zaslavsky G. M. and Klafter J. (1993); Nature **363**, 31.

Skarda C. A. and Freeman W. J. (1987); Brain and Behavioral Science **10**, 161.

Softky W. R. and Kammen D. M. (1991); Neural Networks **4**, 337.

Taylor J. G. and Coombes S. (1993); Neural Networks **6**, 423.

Tedeschini-Lalli L. and Yorke J. A. (1986); Commun. Math. Phys. **106**, 635.

Vilela Mendes R. and Duarte J. T. (1981); J. Math. Phys. **22**, 1420.

Vilela Mendes R. and Duarte J. T. (1992); Complex Systems **6**, 21.

Yao Y. and Freeman W. J. (1990); Neural Networks **3**, 153.

Yao Y., Freeman W. J., Burke B. and Yang Q. (1991); Neural Networks **4**, 103.

Wang X.-J. (1990); Commun. Math. Phys. **131**, 317.

Zumofen G. and Klafter J. (1993); Phys. Rev. E **47**, 851.

Zumofen G. and Klafter J. (1994); Europhys. Lett. **25**, 565.

Lecture Notes in Physics

Monographs

For information about Vols. 1–10
please contact your bookseller or Springer-Verlag

Vol. m 11: A. D. Yaghjian, Relativistic Dynamics of a Charged Sphere. XII, 115 pages. 1992.

Vol. m 12: G. Esposito, Quantum Gravity, Quantum Cosmology and Lorentzian Geometries. Second Corrected and Enlarged Edition. XVIII, 349 pages. 1994.

Vol. m 13: M. Klein, A. Knauf, Classical Planar Scattering by Coulombic Potentials. V, 142 pages. 1992.

Vol. m 14: A. Lerda, Anyons. XI, 138 pages. 1992.

Vol. m 15: N. Peters, B. Rogg (Eds.), Reduced Kinetic Mechanisms for Applications in Combustion Systems. X, 360 pages. 1993.

Vol. m 16: P. Christe, M. Henkel, Introduction to Conformal Invariance and Its Applications to Critical Phenomena. XV, 260 pages. 1993.

Vol. m 17: M. Schoen, Computer Simulation of Condensed Phases in Complex Geometries. X, 136 pages. 1993.

Vol. m 18: H. Carmichael, An Open Systems Approach to Quantum Optics. X, 179 pages. 1993.

Vol. m 19: S. D. Bogan, M. K. Hinders, Interface Effects in Elastic Wave Scattering. XII, 182 pages. 1994.

Vol. m 20: E. Abdalla, M. C. B. Abdalla, D. Dalmazi, A. Zadra, 2D-Gravity in Non-Critical Strings. IX, 319 pages. 1994.

Vol. m 21: G. P. Berman, E. N. Bulgakov, D. D. Holm, Crossover-Time in Quantum Boson and Spin Systems. XI, 268 pages. 1994.

Vol. m 22: M.-O. Hongler, Chaotic and Stochastic Behaviour in Automatic Production Lines. V, 85 pages. 1994.

Vol. m 23: V. S. Viswanath, G. Müller, The Recursion Method. X, 259 pages. 1994.

Vol. m 24: A. Ern, V. Giovangigli, Multicomponent Transport Algorithms. XIV, 427 pages. 1994.

Vol. m 25: A. V. Bogdanov, G. V. Dubrovskiy, M. P. Krutikov, D. V. Kulginov, V. M. Strelchenya, Interaction of Gases with Surfaces. XIV, 132 pages. 1995.

Vol. m 26: M. Dineykhan, G. V. Efimov, G. Ganbold, S. N. Nedelko, Oscillator Representation in Quantum Physics. IX, 279 pages. 1995.

Vol. m 27: J. T. Ottesen, Infinite Dimensional Groups and Algebras in Quantum Physics. IX, 218 pages. 1995.

Vol. m 28: O. Piguet, S. P. Sorella, Algebraic Renormalization. IX, 134 pages. 1995.

Vol. m 29: C. Bendjaballah, Introduction to Photon Communication. VII, 193 pages. 1995.

Vol. m 30: A. J. Greer, W. J. Kossler, Low Magnetic Fields in Anisotropic Superconductors. VII, 161 pages. 1995.

Vol. m 31 (Corr. Second Printing): P. Busch, M. Grabowski, P.J. Lahti, Operational Quantum Physics. XII, 230 pages. 1997.

Vol. m 32: L. de Broglie, Diverses questions de mécanique et de thermodynamique classiques et relativistes. XII, 198 pages. 1995.

Vol. m 33: R. Alkofer, H. Reinhardt, Chiral Quark Dynamics. VIII, 115 pages. 1995.

Vol. m 34: R. Jost, Das Märchen vom Elfenbeinernen Turm. VIII, 286 pages. 1995.

Vol. m 35: E. Elizalde, Ten Physical Applications of Spectral Zeta Functions. XIV, 224 pages. 1995.

Vol. m 36: G. Dunne, Self-Dual Chern-Simons Theories. X, 217 pages. 1995.

Vol. m 37: S. Childress, A.D. Gilbert, Stretch, Twist, Fold: The Fast Dynamo. XI, 406 pages. 1995.

Vol. m 38: J. González, M. A. Martín-Delgado, G. Sierra, A. H. Vozmediano, Quantum Electron Liquids and High-Tc Superconductivity. X, 299 pages. 1995.

Vol. m 39: L. Pittner, Algebraic Foundations of Non-Com-mutative Differential Geometry and Quantum Groups. XII, 469 pages. 1996.

Vol. m 40: H.-J. Borchers, Translation Group and Particle Representations in Quantum Field Theory. VII, 131 pages. 1996.

Vol. m 41: B. K. Chakrabarti, A. Dutta, P. Sen, Quantum Ising Phases and Transitions in Transverse Ising Models. X, 204 pages. 1996.

Vol. m 42: P. Bouwknegt, J. McCarthy, K. Pilch, The W3 Algebra. Modules, Semi-infinite Cohomology and BV Algebras. XI, 204 pages. 1996.

Vol. m 43: M. Schottenloher, A Mathematical Introduction to Conformal Field Theory. VIII, 142 pages. 1997.

Vol. m 44: A. Bach, Indistinguishable Classical Particles. VIII, 157 pages. 1997.

Vol. m 45: M. Ferrari, V. T. Granik, A. Imam, J. C. Nadeau (Eds.), Advances in Doublet Mechanics. XVI, 214 pages. 1997.

Vol. m 46: M. Camenzind, Les noyaux actifs de galaxies. XVIII, 218 pages. 1997.

Vol. m 47: L. M. Zubov, Nonlinear Theory of Dislocations and Disclinations in Elastic Body. VI, 205 pages. 1997.

Vol. m 48: P. Kopietz, Bosonization of Interacting Fermions in Arbitrary Dimensions. XII, 259 pages. 1997.

Vol. m 49: M. Zak, J. B. Zbilut, R. E. Meyers, From Instability to Intelligence. Complexity and Predictability in Nonlinear Dynamics. XIV, 552 pages. 1997.

Vol. m 50: J. Ambjørn, M. Carfora, A. Marzuoli, The Geometry of Dynamical Triangulations. VI, 197 pages. 1997.

Vol. m 51: G. Landi, An Introduction to Noncommutative Spaces and Their Geometries. XI, 200 pages. 1997.

Vol. m 52: M. Hénon, Generating Families in the Restricted Three-Body Problem. XI, 278 pages. 1997.

Vol. m 53: M. Gad-el-Hak, A. Pollard, J.-P. Bonnet (Eds.), Flow Control. Fundamentals and Practices. XII, 527 pages. 1998.

Vol. m 54: Y. Suzuki, K. Varga, Stochastic Variational Approach to Quantum-Mechanical Few-Body Problems. XIV, 324 pages. 1998.

Vol. m 55: F. Busse, S. C. Müller, Evolution of Spontaneous Structures in Dissipative Continuous Systems. X, 559 pages. 1998.